Gina Maranto

Quest *for* Perfection

The Drive to Breed Better Human Beings

A LISA DREW BOOK

Scribner

FOR MARK

A LISA DREW BOOK/SCRIBNER
1230 Avenue of the Americas
New York, NY 10020

SCRIBNER and design are
trademarks of Simon & Schuster Inc.
A LISA DREW BOOK is a trademark
of Simon & Schuster Inc.

Set in Adobe Caslon
Designed by Jenny Dossin

Manufactured in the United States of America
1 3 5 7 9 10 8 6 4 2

Library of Congress Cataloging-in-Publication Data
Maranto, Gina.
Quest for perfection : the drive to breed better human
beings / Gina Maranto.
p. cm.
"A Lisa Drew Book"
Includes bibliographical references and index.
1. Human reproductive technology—History.
2. Genetic engineering—History. I. Title.
RG133.5.M37   1996
176—dc20   96-15795
CIP

ISBN 0-684-80029-2

ACKNOWLEDGMENTS

A number of physicians around the country were generous enough not only to illuminate the fine points of reproductive medicine for me but also to allow me to observe them at work. My thanks go especially to Bernard Cantor, Jamie Grifo, David Meldrum, Richard Paulson, Mark Sauer, and Arthur Wisot, and to those patients who permitted my presence in the examination and operating room as they underwent various procedures. David Meldrum kindly facilitated, and Lisa Beillen helped arrange, my attendance at the UCLA School of Medicine program on in vitro fertilization (IVF) and embryo transfer. Alan Trounson candidly discussed with me the development of IVF in Australia. Howard Jones, Jr., provided me with the text of a speech on the introduction of IVF at the Norfolk Institute. Dorothy Mitchell-Leef spent many hours sharing her expertise and experiences by phone. Also by phone, Wulf Utian recalled the early days of IVF.

Santiago Munné, then at New York Hospital–Cornell Medical Center, elucidated embryonic genetics. W. French Anderson gave liberally of his time and acumen during my visit to his Gene Therapy Laboratories at the University of Southern California School of Medicine; his executive associate director, Sharon Leming, pulled together a comprehensive packet of materials. Steve Broder gave me an informative tour of California Cryobank. Jean Benward, Dorothy Greenfeld, Patricia Mahlstedt, and other members of the Psychology Special Interest Group of the American Society for Reproductive Medicine provided insights into the needs and emotions of couples employing assisted reproduction, as did Marsha Staller in Miami. I owe much of my understanding of the ethical issues involved in reproductive medicine to a stimulating interview with Alexander Capron of USC. Joyce Zeitz of the ASRM was a valuable general resource.

Gabriele Macho, Alan Walker, Randall White, and Mary Ursula Brennan answered questions regarding paleolithic infanticide, and Jeffrey Schwartz talked with me about infanticide during the classical period. David Rindos was a font of anthropological wisdom, and Sarah Keene Meltzoff supplied assorted enthographic details, as well as valued friendship. Hugh Jarvis and the members of Anthro-L discussion group provided much food for thought to this inveterate Internet lurker. Franklin Loew guided me to key sources having to do with animal husbandry. Marla Stone suggested important works regarding women under the Nazis and shared her insights about various fascist family planning schemes.

I am grateful for the encouragement, by E-mail and otherwise, of numerous colleagues, including Sarah Boxer, Allan Chen, Tom Levenson, Marjory Spraycar, Dennis Overbye, and Bruce Schechter. Eager, thoughtful, and reverent conversations with Dick Preston in Miami, and at twenty thousand feet en route to Houston, spurred me onward. Assignments carried out for Bobbie Conlan, Pat Daniels, and Karin Kinney at Time-Life Books helped get me thinking about some of the ideas related to this book. A piece commissioned by *The Atlantic Monthly* allowed me to explore issues surrounding delayed childbearing. Accordingly, my thanks go to William Whitworth, Mike Curtis, Sue Parilla, and the magazine staff. Lynne Swanson of Wilton Library provided a key bit of research assistance.

I am indebted to Elise Hancock, for encouraging my first forays into science writing; to Frank Trippett, Leon Jaroff, and the late Sydnor Vanderschmidt for bringing me to *Discover* those many years ago; to Gil Rogin for the wide latitude he gave me to pursue stories. My respect and gratitude go to Lynne Viti, Darcy O'Brien, the late Don Barthelme, and John Barth for tutelage in the craft of writing.

Barney Karpfinger, my agent, patiently stood by as I groped toward this project, and he and Darren Haber gave me every assistance when I finally figured out what I wanted to do. My editor and publisher, Lisa Drew, provided enthusiasm—an invaluable commodity—and astute guidance in honing the final product. Thanks, too, to Marysue Rucci.

Wendy Kaplan, with several books to her credit, gave particularly valued advice. From Sag Harbor and Naples, Nancy Hooper was a constant source of reassurance and cheer. Bob Singer and Graziella Pozzebon offered intelligent commentary as I began conceiving the book. Karen Cook, Rick Counihan, Sue Earley, Laurie Julia, Denise Luttrell, Thelma and John Luttrell, Garry Rindfuss, Linda Srifuengfung Svennas, and Veronica Uytana have succored and sustained me for years: their fast friendship I can never adequately repay.

My parents, Joseph and Maggie Maranto, introduced me to the pleasures of language, which I consider their great gift to me. Thanks also to my grandmother Lennie Stubblefield and brother Marcus Maranto, and to the entire Derr clan.

To my husband, Mark Derr, I am most deeply grateful. With good heart and true, he walked every step of the way with me, helping me traverse sheer faces and negotiate scree. I couldn't have made it without him.

That which has worth in this volume rightly belongs to all these people; the dross is my own.

CONTENTS

There was never any more inception than there is now,
Nor any more youth or age than there is now,
And will never be any more perfection than there is now,
Nor any more heaven or hell than there is now.
Urge and urge and urge,
Always the procreant urge of the world.

WALT WHITMAN
Song of Myself

Continuous progress in man's knowledge of his environment, which is one of the chief conditions of general Progress, is a hypothesis which may or may not be true. And if it is true, there remains the further hypothesis of man's moral and social "perfectability," which rests on much less impressive evidence. There is nothing to show that he may not reach, in his psychical and social development, a stage at which the conditions of his life will be still far from satisfactory, and beyond which he will find it impossible to progress.

JOHN BAGNELL BURY
The Idea of Progress

I guess that's the question, isn't it? Is "eugenics" a bad word? Eugenics always brings up a negative connotation, but it really shouldn't. I mean, what parent would choose to have an unhealthy child? When you're the physician given the task of producing a child for a couple, how do you do it? It seems to me obvious that of course you choose stock that is as good as it can be. Eugenics is something that's being done every day, with the agreement of couples. Why would anyone faced with our task do it in a haphazard, irresponsible manner that would allow certain diseases and genetic problems to enter the gene pool?

MARK SAUER,
reproductive endocrinologist

INTRODUCTION

ITEM

After Floridian Manny Maresca died in a car crash, his wife of two weeks, the former Pam Williams, asked that sperm be taken from his body and frozen. Accordingly, fertility specialist Mark Jutras dissected the corpse's scrotum and removed the epididymis, a pinched tube in which sperm are processed prior to ejaculation. He minced it with scissors and a scalpel in a lab dish, and captured sufficient sperm to deposit in a sperm bank. Pam later told a reporter for *Vogue* magazine, "I wanted kids with my husband, and I wanted to raise them with him, but if that's not going to be a possibility, then I can still have his kids and raise them knowing who their father was and what a wonderful person he was." Said Manny's mother, "We want this child at all costs. It would be upsetting if she didn't [use the sperm for an artificial insemination attempt]; that was the game plan. If Pam decides never, we will support her. It doesn't rest on her shoulders. I'm not telling her one way or the other. We're hoping she has this baby."[1] According to a January 1994 AP report, at least seven other women in the United States had doctors remove sperm from their husbands during autopsy for possible future use.

ITEM

In New Orleans, on May 30, 1995, a federal district court judge ruled that Judith Christine Hart, a three-year-old, was entitled to Social Security benefits owed to Edward William Hart, who died in 1990 of cancer. Hart had cryopreserved a sample of sperm before undergoing chemotherapy, in case the treatment rendered him sterile. Before dying, he told his wife, Nancy, "There could always be a child for you." A few months after Edward's death, Nancy underwent a form of in vitro fertilization using the thawed sperm, and conceived Judith. Under Louisiana law (as under that of many states and nations), such a child is considered illegitimate. However, the judge agreed that Judith was Edward's biological offspring, borne according to his wishes, and therefore deserved a $9,000 lump-sum payment and approximately $700 monthly thereafter.

ITEM

Circa 1993, a Nigerian man with a Ph.D. flies to southern California from Africa with his wife to visit one of the country's leading assisted-reproduction clinics. His wife has already borne him five daughters; now he wants a son. He blames his wife: It is her fault he has no son. He wants the physicians at the clinic to perform in vitro fertilization using his sperm and an egg taken from another woman, the resulting embryo to be implanted in his wife for gestation. The physician explains to the man that sex is determined by the sperm. Eggs carry only female-determining X chromosomes, whereas sperm are ambidextrous—some carry X chromosomes, while others carry male-determining Y chromosomes. Some men simply produce more X-bearing sperm, ergo all those daughters. The man listens, but cannot be diverted from his purpose. The physicians ultimately refuse to acquiesce to his wishes.

ITEM

Rome fertility specialist Pasquale Bilotta claims that in 1992 a young couple, whom he refuses to identify, came to him seeking in vitro fertilization. Bilotta succeeded in retrieving eight eggs from the twenty-six-year-old wife, "Elisabetta." From these he produced eight embryos using her husband's sperm. Four of the embryos were transferred into Elisabetta over the course of treatment, although none resulted in a pregnancy. The four others were cryopreserved. In December 1992, the woman was killed in a car crash, and her distraught husband came to Bilotta. He wanted one of the remaining embryos transferred into the womb of his thirty-three-year-old sister, "Elena," for gestation. Bilotta complied with the request, and Elena delivered a baby girl. When the story got out, it caused an uproar. Parliament member and granddaughter of Il Duce, Allessandra Mussolini, pronounced the birth "immoral and incestuous," a sentiment shared by more than half of 1,170 Italians polled. A few scientists questioned the veracity of Bilotta's tale. The Catholic Church, however, did not quibble over details, but flatly condemned the entire event.

ITEM

At least three postmenopausal women, a fifty-nine-year-old Englishwoman, a fifty-year-old from Glendale, California, and a sixty-one-year-old in Rome, gave birth in 1994 thanks to fertility specialists, who provided them with embryos crafted with the eggs from younger women and revivified their aging reproductive systems with hormones. The Englishwoman delivered twins, the Glendale woman triplets. The births triggered energetic public debate. Critics pointed out that the women would be in their seventies and eighties when their children graduated college. Defenders claimed there was a double standard at work: No one bothered to raise a ruckus whenever sixty- and seventy-year-old men fathered children. Mark Siegler, director of the Center for Medical Clinical Ethics at the University of Chicago, told a *New York Times* reporter, "The history of medicine has been devoted to overcoming the natural lottery, the hand fate has dealt each one of us. We are already pushing the bounds of attractiveness, sexuality, human well-being. Why draw the line at reproduction?"

ITEM

A couple in the Netherlands went to the Amsterdam daily *Het Parool* with a front-page story about a critical mix-up at the prestigious University Hospital fertility clinic. In March 1993, the couple, pseudonymously known as Wilma and Willem Stuart, had gone to the Utrecht clinic for in vitro fertilization. Wilma conceived twins, both male, who were born in December. One twin, Teun, was fair, while the other, Koen, was much darker. The Stuarts approached the clinic about the matter, but were stonewalled. Later, the clinic agreed to perform DNA tests, which showed that Willem was not Koen's father. After an investigation, the hospital administration blamed the "deeply regrettable mistake" on a technician who it said had neglected laboratory rules and reused a pipette during the fertilization process. The hospital surmised that Koen's genetic father was an Aruban man who had also visited the clinic with his wife for in vitro treatment. The Stuarts told the *Het Parool* reporter that they had been subject to censure from neighbors in the largely white village where they lived. They worried that the biological father might initiate a custody fight for the child. Newspaper editorialists in Holland wondered whether the furor raised by the story had more to do with shaken confidence in "a technique that was considered infallible," than with disapproval "because inadvertently a black child has landed with white parents."

ITEM

In September 1995, agents for the FBI, IRS, and California State raided the offices of physician Ricardo Asch in Laguna Hills and

Santa Ana and confiscated the records of his private practice. Agents also seized material from his Newport Beach home and from his partners, Jose Balmaceda and Sergio Stone. Asch and Balmaceda, two of the world's leading experts in infertility, had been accused in May of having misappropriated the embryos of at least thirty couples who had come to them for in vitro fertilization. As a result of the allegations, the University of California had shut down Asch's in vitro clinic at the school's Irvine branch and instituted legal proceedings against him. Some months later, on the *Oprah Winfrey* show, the story was billed as a "high-tech baby kidnapping," and several patients aired their grievances, claiming that Asch had stolen their embryos and given them to other couples without permission. One couple, John and Debbie Challender, believed that they knew who had received their embryos: an Asch patient who had given birth to twins. Debbie said, "Every twin I see, I can't help thinking these are my kids." Added John, "Those embryos were our *children.*" On the same show, a university spokesperson accused Asch of having committed "biomedical rape." Asch and his colleagues categorically denied all accusations.

O ver the past several years, stories like the ones above have been cropping up in the oddest places. One expects the fantastic and hypothetical to mingle in the pages of the tawdry tabloids, where the likes of babies from Mars or infants riding around in utero for three years are common fodder, but not in those papers of record, the Los Angeles or London or New York *Times.* Yet here they are, these tales of sepulchral romance and embryo kidnapping, of lives owed to laboratory mistakes or medicine as a kind of plastic surgery upon the future—all thanks to the wonders of assisted reproduction, thanks to the interventions of physicians employing in vitro fertilization (IVF), artificial insemination, and a range of other techniques for manipulating sperm, eggs, and embryos. These so-called reproductive technologies have already rocked social norms and promise to redefine the prevailing Western notions regarding kinship and childbearing.

Most immediately, the laboratory scientists and medical researchers who have developed and perfected assisted reproduction have intended to help couples overcome infertility. Indeed, worldwide, over forty thousand women have given birth to children as a result of IVF; over the past fifty years, millions of babies have been delivered because of artificial insemination and fertility drugs.

Until recently, few people who had not encountered infertility problems bothered to undertake assisted reproduction, which is costly and time-consuming and often unsuccessful. Yet, as is apparent from the vignettes above, these technologies open up a host of possibilities which some fertile individuals and couples have begun exploiting. Many observers believe that it is only a matter of time before larger numbers of people able to procreate quite well on their own begin seeking out assisted reproduction with the idea of obtaining a "better" baby. Indeed, infertility patients are already presented with choices by which they may determine in some fashion the quality or "design" of their offspring.

In short, the technologies opened a new phase in the chronicles of human reproduction, one which may or may not be benign. It has, for example, always been possible for women to lose a partner during the nine months of pregnancy, as happened frequently during wartime, and give birth to a father-less child. This was never considered terribly desirable, but cryopreservation of sperm has made it one option which some women, like Pam Maresca, choose. Bizarrely enough, as the Bilotta case showed, a man who longs to engender a child with a dead woman can today also fulfill this desire, as long as the two had previously banked frozen embryos made from his sperm and her egg. He needs merely to find a surrogate mother to carry the embryo to term and relinquish the child to him at birth.

One scenario raised by assisted reproduction has received the greatest amount of media attention: Framed generally as the Frankenstein option, it poses the specter of thousands of cloned look-alikes roaming the land. Unfortunately, such projections distort scientific realities and do a grave dis-service to serious discussions of the moral, ethical, and legal implications of reproductive technologies. They dull the impact of the genuine issues, which are far less Hollywoodish than the hyperbole indicates, yet distinctly troubling.

This book undertakes to correct misapprehensions that may have been fostered by tabloid-style reporting, and also to place these medical innova-tions in context, since so many accounts of them lack any sense of history. To paraphrase American social critic Paul Goodman, the application of technol-ogy is the exercise of a moral philosophy. It goes without saying that it is impossible to view reproductive technologies, or to understand the moral philosophy they stand for, without examining attitudes toward children, con-ception, and heredity over time. As Ashley Montagu, the brilliant American geneticist who so incisively exposed the fallacies of Nazi race hygiene and so-called scientific racism throughout his career, wrote:

> No activity of man, whether it be the making of a book, the con-traction of a muscle, the manufacture of a brick, the expression of an idea, or the writing of a work such as this, can be fully under-stood without a knowledge of the history of that activity insofar as it has been socially determined. For, obviously, any neglect to take into consideration the relations of the social framework can only lead to a defective understanding of such events.[2]

In and of themselves, reproductive technologies can hardly be cast as an unprecedented threat to civilization. Rather, in their eugenic aspects, they represent the extension of an ambition to master the vagaries of reproduction and, pointlessly, "perfect" the species. The fact of the matter is that individuals and societies, particularly in the West, have manifested a long-standing indifference to infants, whom they have frequently viewed as products needful of quality control and an overall improvement. Time and again, humans have endeavored, in however haphazard a fashion, to modify the pool of offspring through control of mating, through sterilization, through selective infanticide. Thus, the aims of reducing birth defects through genetic manipulations and of creating a better baby are not new; but the means to achieve those aims have improved.

Probably for as long as humans have been human, they have attempted to influence birth outcomes. Certainly within historical memory they have done so with an intensity of purpose verging on the religious. To the end of ensuring that they will get the baby they want, people have developed a panoply of rituals. They have worn protective amulets or performed elaborate ceremonies, have incanted and prayed, lit candles, chanted. Couples have made matches on account of bloodlines so that their offspring will be endowed with "superior" characteristics, and made love in prescribed positions to "guarantee" a child's gender. While pregnant, women have avoided gazing upon certain objects and persons (especially, in a remarkable number of societies, midwives) or eating certain foods or encountering certain beasts thought capable of tainting the womb and damaging the fetus. In the hopes of bettering their babies by educating them in utero, mothers-to-be have programmatically exposed themselves to the strains of classical music, attended mentally stimulating lectures, studied the paintings of the Great Masters.

Through methods both superstitious and supposedly scientific, people have sought to make sure that their offspring would be talented and attractive and above all healthy. When such efforts have failed to yield the wished-for outcome, individuals, families, and whole societies have shown themselves capable of ruthlessly direct remedies.

For most of human history, the choice has been less whether to have children than whether to keep them. Upon giving birth to unwanted girls, mothers have smothered or buried alive or drowned the infants. And when they have fathered or borne infants misshapen or missing limbs or unrecognizable as human—the accidents of biology known in classical Rome as *monstri*, literally "monsters"—parents have brought these small, unfortunate lives to an end, sometimes with merciful dispatch and other times through neglect, starvation, or exposure in the wilds (which carried the psychological advantage of uncertainty, so that fantasies of an infant's survival might be, and frequently were, nursed as compensation for its loss). The desire among both individuals and groups to shape the results of pregnancies, to control the nature of the children who are produced, runs so deeply within our species that it could almost qualify as an innate psychological drive.

To survey the history, and the sketchily inferred prehistory, of the relationship of parents with their offspring, is to recognize the truth of French historian Philippe Ariès's statement that "childhood is an age that we disguise by embellishing, by using it to embody our ideals."[3] More than that, children, in their very existence, afford adults the opportunity to act upon convictions concerning the value and meaning of life, and upon expectations for the future. Throughout time, humans have with great regularity chosen to terminate the lives of infants who have not been seen to comport with those varied convictions and expectations; people have literally killed off the possibilities inherent in those newly issued lives, for reasons, when they are ascertainable, mostly having to do with expediency.

It is a characteristic of the historical perspective that actions taken in the immediacy of the moment may also, with the accretion of time, assume the appearance of trends (whether or not these trends are real or mere illusions is a question I happily leave for philosophers to decide). So that infanticide can from the modern purview be seen as a strategy for controlling birth outcomes, which in turn can be viewed as part of an overarching quest by human beings to perfect themselves.

The origins of this quest, which entails in essence a problem of biological engineering, possibly lie with the earliest direct forebears of *Homo sapiens*. Whether by the intentionless force of evolutionary selection or the operation of consciousness upon behavior, members of the *Homo* line proved as capable of exterminating their young as their knuckle-walking primate kin and as members of the animal kingdom in general, vertebrates and invertebrates included. By around seven thousand years ago, with the rise of the earliest cities, infanticide appears to have been practiced by all extant groups of people, and in some instances may have been ceremonially elevated, taking its place alongside other forms of live sacrifice.

During the classical period, people throughout the Mediterranean world deployed political rationales for infanticide. Aristotle and Plato in particular devised theoretical programs for ensuring the vigor and viability of the hypothetical inhabitants of the utopias they imagined. In their ideal states, which we read about in *Politics* and *The Republic*, weak or deformed infants were to be weeded out ruthlessly. The state meanwhile would strictly regulate which men and women would have children, preventing those considered mentally or physically inadequate from procreating. From the Greeks, then, have come down to us the earliest known versions of the idea that human perfection is attainable in part through control of mating, which is supposed to enable the production of a superior form of offspring, possessed of all the best traits of humanity and none of the worst.

Greeks, Romans, and other Western thinkers on up through the Middle Ages readily drew parallels between the husbandry of domesticated animals and of humans, parallels which became ever stronger in the eighteenth century with the rise of scientific breeding. By the early nineteenth century, techniques that had been originally developed for manipulating animal sperm had been adopted for use with humans, and over the years, each ensuing advance on the farm, as it were, found its way into the clinic and the hospital. In a very real sense, the birth in 1978 of Louise Brown, the world's first in vitro fertilization baby, was the crowning achievement of several centuries' worth of animal research. That birth simultaneously launched a new era, in which a range of technologies for manipulating human reproductive processes at the level of gametes (eggs and sperm) and zygotes (embryos) have appeared with stunning rapidity.

It is within the context of thousands of years of documentable efforts to control the quality of offspring in the hopes of improving upon the species' standard model that this book places the latest advances in reproductive medicine. Today, possessed of the ability to tinker with embryos in the laboratory and to alter their very genetic material, we have in hand the tools necessary for a radical overhaul of the human line. To a degree never before possible, we stand able to act upon goals embraced from ancient times. Every engineering project requires specialized tools, and those tools—from sonographs to synthetic hormones to computerized cryopreservation machines—are available at clinics worldwide.

In the West, the goal of engineering human beings has always been intimately wound up with politics. From at least the seventeenth century on, issues having to do with childbearing have been constellated with larger issues regarding population, class, race, and evolution (used in its pre-Darwinian sense of development from a rudimentary to a complete, perfected state). Ironically, as European nation-states strengthened and through force colonized the globe, fears of decline and fall gripped the imagination of educated elites, particularly in England, France, and Germany. The flip side of those animating dreams of world hegemony turned out to be paranoid fantasizing about the destruction of the "white," "Nordic," "superior" race through miscegenation, through interbreeding with the "dark," "southern," "inferior" races. Ultimately emerging from this deranged fixation on bloodlines and supremacy came the supposed science of eugenics, which simultaneously encouraged northern Europeans' fears of "race degeneration" and secured their sense of preeminence with statistics. Moreover, it armed them with theory and a set of social prescriptions based on that theory, and sent them to the barricades eager to beat back the infidel.

"Eugenics," a word adapted from the Greek *eugenes*, meaning "wellborn," was introduced to the English language by the British polymath Francis Galton, a cousin of Charles Darwin. Galton proposed a scheme of selected breeding for humans which he felt would reverse the downward drift of humanity. Galton's ambitious plans for salvaging a devolutionary human race struck a chord with people throughout Europe and the Americas, where dozens of societies devoted to implementing eugenic precepts sprang up. These were not fringe gatherings, convened by kooks or founded by eccentric, moneyed do-gooders; rather, they were serious, mainstream organizations established and joined by scientists.

To be a geneticist in Britain and the United States during the first decade of the twentieth century was almost by default to be a eugenicist. Typically, geneticists in the United States belonged to both the American Genetics Association and the American Eugenics Society. For some decades, the two organizations even shared an official organ, the *Journal of Heredity*. In Germany, geneticists, physicians, and natural historians became ardent eugenicists and laid the groundwork in the 1920s and 1930s for the Third Reich's "murderous science." (This is German geneticist Benno Müller-Hill's phrase for the collusion of virtually the entire university, research, and clinical establishment with Hitler's extermination programs, first of mental patients, then Gypsies, homosexuals, Jews, and Communists.)

After World War II, when everyone learned of the Nazi concentration camps, eugenics ceased to be openly touted, but it continued to draw adherents. Effectively, eugenics went underground, to surface again in the 1960s. In September 1961, writing in the respected journal *Science*, Nobel Prize–winning geneticist Hermann Muller, who sat on the board of the American Eugenics Society in the 1930s, proclaimed that it was time to resuscitate eugenics. He wrote that eugenics had fallen into disrepute "as a result of its spurious use in support of the atrocities committed by those with class and race prejudices." However, Muller insisted,

> the odious perversions of the subject should not blind us longer to a set of hard truths, and of genuine ethical values concerning human evolution, that cannot be permanently ignored or denied without ultimate disaster. On the other hand, if these truths are duly recognized and given expression in suitable policies, they may open the way to an immeasurable extension and enhancement of the potentialities of human existence.[4]

Muller was hardly a voice in the intellectual wilderness. In the early 1960s, the likes of Nobel Prize–winning geneticist Joshua Lederberg and biologist

Julian Huxley (brother of *Brave New World* author Aldous Huxley, and grandson of the famed British scientist T. H. Huxley, who for his tenacious defense of evolutionary theory was dubbed "Darwin's bulldog") also publicly endorsed eugenics, and emphasized the evolutionary benefits that a considered program of biological engineering would surely bring. Nobel chemist Linus Pauling, the Nobel laureate in chemistry and peace, even proposed in 1968 that anyone who was tested for and found to carry the gene for sickle-cell anemia or phenylketonuria disease, a metabolic disorder, should be tattooed, for the practical reason that "two young people carrying the same seriously defective gene in a single dose would recognize this situation at first sight, and would refrain from falling in love with one another."[5]

All of this is to say that the invention and spread of new reproductive technologies did not happen in a vacuum. In some cases, the scientists developing the techniques for handling sperm, eggs, and embryos held eugenic convictions. In many cases, those who promulgated the technology aspired to further the cause of biological engineering. Whatever their individual convictions regarding the possibilities of enhancing the human line, physicians today practicing assisted reproduction operate inextricably within a historical continuum. The development and further uses of these techniques have been and will be shaped by the eugenic precepts which have been embraced for millennia, by people in the West particularly. While each couple seeking to employ any of these new procedures may not begin with conscious intent to fabricate a better baby, they frequently find themselves caught up in making decisions, with the aid and abettal of medical professionals, that can only be construed as eugenic. With each additional fillip to assisted reproduction, we are carried further into the world of human engineering, in which parents will attempt not only to eliminate disease but also to determine which of many other inherited characteristics their children will receive.

1

Group Portrait
with Babies

On a weekly basis, Jamie Grifo, father of three, attempts to impregnate half a dozen women or more.

Grifo is the kind of doctor nurses adore. A compact man with a shock of brown hair and a boyish aspect, he won't shy from shouldering the cleanup and prep work in the operating room if his surgical team has gotten itself running late. He bundles used blue paper sheets and stuffs them into plastic trash bags blazoned with biohazard symbols, strips catheters from their antiseptic packages, performs any of the hundred little tasks usually reserved for the nurses whose job it is to remove the traces of one patient and ready the room for the next. A sometime jogger, Grifo moves with a determined swiftness about the operating room, his brown eyes, behind the veil of his surgical mask, catching everything. Grifo is outspoken (a trait which, in combination with a professed dislike for "micromanagers," has occasionally gotten him into trouble with his academic superiors), and has a repertoire of quips and jokes (some of them slightly ribald) which he deploys throughout the day. (Grifo: What does President Clinton say to his wife after sex? Patient: I don't know, what? Grifo: Hi, honey, I'll be home in twenty minutes.) He can't do embryo transfers without music. During those times when the operating room is under his command, strains of rock and roll filter out of the swinging doors, the nasal twang of Tom Petty singing, "You don't know how it feels to be me," or the plaintive whine of Counting Crows' Adam Duritz drifting down the hallway toward the maternity ward. Grifo has so little of the stereotypical physician's hauteur that he will even stoop to repair a squeaky rollaway stool between surgical procedures. Bending to this task, twiddling the offending caster and dosing its mounting with mineral oil, he pauses, holds the plastic bottle of oil aloft, and notes, "We use this stuff for embryos." (Mineral oil is often an ingredient of the culture medium in which human embryos are sustained outside the female body.)

For all his banter, Grifo is keenly serious about his work. That he is hard driving and ambitious goes without saying. At age forty, he is a leader in his chosen field, a man whose daily routine is deeply embroiled with the mysteries of life, a man with such a fine-tuned knowledge of human genesis that he qualifies, along with several hundred colleagues around the globe who share his expertise, as a latter-day fertility god. They are the ones who perform in vitro fertilization and embryo transfer, and all the other varieties of so-called assisted reproduction, from intracytoplasmic sperm injection to zygote intrafallopian transfer.

Physicians like Grifo, whose specialty is known as reproductive endocrinology, undergo a multistep education, beginning with medical school and a residency in obstetrics and gynecology. After this, they must pass a written and oral exam administered by a panel of experts in that specialty, a process known as board certification. Then, at a time when their peers are establishing their careers (and, not incidentally, earning the money to pay back their medical school loans), those who would be reproductive endocrinologists must enter a fierce competition to win one of only a handful of fellowships around the world. Such fellowships provide two years of intensive training in the intricacies of human procreation, at the end of which time a fellow is required to write a thesis and submit to yet another round of written and oral examination.

Although virtually any doctor in the United States can hang out a shingle claiming to be a fertility specialist, only those who have undergone such an arduous preparation can really make a claim to comprehending the complex orchestration of hormone releases, of cellular changes in the reproductive organs, and of feedback loops to the brain which must all proceed without a hitch if pregnancy is to occur.

Reproductive endocrinologists are heir to a body of knowledge concerning animal and human reproduction amassed over the course of two millennia by philosophers, physicians, gross anatomists, microscopists, botanists, veterinarians, physiologists, cell biologists, geneticists, neurologists, endocrinologists, and molecular biologists. Through observation and experiment, these investigators probed the structure and function of the sex organs. They studied the production and release of germ, or sex, cells—that is, the female oocytes which mature into fertilizable ova, or eggs, and the male spermatozoon, or sperm—and discovered the hormonal and neurological components of these processes. They learned how to manually inseminate females with select sperm, and to freeze sperm for long-term storage; to keep sperm and eggs alive in glass vessels in the lab, to achieve fertilization of them, to grow the resulting embryos, and to transfer them into receptive females. They gained the ability to trigger key events in the female monthly cycle, to amplify or dampen the body's natural rhythms. With the aid of medical engineers, they deployed the laparoscope and ultrasound for viewing the ovaries, fallopian tubes, and uterus, and for monitoring the gestating fetus. With the aid of biochemists and geneticists, they developed amniocentesis and chorionic villus sampling (CVS), which test for congenital defects and inheritable diseases. A few particularly skillful hands acquired proficiency with microsurgery tools and, exhibiting the dexterity of, say, medieval monks able to carve whole scenes replete with dozens of people in a space the size of a walnut—or maybe just kids with a video game—succeeded in removing single cells, or blastomeres, from early embryos, or blastocysts, no bigger than the period at the end of this sentence. In similar fashion, they plucked out chunks of the zona pellucida, or sac surrounding early embryos, traded the entire cytoplasmic contents of one zona pellucida for that of another, and divided embryos to create genetically identical doubles. In the technique called intracytoplasmic sperm injection (ICSI), which has taken the field of assisted reproduction by storm since its introduction in 1992, nimble micromanipulators catch a single spermatozoon and, holding an egg in position with slight suction from a glass pipette, manually poke the sperm through the zona pellucida into the egg's cytoplasm to effect fertilization.

Because the reproductive endocrinology specialty did not even exist until the 1970s, there are today only 584 of these physicians in the entire world. As such, they hold an exalted place within the obstetrical and gynecological community. A handful of them qualify as superstars, and, like topflight baseball players, have a history of moving from city to city, from team to team. The notable franchises in this major league are located in Los Angeles, Houston, Atlanta, Boston, New York City, and, improbably, Norfolk, Virginia. By my tally, at least nine of the top two dozen or so reproductive endocrinologists hopscotched from one program to another during the year and a half that this book was being researched. In addition, an entire group of genetical researchers connected to a university-based assisted-reproduction program decamped for another institution, and a well-known embryologist migrated from one major outpost to another. Wherever they go, the superstars carry their reputations and research. They have the power to draw crack physicians, embryologists, and lab technicians to their team, and to attract clienteles from around the globe. Like most experts, reproductive endocrinologists form a clique in which everyone knows everyone else, friend and foe, and can recite their foibles and strong suits. The field has occasionally been riven by intense rivalries, which some wags have dubbed "the Egg Wars."

Reproductive endocrinologists as a group guard their status jealously. It is not uncommon to hear one of them say, when discussing another physician in the fertility field, something like, "I don't mean to put Dr. XYZ *down*, but he's *not* even a reproductive endocrinologist." In addition, a number of reproductive endocrinologists actively dislike bioethicists—sometimes names are named, other times individuals are merely alluded to by their most recent transgression—who are perceived as having "made their reputations" by playing up the most extreme issues raised by the assisted-reproduction enterprise, for instance, the cloning of humans. Partners in this crime are generally held to be the media. Along with probably a majority of scientists, those involved in the field feel the popular press has presented a distorted portrait of their work and shown itself ever ready to level aspersions against physicians whom it supposes to be engaged in "playing God" or making Faustian bargains or indulging Frankensteinian urges.

Yet inarguably, control is the sine qua non of science, and certainly it is at issue here. The whole point of the assisted-reproduction enterprise, that culmination of those two thousand years of persistence, intuition, and inventiveness, is to exert mastery over procreation, to remove chance from a natural biological process. Already, reproductive endocrinologists and others in the infertility business are participating in eugenic decisions made by couples employing their services. One day, if they have their way, scientists-cum-physicians will replace faulty genes in embryos as if they were bad carburetors.

The treatment of infertility is a booming field. Membership in the American Fertility Society, recently renamed the American Society for Reproductive Medicine (ASRM), went from 3,600 in 1974 to 12,000 in 1995. The organization boasts a gleaming new headquarters in Birmingham, Alabama, and its fiftieth-anniversary conclave in 1994 drew fully a quarter of the membership to San Antonio, Texas. That some of them managed to wangle tickets to the sold-out concert by the Rolling Stones, who happened to be in town, was just an added bonus to the six days of symposia, seminars, roundtable luncheons, and presentations of scientific papers at the city's riverside convention center and the meeting rooms of two adjoining Marriott hotels.

The run-up in the number of physicians billing themselves as fertility specialists coincided with the entry of baby boomers into their prime childbearing years. This was the market economy at work: Physicians saw a need and filled it. Compared with their mothers, baby boom women have been far more prone to delay childbearing, whether because they are pursuing careers or cannot find a suitable mate or want to become more financially or emotionally settled or cannot make up their minds whether they do or do not want to procreate. Whatever their reasons, as of 1992 a significant fraction of the 55 million American women between ages twenty and thirty-four remained childless. Of women twenty to twenty-four, 66.7 percent had not given birth; of those twenty-five to twenty-nine, 43.8 percent had not; of those thirty to thirty-four, 26.1 percent had not.

Partly because women's capacity to become pregnant erodes noticeably beginning around age thirty, pushing the age of childbearing up across the board also increases the likelihood that when these women finally decide to become pregnant they will have difficulty. An analysis of 1988 National Survey of Family Growth data by researchers at the National Center for Health Statistics suggests that childless married women in the United States run an overall one-in-twelve chance of infecundity, that is, of encountering problems conceiving, with the odds worsening the older a woman gets. Not all of this infecundity can be attributed to age, and some of it is owed to the male side of the equation, but in any case, the market for infertility services in the United States alone amounts to some 5 million couples. Add to this large demand the relatively low cost of setting up an infertility clinic and the swelling of the ranks of the ASRM becomes more understandable.

It was also one of those fortunate historical coincidences that IVF and embryo transfer happened to come along right when they did, adding a certain glamour and reputability to an otherwise lackluster subspecialty. Prior to the transfer of reproductive technologies from the basic science lab to the clinic, physicians who wished to help subfertile or infertile men, those who for whatever reason could not manufacture or deliver sperm capable of fertilizing an egg, basically had recourse to one reliable remedy: artificial insemination. In some cases, the man's sperm might prove adequate to the task when concentrated beforehand and then launched into the vaginal canal by syringe or placed in a small cup over the cervix or inserted into the uterus. If that failed, the next step was to use competent sperm provided by another man, an expedient known as artificial insemination by donor (AID). Although estimates vary, in modern populations, about 40 percent of infecundity can be pinned on so-called male factor infertility.

Another 40 percent of the time, the incapacity is attributable to faulty functioning of the female organs. (About 20 percent of couples experience infertility that is idiopathic, or of no known cause.) Traditionally, women with reproductive difficulties had not many options either. From the late 1800s, gynecological surgeons were able to correct some anatomical difficulties. But surgery to correct blockages of the fallopian tubes caused by disease or surgery or prior pregnancies often—as it does even today with improved tools and methods—provoked the formation of additional scar tissue which further impeded the passage of eggs from the ovaries to the uterus.

Doctors could also, beginning in the 1960s, dose women with fertility drugs. If this tactic succeeded, the lucky woman stood a sizable chance of conceiving not one baby, but two, three, four, five, or more. The manufacturer of one fertility drug, clomiphene citrate, sold under the trade name Clomid by Merrell, puts the added risk of a multiple pregnancy at 6 to 8 percent.

A 1990 study carried out by the British Department of Health found that the rate of triplet deliveries in England and Wales alone went from a steady 10 per 100,000 before fertility drugs were introduced in the late 1960s to nearly 13 per 100,000 a decade or so later. After the introduction of the variants on IVF called gamete intrafallopian transfer (GIFT) and zygote intrafallopian transfer (ZIFT), the rate climbed to 28.6 per 100,000 deliveries in 1989. That same year, British obstetricians also delivered 183 women of triplets, 11 of quadruplets, and 1 of quintuplets.

In GIFT, physicians place eggs and sperm—in ZIFT, embryos—directly into a woman's fallopian tubes. Because the chance of a pregnancy rises when multiple eggs or embryos are transferred, British physicians, along with everyone else performing these procedures (which have now fallen somewhat out of fashion), initially opted for the more-is-better philosophy, and in 1987 reportedly were transferring an average of 6.3 ova in each GIFT attempt. The problem with multiple conceptions is not simply that of an embarrassment of riches: Women who carry more than one child often experience complications that may land them in the hospital and, in extreme cases, threaten their lives. The growth of the fetuses, crowded into a uterus best suited to a solo occupant, is stunted, and multiples are rarely carried to term. Born prematurely, many such infants perish immediately or shortly afterward, and if they survive stand a greater chance than normal of cerebral palsy and other ailments.

Offering little opportunity for doing innovative medicine, the fertility field did not hold a great deal of allure for many young, ambitious interns rotating through departments in search of a life's calling. But in 1978, all that changed, as baby Louise Brown's visage was splashed across newspapers and magazines worldwide. The impact of that first successful IVF and embryo transfer, carried out by Britain's Patrick Steptoe, a gynecologist, and Robert Edwards, a physiologist, was twofold: First, it drew the attention of hot young premeds and medical students like Jamie Grifo. Grifo and his forty-something peers, including the likes of Richard Paulson of the University of Southern California School of Medicine, Mark Sauer of Columbia University medical school, Bill Yee of the University of California at Irvine, Jeffrey Sher of Pacific Fertility Medical Center in San Francisco, Dorothy Mitchell-Leef of Reproductive Biology Associates in Atlanta, Gianpiero Palermo of Cornell Medical College, and Kwang-Yul Cha of Younsei University Medical College in Seoul, South Korea, have amplified the work of those immediately responsible for the introduction of IVF and embryo transfer, the pioneers who continue to dominate the field. Although Steptoe died in 1988, his partner Robert Edwards persists in his research. Most of the other early notables also remain active, including Australia's Carl Wood and Alan Trounson and America's husband-and-wife team, Howard Jones and Georgeanna Seeger Jones; John Buster; Maria Bustillo; Richard Marrs; Jacques Cohen; and David Meldrum. At professional gatherings of fertility specialists, hallway conversations are frequently interrupted by remarks like, "Oh, look, there's so-and-so, he/she had the first frozen embryo [or GIFT or ZIFT or ICSI] baby." (Odd as it sounds, this is the preferred shorthand: Physicians are said to have "had" the first such-and-such baby. Whether this borrowing of the term normally reserved for mothers signals a healthy psychological identification or an unfitting proprietary claim on the part of the profession is open to interpretation.) This chronic dropping of physicians' names—as opposed to the names of the women and babies who literally embody these accomplishments—serves to remind one that, whatever else might be involved, a main point of the assisted-reproduction exercise for top fertility specialists is professional recognition, advancement, and remuneration. This is not to say that compassion and altruism cannot be found in their black bags as well, but simply that obtaining a full portrait requires contemplating the benefits fertility specialists accrue as well as those they may endow.

The second thing Louise Brown's birth did was to galvanize infertile couples, many of whom had given up on the possibility of bearing children. Almost immediately, Steptoe's mailbox was overflowing with letters from the curious, eager, or desperate, all of whom wanted access to the procedures he and Edwards had refined.

Infertility has, in many societies in many eras, been construed as a shameful state, a reflection of moral turpitude or inadequacy. Throughout premodern Europe, men who could not beget children often were seen as weak, lacking in virility; women who could not bear them, as emotionally cold, physically dried-up, tainted. Women who did not breed were thought to be bewitched, or capable of witchery—capable, in fact, of rendering other women or female animals barren at will. This association of childlessness and devilry appears in part to have been related to attempts to cast opprobrium on birth control: Recipes for ancient contraceptives and abortifacients tended to be the province of female herbalists, some of whom would surely have employed their own devices to remain childless.

At any rate, the negative aura surrounding infertility persisted into the twentieth century. While it appears to be abating in some segments of some societies, infertility continues to be viewed by many as a horrible sentence. Psychologist Patricia Mahlstedt, who has been associated with the infertility program at Baylor College of Medicine and maintains a private practice in Houston, Texas, has produced several scholarly papers which give a harrowing overview of the damage infertility wreaks upon individuals, couples, and families.[1]

Most infertile couples suffer from emotional distress. Initially, they may

feel inadequate due to their inability to conceive after a year or more of trying. They may deny the problem or blame each other. Their sex life may
become fraught with tension. Under the pressure, men may watch their self-
esteem dwindle. They may become impotent or have affairs in an attempt to
prove their manliness. Women experiencing difficulty conceiving have told
counselors that they feel deeply inadequate and failed. Some infertile couples
become anxious or self-conscious in the presence of family members or
friends who already have children. One woman told Mahlstedt, "I feel like I
don't belong, like a second-class citizen with no place to go. Without a child,
I don't belong in the group with kids who play in the park. Without a child,
my husband and I don't fit in with our friends who do."

Couples feel stigmatized, freakish. Some men and women offer their
spouses an out: They offer to leave so that they can find a fertile partner.
Especially in cultures which emphasize the religious aspects of procreation or
revere the family above almost all else, infertility can be read as a sign of a
curse or of divine displeasure. From time immemorial, women have been
blamed for infertility almost reflexively, and in many places today, a woman
who cannot produce an heir will be jettisoned, so that her anxieties over the
matter are more than understandable: Childlessness threatens not only her
standing in the community, but also her marriage and therefore her economic
well-being.

Unfortunately, the turmoil does not necessarily cease for those who seek
medical aid. Mahlstedt explains that many couples who undertake infertility
treatment get caught up in an obsessive routine:

> They become patients, and their lives begin to revolve around con-
> ceiving by following a physician's game plan. They focus their
> attention on what they have failed to accomplish [parenthood] and
> soon start neglecting other goals and needs in their lives. In most
> cases, this type of focusing leads to losses of self-esteem, confi-
> dence, security, health, close relationships, and even hope. Some
> patients lose sexual potency or interest in sexual intercourse, losses
> that interfere with treatment goals and put significant stress on
> marriages. There is no balance in their lives. There is, instead, hope
> one week, grief the next. This cycle creates a very confusing roller
> coaster, with depression, anger, and guilt as part of its down side.[2]

In other words, the very process of undergoing assisted reproduction tends to
worsen the pathology, not the least because it is physically onerous, requires
an enormous investment of time on the part of women, and by dint of its
complexity demands that women pay an excessive amount of attention to
what their bodies are doing. Being an infertility patient is about as stressful as
being an air-traffic controller. (In a certain sense the word "patient" is inapt,

since infertility is not an illness; but insofar as the client of a physician is a patient, those seeking assisted reproduction are patients.) Meanwhile, caring friends advise, "You would probably get pregnant if you would just *relax*."

All the drugs contribute to the unpleasantness of the ride. A standard regimen for an IVF attempt, or "cycle," lasts over a month, during which time a woman has to give herself daily subcutaneous injections of leuprolide, a synthetic hormone which, in effect, entrains her reproductive system for the other drugs she will take. On day ten, if she is menstruating, she adds to this a series of intramuscular injections—generally deep into the buttock—of another hormone called human menopausal gonadotropin. These painful shots are usually given at home by her husband or a willing friend or family member. About seventeen to nineteen days into this routine, she receives a shot of yet another hormone, human chorionic gonadotropin, to induce ovulation. Afterward, she continues with the human menopausal gonadotropin for two more days. A few days later, she begins taking shots of progesterone. These hormones can leave her weepy or on edge, depressed or unable to concentrate; she may experience hot flashes, exhaustion: all on top of an already precarious emotional state.

During this entire period, she will also have made multiple trips to the clinic for urine and blood tests, and for ultrasound scans of her ovaries, perhaps accompanied by her husband, perhaps not, depending on how involved or available he is. Since all this to-and-froing practically constitutes a full-time endeavor, a woman who works has a lot of juggling and apologizing to do and, if she is trying to hide the fact that she is undergoing infertility treatment, a lot of lying.

As close as possible to 34.5 hours after receiving her dose of human chorionic gonadotropin, the woman goes in to have her eggs "retrieved" under general anesthesia. If the physician succeeds in sucking any eggs from her ovaries, they are then fertilized in a petri dish, allowed to grow for three days, then transferred in yet another procedure, which is quick and relatively painless and done without anesthetic, but which a woman may approach with more trepidation than anything else: Here is the ultimate test of her fertility, and the (possibly fruitless) culmination of all her efforts. Two weeks later, the woman goes to the clinic for a pregnancy test, and if the answer is negative, she will in all likelihood be thrown into a tailspin of sorrow, frustration ("Why won't my body do what it's supposed to do?"), and anger. Confessed one patient to Mahlstedt, "My husband told me he hated what the past few years had done to me. He said he watched me turn into an angry, bitter, hateful person."[3] Sometimes the anger is turned on the physician, sometimes on that other convenient target, the spouse.

Marriages can be destroyed by the rage and recriminations of one or both partners. Since 80 percent of those undergoing assisted reproduction fail to go home with a baby, opportunities for blaming are rife. Psychologists have also found that coercion among couples is common. Linda Hammer Burns, a psychologist and assistant professor at the University of Minnesota Medical School, tells the story of one husband, whose wife balked at going through a ninth IVF cycle. Says Burns, "The husband said, 'I'll divorce you and get someone younger.'"

Other couples negotiate rather than deliver ultimatums. Sociologists Judith Lorber and Lakshmi Bandlamudi did a fascinating study in 1993 of how this dynamic works when men are infertile. Although they interviewed only nine couples and three wives who were undergoing IVF for male infertility, their findings tend to jibe with tales told by a broad spectrum of counselors. Because many men want a biological heir, they resist artificial insemination by donor, despite the fact that it is relatively simple, comparatively inexpensive, and requires no heroic effort on the part of the woman. They prefer IVF because it "provides a technological means for a man who has low sperm count, poor sperm motility, or badly shaped sperm to impregnate and for his fertile partner to have *his* child."[4]

When women married to such men are themselves subfertile, they may volunteer to undergo IVF, in effect shifting the physical and emotional burden entirely onto themselves. Oftentimes, they view themselves as infertile, when in fact they might conceive easily were their husbands' sperm up to par. Three of the women in the group Lorber and Bandlamudi interviewed had already borne children with other men, and undertook IVF for altruistic reasons. One husband, explaining why the couple had sought IVF, said, "Well, for myself—to have a biological child. I do not have one. I would really enjoy that—the whole experience of going through the pregnancy, . . . the whole path of it, which is different from adoption."

Three men had pressured their wives into IVF. Reported one wife who had gone through five years of unsuccessful treatments, "It is an egocentric thing—he wants to have a child of his own genetic make-up with his own DNA. I think he wants to have a child with his ears or whatever, and I would love to have a child with his ears and his freckles . . . you know, my eyes . . . whatever."[5] Still, she was exhausted and had lost the will to go any further.

One of the most perverse aspects of the assisted-reproduction ride is that no matter the pain, aggravation, and distress involved, most people who climb on have a hard time getting off. One reproductive endocrinologist told me about a fifty-one-year-old woman who had been through multiple IVF attempts and refused to give up, even though he had repeatedly told her that she stood little chance of conceiving unless she opted for what is known in the trade as "donor egg." This procedure employs straight IVF technology, but instead of a woman's own eggs, those of a "donor" are used—sometimes a family member, sometimes a woman recruited by the clinic. Donors generally receive a payment of about two thousand dollars. Fertility specialists maintain, because the issue of commodifying body parts is so incendiary, that this is not the purchase price of eggs, rather compensation for the women's trouble: Indeed, egg donors submit to most of the drug rigmarole that an IVF patient must go through. Egg donors enlisted by clinics tend to be younger than recipients, and for women over forty, going this route increases their chances of conceiving, which are otherwise virtually nil. The fifty-one-year-old, though, insisted on IVF with her own eggs. Eventually, physicians with ethical compunctions simply deny treatment to such patients, but it may take a while.

Given the wide array of psychological complications that accompany assisted reproduction, one would think that counseling would constitute a major element of any infertility practice. However, physicians in the field have on the whole exhibited reluctance to involve mental health professionals. The first inroads were made by academics interested in profiling those couples who opted to pursue IVF when it was still highly experimental (a good case can be made that it remains experimental, only somewhat less so than originally). This broad curiosity gave way to efforts to assess the psychological impacts of assisted reproduction. In 1984, a small, self-selected cadre of counselors began attending the annual American Fertility Society meetings, and eventually lobbied for standing in the organization. Today, while Britain requires, through its Human Fertilisation and Embryology Act, that infertility centers provide prospective and active patients with counseling, the United States mandates only that physicians obtain informed consent from those undergoing assisted reproduction. (Informed consent is a legacy of the Nuremberg Tribunal, and its essence is the moral precept that people should not be involuntarily subjected to medical treatment.) Many physicians and clinics practicing assisted reproduction do not, as a consequence, offer counseling on a routine basis, but merely on an "as needed" basis: which is to say that someone has first to notice that a woman, or a couple, is having problems. Whether or not this happens would seem to depend in large measure on the size of the infertility program. At the largest centers in the country, the overall level of anxiety among patients is so high that it's hard to know how anyone could be expected to notice when patients cross that line into neediness, especially since they may, indeed, have needed counseling from the start. Some critics of assisted reproduction have argued that far more helpful to the worst-case couples who enter clinics in search of technological solutions would be to help them deal constructively with the profound sense of loss they are experiencing, to help them grieve and get on with their lives, instead of spending two, five, or ten years in a pursuit almost guaranteed to exacerbate their woes.

In the United States, the demographic profile of patients who resort to assisted reproduction is of the sort that makes marketers drool. Whereas the national health programs of, say, Britain and Australia cover a certain number of IVF attempts, only six states in the United States—Arkansas, Hawaii, Illinois, Maryland, Massachusetts, and Rhode Island—require insurance companies to cover this and other "high-tech" procedures. Consequently, only those able to afford the $6,000 to $12,000 per cycle (prices for the more elaborate procedures, like ICSI, are highest), payable in cash ahead of time, have access. A 1994 analysis by statisticians Anjani Chandra and William Mosher of the National Center for Health Statistics revealed that "users of any infertility service are more likely to be non-Hispanic white, college educated, higher income, older than thirty," without children, and married at least once.[6] Other figures show that of the roughly 2.5 million couples seeking treatment in 1994, only about 40,000 attempted IVF, ZIFT, or GIFT. According to the Society for Assisted Reproductive Technology, an arm of the ASRM to which some 300 American and Canadian clinics voluntarily report, 41,209 women underwent these procedures in 1993, and 8,741 gave birth. In baseball, a .212 batting average could get you benched, cut, or sent back to the minors. In any other area of medicine, a one in five success rate for a new therapy or surgical procedure would generally not be considered good enough to warrant its widespread adoption.[7]

The savviest shoppers for reproductive services know the failure rates. They can recite them to you, chapter and verse. They know, too, the players; have researched the programs, sometimes by plowing through the foot-high "Clinic-Specific Outcome Assessment" compiled by the Society for Assisted Reproductive Technology. They have read the consumer guides, like *New Options for Fertility,* the comprehensive handbook by physicians Arthur Wisot and David Meldrum, and even some of the ancillary literature, like journalist Anne Taylor Fleming's *Motherhood Deferred,* the sad story of a decade spent in and out of fertility clinics, interwoven with a personal history of the women's movement. Frequently, women second-guess their physicians. Shouldn't the hormone dose be lower? Didn't this or that sign seem to require this action as opposed to that?

If these patients are so smart, why do they hazard an enterprise so bound to fail? There seems to be one primary reason: They do not simply want to have children, they want to have children who have some biological attachment to them. That is, women want a child who has spent nine months in their womb, and couples want a child who is the product of their own gametes—that is, their own sperm and eggs.

Ask couples undergoing assisted reproduction why they simply do not adopt, and you will hear plenty of excuses: Babies are hard to find; adopting over forty is impossible; adopted children never really adjust; the birth mother might demand the child back. One couple told me, "We don't want some eighteen-year-old changing her mind." Indeed, any or all of these may be true, but listen long enough and you will hear another reason which amounts to a refrain: We want our *own* children.

On the face of it, this might seem an atavistic need, and sociobiologists especially will tell you that everything humans do is designed to address this need. But that argument doesn't bear up under the most rudimentary scrutiny. People have, throughout history, shown themselves more than able—prone, in fact—to engage in actions which do little to maximize their evolutionary fitness, that is, to get as many of their genes into the next generation as possible. War, for example, which in modern times has consumed hundreds of millions of lives, is hardly a good way of accomplishing that supposedly paramount goal, and yet whole societies have been centered on the enterprise of war, and generations of men have viewed soldiering, especially soldiering in battle, as the sine qua non of manhood. Take it in the chest with a mortar at nineteen and you are not going to sow many genes into perpetuity. Fitness, except in the treadmill and weight-room sense, is something few people consciously aim for. Like a eulogy, it is what can be said of them after they are dead. Fitness in evolutionary terms really only has meaning over the span of multiple generations, not in the span that can be captured in a family portrait.

In many aboriginal societies, far less emphasis was placed on biological links to children. Especially in polyandrous societies, women may not even have known who the father of a child was; the child might have been raised largely by her brother. Even today, certain Pacific Basin peoples readily give children away to siblings or friends who have none, or to strangers who might take the child to Canada or Europe or the United States. They consider farming a child out beneficial not only to the child but also to the entire family or clan. As the highly respected British anthropologist Marilyn Strathern has written,

> In many cultures of the world, a child is thought to embody the relationship between its parents and the relationships its parents have with other kin. The child is thus regarded as a social being, and what is reproduced is a set of social relations.[8]

What Western couples tend to focus on, instead of a larger network of people with whom one interacts and to whom one has obligations, is "the making of persons, and specifically individual persons." Writes Strathern, "What is apparently at stake is the fate of human tissue, and what these [assisted-reproduction] techniques reproduce is parental *choice*. The child will embody the desire of its parents to have a child."[9]

Once children have been commodified in this manner, it is a short step to incorporating additional aspects of parental desire. That is, they want not just any baby (of theirs), but the best baby science can provide. A July 1992 March of Dimes Birth Defects Foundation poll of 1,000 Americans found that 43 percent approved of hypothetical prenatal genetic manipulations that would improve a baby's looks, while 42 percent liked the idea of manipulations that would boost his or her intelligence. Over a third believed that the goal of designing the "perfect" baby was a desirable one, although the poll itself did not specify what that perfection might entail.

Patrick Steptoe and Robert Edwards, the men who engineered the conception of Louise Brown in a petri dish, repeatedly told reporters that they wanted only to enable women to fulfill the simple dream of having a child. But when they elaborated their intentions in scientific forums, they revealed much more. Steptoe and Edwards and the many other researchers who raced to be among the first to achieve IVF and embryo transfer in humans had other motivations. They wanted, for the sheer mastery of it, to be able to make sperm, eggs, and embryos in the lab carry out their business as they did inside the body. To acquire knowledge about embryogeny, the origin and growth of embryos, that most essential of physiological processes, they wanted those skills. And they wanted, having initiated fertilization and maintained early embryonic growth in the lab, to sort and select embryos—for sex and for quality.

Eugenics is inherent in reproductive medicine. Those opting for artificial insemination by donor literally shop for the future genetic father of their children in catalogs, and, of course, they want the "best." Those who rely on clinics to provide them with eggs for IVF with egg donation accede to eugenic selections made by physicians, who choose young, healthy women, who frequently have already proven the "quality" of their eggs by having had children, and turn down volunteers who are overweight or reveal a history of emotional problems, including depression.

The next wave of assisted reproduction is already upon us. Jamie Grifo and others are paving the way with a technique known as preimplantation genetic screening, in which embryos are tested for defects before even being transferred. (Grifo is not only a reproductive endocrinologist but also a geneticist.) At New York Hospital–Cornell Medical Center, among the places where this method is being tried experimentally, several couples at risk of heritable genetic defects have already opted to have their embryos analyzed in this fashion prior to having them transferred, to avoid the possibility of having to abort a fetus whose CVS or amniocentesis showed that it would be affected.

Scientists want to wipe out heritable genetic disease. Preimplantation genetic diagnosis affords one approach to doing this. Germ-line gene therapy, or making direct revisions to the genes of embryos, is another. This appears to be a laudable goal, to correct diseases in vitro. Those born with the nastiest of the genetic killers—such as Huntington's chorea, alpha- and beta-thalassemia, Tay-Sachs disease, cystic fibrosis, and sickle-cell anemia, whose infamy far exceeds their incidence—suffer awfully, as do those who love and care for them. No one would willingly inflict such agony on an infant. Yet the impulse to eliminate genetic disease also partakes of eugenics, and consequently raises complicated issues concerning the agents and nature of such alterations. Who is deciding which genes to change? What might the long-term effects of such changes be? And why, when so many other debilitating ailments which affect the health and well-being of millions of children around the world can be prevented with so little investment, is medicine focusing so much talent and money on dealing with diseases that strike a (terribly unfortunate) few? As is usual in science and medicine, decisions on this score are being driven as much by ego, money, and power as by beneficence.

When we take charge of the human genome, the chromosomal archive of our species, what exactly are we taking charge of? What do we intend to do with it? And why? What guarantees that medical geneticists' efforts to tinker with the species will prove any more successful than, say, those by biologists to concoct artificial wetlands or self-regulating biospheres intended to function as smoothly as nature itself? To date they have been botches all, our attempts to wield the wand of creation: Faked ecosystems invariably fail to perform as intended, requiring inputs of energy to maintain equilibrium or lapsing into irreversible decline. How then will it go with the genes?

There is, in truth, nothing new in the desire to perfect the species. Its roots lie, seemingly, deep in prehistory. Improbably enough, the manipulations going on in the embryology lab and medical clinic today can be seen as the culmination of a long sweep of events beginning some ten thousand years ago, when one of our hominid forebears abandoned a newborn child somewhere in the wilds and left it to die.

2
They Kill
Their Young

I f you were to take a map of New York and environs and draw a line east to
west, roughly dividing the greater metropolitan area in two, the number of
people living in each half would total about 10 million. (Los Angeles or Mexico City or Tokyo-Yokohama would serve equally well for this exercise, since all
have populations approaching 20 million.) Now, if you were to scatter those 10
million people from our demi–New York (or L.A. or Mexico City or Tokyo-
Yokohama) around the emptied-out continents of Africa, Eurasia, Australia,
and North and South America, distributing them in groups of no more than
fifty in any one place, you would have a fair idea of the sparse character of the
human presence on earth some ten thousand years ago, when, in certain
areas, farming as a way of life had begun to supplant the foraging and hunting
which had sustained the *Homo* line for a couple of million years.

If by some quirk of the time-space continuum you were able to tour
around this thinly populated planet of the Neolithic period and visit an
assortment of its two hundred thousand or more bands of *Homo sapiens,* you
would find that these peoples displayed a variety of settlement patterns, social
strictures, rituals, and yearly rounds of activity, but shared a few simple technologies—the manufacture of bladed stone tools prime among them. Having
compared and contrasted the habits of a substantial portion of these subjects,
you would likely be forced to conclude that they also bore something else in
common: a proclivity for killing off their young, for committing infanticide.

Virtually all animals, from spiders to spider monkeys, are known to kill
their offspring. Invertebrates, often lacking the faculty to distinguish kith and
kin, may wreak as much or more havoc upon their broods as predators or the

harrowing elements. Among creatures with spines, adult fish regularly gobble up their own spawn, and birds of many species routinely ignore the youngest hatchling at feeding time and will not usually interfere when older siblings push weaker nest mates out to certain death.

Profoundly social creatures that most mammals are, they engage in infanticide within a communal setting, under the rubric of dominance and submission that so thoroughly characterizes their sexual relationships. Male and female rivalries flair, putting youngsters' lives at risk. It is not uncommon for prides of lions to be raided by outside males, who, if they succeed in their coup, promptly slaughter cubs sired by their predecessors. A young silverback gorilla who moves in on an older male's troop and succeeds in wresting away control will often go after the departed silverback's youngest offspring and kill them—if he can pry them from their mothers. Wild canids, males and females alike, stage hit-and-run attacks on other packs, cutting down pups. And within groups of both canines and primates, dominant females have been known to destroy a lesser female's newborn litter, especially if the dominant female has herself recently whelped.

Humans, too, have a heritage of exterminating their own young. The historian William Langer has defined infanticide as "the wilful destruction of an infant through exposure, starvation, strangulation, smothering, poisoning, or through the use of some lethal instrument or weapon," and it is probably true that the usual victims have been newborns.[1] But under certain circumstances toddlers have been killed, and older children have effectively been murdered through repeated physical abuse as well as by insufficient feeding and other forms of profound neglect. Justifications for these acts have differed according to time and place but have often hinged upon worries about food shortages; or fears of bad luck brought on, say, by the birth of twins; or sheer incapacity to care for another child when family size has already overtaxed parents' economic and energetic means.

Although frequently committed in secret, sometimes indistinguishable from natural attrition, and often discussed only within the confines of the family or immediate social group, infanticide has been recognized as universal. In 1922, British historian Alexander Carr-Saunders made a thorough survey of anthropological literature and discovered accounts of infanticide for a significant fraction of what were then termed the "primitive races." Several subsequent studies, carried out after World War II and drawing on a comprehensive archive of ethnographic material known as the Human Resources Area Files, have fully supported Carr-Saunders's findings. Although infanticide was unusual in Africa and appears to have been strictly proscribed by the Vedda of Sri Lanka, the Andamanese islanders living off the northern tip of Sumatra, and by certain California tribes, it otherwise played a role in the life of all peoples who, at the time of first European contact or thereafter, maintained ancient patterns of existence as hunter-gatherers, nomads, pastoralists, or subsistence farmers. The custom prevailed in Australia and Tasmania, throughout Oceania, in India, China, and Japan, across Siberia and the central Asian steppes, and from the farthest reaches of North America on down into South America. Some Victorian observers registered this as yet another reason to morally condemn the savagery and barbarism of those "primitive races," conveniently forgetting that Europe and the Mediterranean region had their own legacy of baby killing with which to reckon. In short, infanticide has been practiced "on every continent and by people on every level of cultural complexity, from hunters and gatherers to high civilizations, including our own ancestors. Rather than being an exception . . . it has been the rule," writes anthropologist Laila Williamson.[2] Indeed, only within the last century have the rise of more effective contraceptives and medically safe abortions helped minimize unwanted births and probably thereby lessened the occurrence of infanticide around the world.

The prevalence of infanticide within historical memory points to its ancient roots. Archaeologists have concluded that our species has engaged in this practice at least since the Neolithic period (formerly referred to as the New Stone Age, and stretching from 8000 B.C. at the earliest to 3000 B.C.). Carr-Saunders was among the earliest to remark upon the implications of digs in his own country. He wrote:

> There is evidence from graves that infanticide was practised in Neolithic times in England. It is thus of particular interest to notice that, where we can catch sight of peoples emerging out of the prehistoric period, we find infanticide established as a practice.[3]

Anthropologists who have taken aboriginal groups as a window both into the Neolithic period and into the preceding Paleolithic period (Old Stone Age) have argued that infanticide constitutes the oldest form of birth control and that it served to limit the growth of hominid populations for hundreds of thousands of years. To be sure, scholars are at a loss to glean solid proof of infanticide from the farthest reaches of prehistory, but many assume that the custom of culling the youngest, weakest members from a band has extremely remote origins.

The Paleolithic era spanned some 2.5 million years. Although it is difficult to comprehend given the incessant innovation which is our contemporary norm, technological change was effectively nil throughout almost this entire time. At the beginning of the period, the earliest exemplars of the *Homo* line roamed eastern Africa and, starting around 1 million years ago, our immediate ancestors, classified as *Homo erectus*, fanned out into Eurasia.

These protohumans survived by gathering nuts, roots, seeds, leafy plants, fruits, and berries, and by hunting and fishing. Their basic shelter was an open-air camp or a cave. Their basic tool kit consisted of crude stone choppers, scrapers, and mashers; not until around forty thousand years ago, with the advent of *Homo sapiens sapiens,* did bone implements, shafted spears, and fishhooks supplement these.

Archaeologists have discovered skeletal remains of Paleolithic hominids in far-flung spots, away from any camp, as if death had come during a hunting or trading expedition. Occasionally, the bones of as many as a dozen individuals have been found grouped together. Sometimes human and animal bones are strewn randomly together, cut and broken open in such similar fashion that archaeologists suspect cannibalism took place, as, for example, at the Krapina rock shelter in northern Croatia. Archaeologist John Wymer has described what was unearthed there:

> In one archaeological layer . . . were the remains of a large fire in which were found almost nothing but human skulls, broken and burnt. Deposits of ash, charcoal and other burnt human bones show that such feasts took place on many occasions. Long bones were found . . . split for their entire length, presumably for the extraction of marrow.[4]

Although such interpretations, based on careful microscopic studies of marks on the bones, have been disputed, they are still supported by experts in this type of analysis, including Pat Shipman, formerly of Johns Hopkins University.

Throughout most of the Paleolithic period, the dead received no more than a cursory burial, if they were buried at all. True interments do not begin to show up until around 50,000 B.C., among *Homo sapiens neanderthalensis.* This, in part, accounts for the paucity of whole skeletons dating from the Paleolithic. In most climates, microbes begin to do their decompositional work on dead bodies almost immediately. Generally, only where bones are covered, intentionally or by accident—as, say, when the roof of a cave collapses—do they stand a chance of being preserved. Encased in soil, bones are exposed to groundwater, which leaches minerals out, leaving a fine, friable web. Then, in certain soils under certain conditions, the chemical magic occurs whereby these delicate structures gain greater permanence.[5] Overwhelmingly more bones sit in the ground in a precarious, highly perishable state—or vanish altogether—than are transformed into fossils.

As a consequence, archaeologists generally assume in dealing with the cache of remains from the Paleolithic period that they are not looking at a representative cross section of the primordial population. Nevertheless, some curious patterns, which hint at truths concerning that population, emerge. For one thing, there is the presence of so many babies and young children in the fossil record. By rights, their slender frames and delicate skulls, which turn to dust far more readily than those of thicker-boned adults, should show up hardly at all. Yet the bones of infants make up a relatively large fraction of the skeletal material dating from this time. In 1961, Henri Vallois, then an éminence grise of paleontology and director of the Institut de Paléontologie Humaine in Paris, surveyed a large sample of Paleolithic and Mesolithic skeletons unearthed at locations in Europe, around the Mediterranean, and in central Asia. Of 175 individuals found in group burials, about 35 percent were infants.[6] That such a proportion of delicate remains litter the Paleolithic record suggests that infants perished in rather large numbers.

Could their deaths be put down to the rigors of prehistoric existence? In the modern world, infant mortality is typically highest where women are poorly nourished both during pregnancy and after giving birth, and where unsanitary conditions contribute to the spread of diarrheal diseases, which are brutally efficient at killing infants through dehydration. But many anthropologists are convinced that while Paleolithic lives were indeed short, they were not necessarily nasty or brutish. Apparently, early hominids led a rather healthy existence, plagued by few of the infectious diseases, diet-related maladies, and chronic ailments common to postagricultural life. According to demographer Thomas McKeown,

> the rise of airborne disease, including notably the respiratory infections, probably dates from the time when human populations first aggregated in groups of substantial size. This explains why infectious diseases became predominant as causes of death from the time of the first agricultural revolution.[7]

Analysis of bones reveals that prehistoric peoples ate well and had ample variety in their foodstuffs. Among hunter-gatherer tribes who still maintained their traditional ways of life into the nineteenth and early twentieth centuries, malnourishment was uncommon and starvation rare, and often, infant mortality was low. In the absence of other confounding factors, healthy women bear healthy babies. If this was also the case for Paleolithic peoples, as it likely was, then maybe at least some portion of the infants crowding those group graves were victims of infanticide.

Another key bit of support for the infanticide hypothesis comes from the observation that males make up the preponderance of skeletons dating from the Lower and Upper Paleolithic periods ("lower" because, in principal, artifacts from it lie in soil layers below those of the later, or "upper," era) and from the Mesolithic period (a term applied only to northwest Europe and ranging from about 8000 to 3000 B.C.). This disparity was of particular interest to Vallois, who also looked at the ratio of men to women in a large sequence of early fossils. Of 307 Paleolithic and Mesolithic individuals included in his study, males outnumbered females 172 to 135. In the subset of strictly Paleolithic remains, there were 73 men and 49 women.

Using this pair of figures, Vallois calculated a Paleolithic sex ratio of 148. The sex ratio is the product of the number of males divided by the number of females, and is conventionally stated as an integer—so, for example, Vallois divided 73 men by 49 women to get a result of 1.48, or a sex ratio of 148. This handy measure features in many scientific studies of populations, Charles Darwin being among the first to devote consideration to it. In *The Descent of Man*, he made a cursory survey of aboriginal sex ratios, and wondered whether the effects of a prolonged practice of infanticide might have contributed in some populations to pronounced imbalance in the number of males born, presuming that "the tendency to produce either sex would be inherited like almost every other peculiarity, for instance, that of producing twins." However, Darwin concluded that "the whole problem is so intricate that it is safer to leave its solution for the future."[8]

As it turns out, obtaining the sex ratio directly from birth data does give a baseline reading of a species' natural tendency to produce greater or fewer numbers of one or the other gender. Among contemporary humans, *Homo sapiens sapiens*, who are taken as representative of *Homo sapiens* in general, the sex ratio has been shown to hover consistently around 105 at birth, meaning that 105 boys are born for every 100 girls.

If the sex ratio is recalculated a few years later, the balance between boys and girls evens out due to the inherently higher mortality rate among male infants (who appear to have a tougher time of it even from conception, since a greater percentage of male embryos are also spontaneously aborted or delivered stillborn). Wherever the childhood sex ratio does not drop but instead slides significantly above 105, it piques interest, because this means that females are dying at a greater rate than males, an anomaly which invariably provokes a suspicion that girls are being systematically eliminated. Vallois's calculation of a primordial sex ratio of 148 was thus more than adequate to raise eyebrows.

Did Vallois discover a genuine characteristic of Paleolithic populations? He himself fretted over the validity of his finding, wondering whether miscategorizations had occurred, whether "some female individuals may have been considered as male." But he immediately dismissed his doubts, writing, "It is hard to believe that such an error may have been made often enough by the anthropologists to produce a sex ratio of 148!"[9] Moreover, the collections of bones he had analyzed were heterogeneous and had been deposited at far-flung places over a wide range of time under very different conditions, "the earliest of [the finds] and at least a part of the others not even being burials."[10] With such a mixed bag, there was no reason to expect consistency. Yet the skewed sex ratio showed up even when Vallois considered the skeletons from each site separately.

Shortly after Vallois published his ponderings, additional clues concerning prehistoric demographics emerged. In April 1966, a symposium on primitive humans was held in Chicago. The four-day meeting, which in retrospect would prove epochal and influence research for decades, drew seventy-five scholars, among them Joseph Birdsell, an American anthropologist who had carried out fieldwork in Australia beginning in the late 1930s. Early on in his studies, Birdsell had noted a direct relationship between annual rainfall, which can be measured in fractions of an inch in some regions of Australia's vast interior desert, and population densities of Aborigines.

Aborigines descend from bands of humans that reached the continent by water as early as 30,000 B.C. By the time Birdsell visited Australia, they comprised some fifteen major, and hundreds of subsidiary, tribes concentrated along the northern and southeastern coasts and in the vicinity of the Macdonnell Ranges at the center of the continent. Exploring the hypothesis that the ranges of hunter-gatherer tribes correlated inversely to mean annual rainfall, Birdsell confirmed that Aboriginal territories were smallest along the coasts, where rain was most plentiful, and expanded in those regions where rainfall was scarcest. Even so, Birdsell supposed that the number of Aborigines probably never topped one person per square mile, and postulated that throughout most of the time they had occupied Australia, Aborigines had lived in bands of no more than twenty-five.

Birdsell's work on population densities led him to attempt to reconstruct the sex ratio for Aborigines prior to British colonization in 1788. Although possessing a rich array of symbolic visual images, the Aborigines employed no alphabet and were instead supreme practitioners of the oral tradition. Until the past decade or so, when creeping Westernization all but obliterated their venerable ways, adult men could recite tribal genealogies, extending back for many generations. Recording some 194 of these, Birdsell was able to calculate from them a precolonial sex ratio, which at adulthood turned out to be 150.

To Birdsell, this figure broached only one interpretation: Aborigines must have routinely practiced female infanticide. Presuming that a small fraction of girls were lost to disease or accident, he estimated that year in and year out, 15 percent of female infants must have been killed. This projection into the recent past of his subjects hardly seemed far-fetched, given that Aborigine parents of the modern day admitted that they would slay infants who were born too soon after a sibling, offering as a rationale that a woman saddled—literally—with two children could not grub for roots or gather plants very efficiently. Aborigines had been reported to destroy infants if their mothers died in childbirth or if food was in short supply, as, for example, after prolonged drought. They may have, on rare occasions, under extreme duress, exploited infants as a food source.[11]

Taking his analysis one step further, Birdsell extrapolated from the data to the distant past. Just as Aborigines had maintained themselves at a fairly stable population level for thousands of years, so, too, the anthropologist suggested, had hominids throughout the Pleistocene epoch. (This term, borrowed from geology, refers to a period commencing 1.6 million years ago. During the Pleistocene, ice sheets repeatedly advanced across the continents and shrank, until glaciers finally retreated to the polar and alpine regions at the end of the last ice age, ten thousand years ago, which is designated the beginning of the Holocene epoch in which we live today.) In a paper titled "Some Predictions for the Pleistocene Based on Equilibrium Systems among Recent Hunter-Gatherers," which was published in *Man the Hunter*, the volume that emerged from the Chicago symposium, Birdsell argued that

> systematic infanticide has been a necessary procedure for spacing human children, presumably beginning after man's entry into the niche of bipedalism, and lasting until the development of advanced agriculture. It involved between 15 and 50 per cent of the total number of births. Among recent hunters it tends to be preferentially female in character and probably was in the Pleistocene.[12]

Birdsell believed that archaic humans were well aware of the limitations of the environments in which they lived, and consequently practiced population control to ensure that their numbers did not overtax the resource base. They were sophisticated enough, he thought, to have recognized that eliminating female infants was vital to keeping population down. The mathematical logic is not difficult to grasp: If there are ten females, they can simultaneously produce ten children through the agency of just one male; however, ten males cannot father ten children off one woman, except over a period of almost ten years. Thus, a society desiring to limit its numbers might well settle upon the strategy of routinely killing a certain proportion of females at birth.[13]

Birdsell had thus arrived at that great puzzle of prehistory, that is, why the *Homo* line—the modern representatives of which have demonstrated an enormous capacity for reproducing themselves, increasing their numbers 500-fold in just 10,000 years—maintained such a low rate of population growth, perhaps as low as .001 percent, for 100 or so millennia. Birdsell estimated that it would have taken a mere 2,202 years for ancestral Aborigines to fill their continent, attaining a population of 300,000.[14] Thus, it would seem that the world should have grown far more crowded in a shorter time than it apparently did.

What kept the human race in check? Birdsell argued that a proclivity for infanticide was responsible. Other anthropologists have contended that in addition to environmental onslaughts like disease, famine, and predation by wild animals, other factors, including intertribal clashes, murder within groups, and various intercourse taboos, would have dampened population growth. Still other investigators have suggested that among some prehistoric peoples, long periods of nursing—up to two or three years, as in some aboriginal cultures—would have prevented conception and kept fertility, and thus population increases, low. However, prolonged lactation only diminishes a woman's ability to conceive for so long, and in no way serves as an absolute bar to pregnancy.

A few researchers have postulated that another physiological regulator, low body fat, might more efficiently retard births. In certain parts of the prehistoric world, dietary intake, while adequate for survival, may have been insufficient for women to build up the fat stores needed to bring on and sustain menstruation. Thus, births would have been tied closely to nutrition, with women conceiving only when their percentage of body fat passed a certain critical level. Some scholars have suggested that a prehistoric recognition of the link between fat and fecundity might explain those fleshy prehistoric fetishes like the so-called Venuses of Willendorf, Lespugne, and Laussel, with their huge, pouching bellies and breasts, and rotund buttocks.[15] As has often been assumed, these figurines may well have been fertility goddesses. If, indeed, they were, the fact that they have been recovered at sites along the entire spine of Eurasia, from the Pyrenees in the west to Lake Baikal in the east, could indicate that birth was problematic enough that the desire for offspring warranted pleas for the intervention of higher powers.

Even considering such alternative explanations, a strong faction within the anthropological community continues to support the view that infanticide played a role in moderating primitive growth patterns. A final bit of evidence comes from the fact that at a certain point, coinciding with the spread of the agricultural way of life, population began spiraling upward, and has not yet stopped. In his popular 1977 inquiry into the origins of cultures, *Cannibals and Kings*, anthropologist Marvin Harris speculated that with the shift to agriculture, food production went up, rendering infanticide unnecessary. This assumes both that the primary cause of infanticide was lack of food and that agriculture bettered nutritional standards of living across the board, when it is now clear that it may not have, at least not for all peoples.

Other scholars, while also tying agriculture to a diminishment in infanticide, suggest that the practice faded due to changes in patterns of labor. Hunter-gatherers migrated seasonally in search of game or other food, often splitting up into smaller units—say, a husband and wife and their children, who spent months on their own foraging over a wide area before returning to a home base and reuniting with the rest of the band. Agriculturalists, on the other hand, tended to stay put, and might not have found it as difficult to care for extra young ones. They might accordingly have dispensed to a degree with the old lethal methods of ensuring adequate spacing between siblings, allowing more children to live.[16]

To gain insights into how prehistoric peoples might have regarded the practice of infanticide, anthropologists have often looked to hunter-gatherer groups. This is a dangerous intellectual enterprise, as British archaeologist R. J. C. Atkinson, known most widely for his work on Stonehenge, was wont to warn. Atkinson long made a habit of criticizing researchers who attempted to delve into the motives and mentality of Neolithic man. Most of them were, he declared, "just bombinating around in the dark."[17] This because the book of prehistory is, by definition, written in bones and stone, in tools and caves and campsites, not in words. And without language, there is just so much we are able to divine before running up against an impenetrable barrier of silence, of blankness. What Atkinson objected to was the tendency, in attempting to fill in the gaps, to spin fantasy images of early hominids. We likely err at both extremes, when we project onto them our perceptions, capacities, fears, and ambitions, and when we see them as incontrovertibly alien or "lower."[18]

The best anthropologists bear this in mind. As long as the habits of the Tikopia or Yanomano are not taken as representing an absolute correlative for those of archaic peoples, all but the most doctrinally pure feel fairly comfortable with this approach. Oblique though it may be, it has yielded evocative information concerning why and wherefore infanticide occurred.

What of the notion that a primary reason for prehistoric infanticide would have been to achieve a desired spacing between siblings? It does turn out that many hunter-gatherers, like Australian Aborigines, relied on infanticide for this purpose, using it as an ex post facto form of birth control.[19] Some groups, as Birdsell suggested, may have had a fine-tuned awareness of the carrying capacity of the land in which they lived, and may have understood from long experience that if their numbers exceeded a certain level, hard times would follow. More likely, the practice arose out of the need to cope with straitened circumstances and possibly oftentimes involved a perceived trade-off between eliminating the infant and keeping one or more older children alive.[20]

In *Anjea*, his classic work on the subject of infanticide, anthropologist Herbert Aptekar argued that the decision to snuff out a new life must generally have been "a reaction by the individual to an immediate, pressing problem, namely, shortage of food," rather than the result of an "abstract formulation of a population policy."[21]

This seems usually to have been the case; however, it appears that some aboriginal peoples, particularly in Oceania, did set an absolute cutoff and dispense with all infants over and above that amount. Certainly, many groups adopted specific policies regarding excess females.[22] If a band wished to ensure that every one of its members could find a mate without looking outside the group, it might selectively cull females. But probably more often in play was the perception that girls carried less intrinsic worth—economically and otherwise—than boys. Darwin, in *The Descent of Man*, quoted the "apparently trustworthy writer, Mr. Jarves," who published a history of the Sandwich Islands in 1843. Jarves remarked, "Numbers of women are to be found, who confess to the murder of from three to six or eight children," and adds that "females from being considered less useful than males were more often destroyed."[23] Most Aleut and Eskimo peoples also held this bias. The Danish voyager Knud Rasmussen, who made many accurate ethnographic observations during a swing from Greenland to Alaska in the 1920s, reported that the Netsilik, who lived above the Arctic Circle west of Baffin Bay, wiped out more than one-third of girls delivered. A Netsilik elder told him, "Parents often consider that they cannot afford to waste several years nursing a girl. We get old so quickly and so we must be quick and get a son."[24]

Other, subtler forms of discrimination operated as well, such as the failure to nurse girls as religiously or for as many months as boys. In many cultures, girls, once weaned, received a poorer diet, lacking in body-building proteins and fat. Neglect and careless handling, too, were the sort of sins of omission that would have taken a toll on female infants.[25]

Obviously, even aboriginal groups ill-disposed toward females did not kill all girls—that would have been collective suicide. The willingness of a mother to fight for the child could have made a critical difference. The anthropologist Mildred Dickemann tells the story of the Papua New Guinea mother who, instructed by her husband to quash the life of their newborn if it were a girl, refused, and gave her daughter a name that would be a constant reminder of that defiance, Letahulozo, which Dickemann translates as "Break It and Throw It Away."[26] But in any case, whether a few women succeeded in contravening tribal customs or whether a tribe put only a small fraction or as many as half of all girl babies to death appears rather immaterial, given that no culture ever reported that it preferentially eradicated *boys*.

Of course, under extreme circumstances, boys might be slain or abandoned along with girls, as when, for instance, an invasion by enemies forced a tribe to beat a retreat. But such exceptional events are said to have caused regrets, as among the Yaudapu Enga of New Guinea, who memorialized children lost in such fashion by imagining that they had been adopted by supernatural beings.

Infanticide, it appears, was prompted by a whole range of concerns, and served many ends. As Aptekar pointed out, infanticide and abortion (which was also practiced by aboriginal peoples, although it posed great risks to the pregnant woman), "are at once something more and something less than 'population controls.' They are moving cultural forces capable of being sustained by diverse cultural elements, and adaptable to an infinite variety of uses."[27] Rather than pathological, infanticide was expedient. Above all, it enabled people to choose how many and which offspring would, having entered the world, be granted the possibility of living out their appointed round of years.

Almost without exception, aboriginal peoples destroyed newborns displaying any signs of physical abnormality. This would, in the main, have been compelled by practical considerations and possibly also by compassion. Aboriginal peoples had a sophisticated understanding of plants and their medicinal properties, but lacked advanced surgical skills. Experience must have taught them that severely impaired newborns did not live very long, and faced with such a birth, they acted swiftly. This was true from the Pacific to the Arctic. During a South Seas sojourn, the author Robert Louis Stevenson remarked that Polynesians routinely killed off disabled or diseased newborns, while early twentieth-century explorers noted that Greenlanders consigned flawed infants to the scythes of wind and ice. Anthropologist Williamson has written that "the reasons for eugenic infanticide seem obvious: unwillingness or inability to assume the burden of caring for such an infant, whose future at best would be unsure."[28]

Whether revulsion, superstition, or a programmatic philosophy also played a role we cannot be certain. However, many tribes did freight birth with heavy symbolism, so that the season or day upon which a child was born as well as the details of his or her birth (such as whether the mother's labor was long or short, and whether the baby's presentation was breech or normal) were thought to determine in part or whole the child's nature or fate. Frequently, mothers drew the blame if a baby was abnormal, and among some groups, if a woman had herself come to exhibit mental or physical aberrations, her infants would be destroyed.

A birth that was out of the ordinary, especially of twins, generally raised fears. A handful of early peoples embraced twins as potent talismans of good luck, including the Mojave Indians of the American Southwest, the Ashanti and Shilluk of Africa, and the Jibaro of Ecuador; the Papago Indians even believed them to carry shamanic powers. But far more cultures viewed twins with dismay or disgust, and obliterated them. A few groups spared one, while choosing to eliminate the weaker, younger, or female twin. Some tribes believed that only intercourse with evil spirits—or with a man not her husband—could fill a woman's womb with two offspring. Williamson observed that Trobriand Islanders likened the mother of twins to the animals and ridiculed her, as did the African Ibo, who found such births repugnant.

Indeed, any conception following intercourse not deemed acceptable by the group might be treated with severity. Premarital, adulterous, or incestuous unions—or even stormy marriages—could guarantee a death sentence for any resulting offspring, as could relations with members of outlying or rival clans or tribes.[29] Attempts might be made to abort the pregnancy, as among outmarrying Fijian women, who wished to eliminate possible claims to land by their husbands' tribes. Among many groups, what amounted to class or race barriers existed, so that a child borne out of a coupling of a low-ranking Tahitian with a high-ranking one would automatically be slain, and, among the Apache, a mixed-breed infant would be choked or discarded in the desert. In the rain forest along the Amazon River, Yanomamo men taking a previously married woman for a wife could, as one of his conditions, require her to kill any small children she might have.

Infanticide, whether undertaken as a survival tactic or as a method for affecting desired social aims, may not have raised great emotions and instead may have been looked upon with complacency. Compassion may have been subordinated to hard-nosed pragmatics, as it was on a general basis among the Hadza of Tanzania. Anthropologist James Woodburn wrote of the Hadza that they were "strikingly uncommitted to each other; what happens to other individual Hadza, even close relatives, does not really matter very much. People are often very affectionate to each other, but the affection is generally not accompanied by much sense of responsibility. If someone becomes ill, he is likely to be tended only so long as this is convenient."[30]

When it came to infants, emotional distance such as that displayed by the Hadza was made easier through categorical estrangement: Often, babies were assigned a nonhuman or subhuman status for a period of days, or years. Several Andean tribes, for instance, would not formally acknowledge infants until they had survived past a certain stage. Among the Machigenga, babies became human after their first feeding, but the Amahuaca of Peru waited three years before accepting a child. Among many aboriginal peoples, a ceremony in which a baby received its name often formally signaled its transformation from a mere life-form to a person; before then, it could be killed with impunity, because it had no standing in the ranks of the human community.

On the other hand, as Marvin Harris has implied, there must have been psychological costs to the killing or fatal neglect of infants. Furthermore, Harris reasons, the "material costs of nine months of pregnancy are not so easily written off. It is safe to assume that most people who practice infanticide would rather not see their infants die. But the alternatives—drastically lowering the nutritional, sexual, and health standards of the entire group—have usually been judged to be even more undesirable."[31]

B y exerting control over reproduction, ancient humans, unconsciously or not, served as agents of their own evolution. Reduced to its essence, evolution constitutes the process whereby changes occur in a gene pool over time. The mechanism producing these changes is natural selection, in which some genes (and therefore some physical forms and some behaviors) persist and others die out. As the term implies, natural selection has to do with sorting. It can be thought of as a gate erected between generations, through which only certain genes—packaged conveniently as a person or a panda—pass. Those genes moving from one generation into the next are the winners, and the individuals bearing them are, by the terms of the theory, fit.

Fitness has been frequently misinterpreted as referring to the quality of the organism in question, the assumption being that it is somehow better (stronger, more worthy) than any other organism on the block. However, fitness actually refers not to quality but instead to fertility: an organism's success in having passed its genes through the gate to the next generation. Those individuals who succeed in producing the greatest number of offspring—in putting the greatest number of individuals bearing their genes on the other side of the natural selection gate—are fittest. A line of organisms that wins the fitness sweepstakes by spawning large numbers of offspring, who in turn spawn large numbers of offspring, will, after multiple generations, have determined the overall character of the gene pool more than one that has failed to do so.

As Darwin realized, humans themselves may participate in the evolutionary process. They do so when breeding animals or plants, whereby they make choices about which individuals will be mated together or propagated. Contemplating the origins of domestication, Darwin wrote, "Unconscious selection in the strictest sense of the word, that is the saving of the most useful animals and the neglect or slaughter of the less useful, without any thought of the future, must have gone on occasionally from the remotest period and among the most barbarous nations."[32] Infanticide, too, can be seen as having functioned as a type of selection. While people eliminated infants for particular reasons having to do with their immediate situation, their actions over the long haul yielded changes in the gene pool. Indeed, modern humans are likely the net result of aeons of prehistoric winnowing of babies.

It is commonly argued that only with the advent of modern contraceptive devices and drugs did humans gain the ability to engage in family planning and to control population. However, aboriginal groups certainly, and primitive humans likely, employed very effective, albeit cold-blooded, means of achieving those ends. By the time that people began to gather in the first cities in Mesopotamia and devote themselves intensively to agriculture, *Homo sapiens* had long since unloosed its perfecting ambition, had already begun making eugenic decisions. In the hands of the Greeks of the classical era, that ambition would come to full intellectual flower.

3

"This Plot of Good Ground"

Sparta was a community that took sex seriously. While the strivers in Athens, to the northeast, over on the Greek mainland, erected their artful paeans to universal harmony in stone upon the Acropolis, Sparta set out to construct the perfect military society. A wall enclosing a dusty patch of ground about a half mile square, a perfunctory temple, communal barracks—these the Spartans considered sufficient to answer their immediate needs. Aside from that, architecture in and of itself held no fascination for them. Their city was spare and stripped down, and so, too, their political philosophy. Sparta had war on its mind, and sex was a weapon of war. In the midst of campaigns, several classical authors report, Spartan men would quit the field at opportune times and hurry back home to couple with their wives, convinced that the heat stirred in their blood by the fight would be imparted to their offspring, who would grow up lean and strong and battle-ready, capable of assuming the mantle of Spartan citizenship.

Lycurgus, the lawgiver and king, loomed over Sparta like Mount Taygetus there to the west, across the valley of the Evrótas River, which courses down the Peloponnesian peninsula to empty into the Mediterranean at the Gulf of Lakonikós. Although his name is all but coeval with Spartan discipline, Lycurgus is for us a figure part ghost, known only from the writings of classical Greek historians. But if the ancient accounts are to be believed, it was he who laid down the Spartan code, following oracular dictates. It was he who engineered a major redistribution of land, appropriating property for thousands of acres around and then parceling out shares to each male of the city; he who commanded that Spartan men live as *homoioi*, as equals before the law who owed, equally, fealty to the state. (Sparta was not an entirely egalitarian society: The ranks of the *homoioi* were filled by ethnic Dorians, who had invaded the Peloponnesus centuries before. The peninsula's original inhabitants, the Helots, still lived in and around Sparta, but were excluded from citizenship and functioned as laborers or slaves.)

Whether Lycurgus was a real entity or merely a legendary founder whose authority it was at times psychologically and politically convenient to invoke, Sparta by about 480 B.C. functioned akin to a tightly run army base, with its eight thousand male citizens (and probably a slightly larger number of women and children) dedicated to the pursuit of physical prowess. The system known as the *agōgē*, firmly established by then, took boys from their mothers at age six and submitted them to a fourteen-year-long regime designed to groom them into elite fighting cadres. Because today's infants would be tomorrow's soldiers, the state took a particular interest in childbearing.

Plutarch, profiling Lycurgus five or six hundred years after the fact, reported that "in order to the good education of [Spartan] youth (which, as I said before, he thought the most important and noble work of a lawgiver), he went so far back as to take into consideration their very conception and birth, by regulating marriages."[1] Plutarch claimed that Lycurgus considered children

> not so much the property of their parents as of the whole community, and, therefore, would not have his citizens beget by the first-comers, but by the best men that could be found; the laws of other nations seemed to him very absurd and inconsistent, where people would be so solicitous for their dogs and horses as to exert interest and to pay money to procure fine breeding, and yet they kept their wives shut up, to be made mothers only by themselves, who might be foolish, infirm, or diseased; as if it were not apparent that children of a bad breed would prove their bad qualities first upon those who kept and were rearing them, and well-born children, in like manner, their good qualities.[2]

Plutarch showed remarkable insight into the reasoning of a man he never interviewed, and who left no written statements behind, and his elaboration of Lycurgan policies must qualify as partly fictive.[3] But historians on the whole agree that state control of reproduction did exist in Sparta and that social relations were in large measure dictated by the desire to produce worthy heirs. Great pressure was brought to bear upon men to mate, and censure heaped upon any who reached a certain age without marrying. Wooing was dispensed with; Spartan men by tradition gained wives through force, abducting them (just as Zeus had a habit of doing with women he fancied, as, for example, Europa, whom he carried off, having assumed the form of a white bull). Afterward, domestic comforts were minimal, with married men continuing to bunk in with their fellows and connubial congress somewhat in the line of a municipal duty. Xenophon, in the fourth century B.C., noted that Lycurgus had observed how, in many cultures, newly wed men "had unlimited intercourse with their wives." This, Lycurgus realized, was not as it should be; accordingly, he declared it a disgrace for husbands to be seen frequenting their wives' quarters. Xenophon wrote that the net effect of thwarting men's access to their wives was to increase their level of desire, and their heightened passion ensured that any offspring born would be "stronger than if the couple were tired of each other."[4]

Sparta being a small town founded on the notion of communitarianism, men reportedly shared all possessions, including wives. The interests of the state apparently prevailed to such a degree that, as Plutarch intimated, husbands might cede their marital rights to facilitate the production of "well-born children." Spartans are said to have bred their wives much in the way that they might breed a prized bitch or a fine mare, selecting a stud from the ranks of the fittest young men, or handing a wife over to a valued and able-bodied friend so that she might conceive by him. Plutarch claimed that an honest man "who had love for a married woman upon account of her modesty and the well-favouredness of her children, might, without formality, beg her company of her husband, that he might raise, as it were, from this plot of good ground, worthy and well-allied children for himself."[5] (One might argue that such practices represent the earliest-known example of donor insemination—albeit effected with considerably more physical directness than in its modern version.)

While boys were being honed into fighting machines, girls submitted to state-mandated aerobics for the sake of their future matronly obligations. According to Plutarch, Lycurgus mandated a regime of exercise for Spartan maidens which included "wrestling, running, throwing the quoit, and casting the dart, to the end that the fruit they conceived might, in strong and healthy bodies, take firmer root and find better growth." In addition, the "greater vigour" they gained might enable women to better "undergo the pains of child-bearing."[6]

For girls, the compulsion to be fruitful and multiply must have been intensely felt. Certainly, women would not have welcomed the certain rejection that accompanied sterility. A woman who did not bear children would be summarily discarded, so that her former husband could take on a new, it was hoped fecund, spouse. Even a woman who succeeded in bearing children was subject to implicit judgment of her reproductive merits: No child survived in Sparta without a go-ahead from the phratry, a body whose word was law. Every newborn was carried by its father before this tribunal of military and community leaders, who inspected it for flaws. Plutarch wrote of these elders that

> their business it was carefully to view the infant, and, if they found it stout and well made, they gave order for its rearing, and allotted

to it one of the nine thousand shares of [community] land . . . for its maintenance, but, if they found it puny and ill-shaped, ordered it to be taken to what was called the Apothetae, a sort of chasm under [Mount] Taygetus; as thinking it neither for the good of the child itself, nor for the public interest, that it should be brought up, if it did not, from the outset, appear made to be healthy and vigorous.[7]

For "the good of the child itself," and in "the public interest," how many infants were tipped from cradling arms out into that "sort of chasm" at the base of Taygetus to plummet to their death?

By the fifth century B.C., thinkers throughout the classical world held that humans could be improved through breeding. Inhabitants of the northeastern Mediterranean region had accrued some five thousand years of experience with husbandry, and from generations of observation, had discerned that like breeds like; that the egg of a hen produces a chick, whereas the egg of a lizard yields a lizard. Furthermore, long experimentation with plants and animals had convinced the Greeks that qualities belonging to parents, such as looks, strength, and endurance, could be inherited. Animals conveyed physical traits to their spawn, and so, too, human parents imparted their attributes to their children. It was commonly believed that a father's "stamp" could be seen in the features of his sons and daughters; doubts about his paternity could be raised if his children's looks departed too greatly from his. Many thinkers contended that the quality of offspring was of paramount concern. If keeping the stocks of animals vital through careful regulation of mating was important, how much more so was protecting the vigor of the human line? For the Greeks, it followed axiomatically that the body politic would only be as strong as the actual bodies of its citizens. A weak people meant a weak state.

Thus, Sparta was not alone in condemning infants to death, merely more rigorous about it. Throughout ancient Greece, infanticide was seen as amply defensible. A father had an absolute prerogative to decide whether his children would live or die, expressed formally in the ceremony known as *amphidromia*. On the fifth, seventh, or tenth day after birth, a nurse carried the infant around the hearth in the presence of the father, who gave it a name or consigned it to death. Everywhere except Thebes, where infanticide apparently was a capital crime, parents took unwanted babies out into the countryside and abandoned them to the elements. This practice, called exposure, shows up in numerous Greek dramas. Characters in Menander's comedies *The Girl from Samos* and *The Arbitration* joke about exposing infants. And the plot of Sophocles' *Oedipus* turns upon an act of exposure which qualifies as the most famous literary example of the deed.

The event setting in motion the tragedy of Oedipus occurs many years before the time frame covered by the play, when an oracle warns King Laius of Thebes that his newborn son is an incipient parricide. Rather than waiting around to test the oracle's accuracy, the king opts for a preemptive strike. "When his child was but three days old, / Laius bound its feet together / and had it thrown by sure hands upon a trackless mountain," or so everyone in the royal household thinks. Those lines come from the William Butler Yeats translation of the play; another, more recent version has Laius piercing and yoking the infant's ankles together, whereupon he is given over to a royal servant to be taken out and "cast away on pathless hills."[8] But the soft-hearted servant subverts the king and secretly finds the infant a berth with a lowly shepherd family. Oedipus lives.

Sophocles picks up the action years later, when Oedipus encounters Laius at a crossroads outside Thebes. At that fateful juncture, their two chariots onrushing, the old man fails to give way and the son lashes out, exacting a lethal punishment for the old man's rude driving. But after Oedipus—having inadvertently married his mother, Queen Jocasta, and discovered his true origins—gouges out his eyes in a fit of self-blame, what classical audience could have missed the point that his tragedy followed on inevitably from his having been exposed, and thereby rendered ignorant of his parentage. Might Sophocles have been suggesting that the murder at the rural intersection had a deeper, cosmic purpose, as retribution for a primal wrong?

Certainly, this theme comes up time and again in Greek myth.[9] Here, gods and monsters are repeatedly portrayed carrying out the job of dispatching with infants and children, a task they execute with gusto until their victims turn the tables on them. The god Cronus, towering over his Titanic domains, routinely devours his children, until his son Zeus grows bold enough to destroy him. The Cretan Minotaur, half man, half bull, annually consumes an allotment of Athenian youth sent into his labyrinthine lair as tribute, until clever Theseus defeats him. One can speculate that such tales served a necessary psychological purpose, expiating to some degree the guilt society felt about its highly contingent relationship with children. In reality, the young had so little power that they could be summarily killed; in myth, they rose up to smite their parents.

But then again, just as women today may undergo abortion with reluctance and in its aftermath feel a mixture of regret and relief, so the Greeks seem to have exposed infants with some disquiet yet viewed the act as indispensable under certain circumstances, particularly in the case of deformity. Exposure partly served, along with herbal preparations to prevent conception or induce abortion, as a legitimized means of birth control.

The skilled Greek physician Hippocrates devoted a good deal of attention to women's medicine, although his knowledge of female anatomy and of the mysteries of reproduction owed as much to folklore as to careful observation. For instance, he and his followers contended that women might suffer from "hysterical suffocation" caused by shiftings of the womb, a misdiagnosis not corrected until Herophilus made careful dissections of female corpses in Alexandria around 300 B.C. Physicians in antiquity also believed that conception occurred most readily directly after menstrual bleeding. American historian John Riddle, in an exhaustively researched study of ancient contraceptive practices, has recently argued that despite such obvious gynecological misapprehensions, birth control in the ancient world was probably more reliable than has commonly been believed, and that when contraception failed, women had ready recourse to abortion. Hippocrates himself, who later commentators claimed had issued a blanket condemnation of abortion, willingly terminated pregnancies by several different means. (Hippocrates objected only to abortions that were brought about by the use of a vaginal suppository, or pessary, which could endanger the mother's health.)[10]

According to Riddle, classical midwives and physicians knew of dozens of concoctions, some derived from recipes thousands of years old, that would act as spermicides or morning-after drugs to prevent conception or to trigger miscarriages early in pregnancy. For example, prior to having intercourse, women might place a piece of pomegranate peel or a mash containing honey and certain herbs directly into the vagina. Pessaries like these altered the vaginal pH sufficiently to kill sperm. One plant, referred to in ancient medical texts as *silphium*, was used throughout the Hellenic world as a contraceptive. Probably a type of giant fennel, it grew wild in North Africa and was heavily exploited for trade. Contemporary authors report that attempts to cultivate the plant failed and that it was eventually harvested to extinction.[11]

A woman already with child had recourse to a number of methods designed to induce abortion. She might lace bathwater with linseed oil, fenugreek, and artemisia, then apply one of several pasty mixtures that would "excite" the uterus and cause it to expel its contents (the sort of procedure enjoined by Hippocrates). Various elaborate brews and teas—prime ingredients included willow, juniper berries, cabbage seeds or flowers, pepper, and ivy—appear to have been reliable abortifacients, the secrets of their preparation handed down from mother to daughter or obtained from male physicians and midwives. (Female physicians were rare throughout the classical world, sometimes banned by law. Athens, for example, prevented women from entering the field until, in 330 B.C., a woman named Agnodike mounted a successful challenge to the exclusionary rule.) Apart from chemical means, a technique called the Lacedaemonian leap, reputedly perfected by Spartan women, required a pregnant woman to jump up and touch her heels to her buttocks, which was supposed to jar the fetus loose. In some places, abortions performed after the fetus had developed recognizably human features could, if discovered, draw fines or other sanctions. But generally, the ancients accepted any method of preventing conception and expressed no moral qualms about eliminating unwanted pregnancies as long as the fetus had not quickened (which usually happens between sixteen and twenty weeks).

Sitting in the shade of a colonnade as Athenians on their daily rounds crisscrossed the agora, or marketplace, Plato must have heard a great deal about the enterprise of human breeding from his mentor Socrates. After his teacher, condemned by the city, quaffed his deadly draft of hemlock in 399 B.C., Plato retreated to the countryside, where amid silvery olive trees he set up his school for young men, the Academy, and distilled what he had learned. In the dialogue known as *The Republic,* Plato, a firm admirer of Spartan customs in regard to reproduction, elaborated upon his conviction that the state had the right and responsibility to oversee what went on in the bedrooms of its citizens.[12]

Book V of *The Republic* finds Socrates holding forth as usual among a group of obliging friends and foils. The topic of the hour is the state—in its ideal manifestation, the perfect city. Now Socrates embarks upon a long discussion about the place of women and children in his utopian scheme.

First off, he says, women and children should be held in common, and marriages should be arranged by the state, which would endeavor to make matches on the basis of similar character. These unions would be declared sacred, at least for purposes of public relations. However, their true value, from the state's point of view, would lie somewhere else altogether. Addressing one of his companions, Socrates rhetorically inquires,

> "How will [these marriages] be most useful? Tell me that, Glaucon. For I see in your house hunting dogs and numbers of pedigree game birds. Pray, have you paid any attention to their matings and breedings?"
>
> "What sort of attention?" [Glaucon] asked.
>
> "First of all, in this admittedly pedigree stock there are some, aren't there, which turn out to be the best?"
>
> "There are."
>
> "Then do you breed from all alike, or do you take the greatest care to choose the best?"
>
> "I choose the best."
>
> "What of their age—do you take the youngest or the oldest, or as far as possible those in their prime?"
>
> "Those in their prime."

"And if the breeding should not be done in this way, you consider the stock will be much worse both in bird and dog?"

"I do," said he.

"What of horses," said [Socrates], "and other animals? Is it different in them?"

"That would be odd indeed," he said.

"Bless my soul!" said [Socrates]. "My friend, what simply tiptop rulers we need to have, if the same is true of mankind!"[13]

Plato draws the parallel emphatically. Humans, like animals, are "stock." And the imputation is clear: Willy-nilly intercourse will degrade the stock's quality. To function well the state needs the best type of man possible. Ergo, the state has a major interest in governing sex.

Rulers, Socrates says, need to secure the pedigrees of their subjects by steering the "best men" to the "best women." He offers a method of accomplishing this through holidays and festivals in which men and women are married en masse. State propaganda, including the ancient equivalent of pop love songs, will be geared toward supporting those unions deemed opportune. "[P]riestesses and priests will intone for each [state-approved] wedding while the whole city prays that the children born will be better children of good parents, and more useful children of useful parents, from generation to generation."[14] The rulers, meanwhile, will carefully adjust the number of marriages, "taking into account war and disease and so forth, in order to keep the city from becoming either too large or too small as far as possible."[15] To remain viable, the state must keep its population at an optimal size, offering both inducements and penalties for childbirth as needed. For example, unmarried young men who excel in war, sports, or other activities will win free sexual access to women, with the design that they will father many children.

Socrates figures that men should be allowed to "beget for the state" from the time their "quickest racing speed is past" (about age twenty-five for the fleet-footed Greeks) to age fifty-five. Women, with a shorter optimal period of fertility, "shall bear for the state from the age of twenty to forty," he says. As soon as men and women exit their childbearing years, they will be permitted to consort with anyone but their children or parents. They will be admonished, though, to prevent pregnancies. If all precautions fail, "if a child is born, if one forces its way through, they must dispose of it on the understanding that there is no food or nurture for such a one."[16]

In fact, anyone who reproduces in disregard of state strictures will call down obloquy upon himself and his progeny. If a man fathers a child illicitly, that child will be excluded from the sacred fellowship of society. It will not be blessed by "holy rite or prayers . . . instead it [will be said that the child] was begotten in darkness with incontinence to the common danger."[17] "To the common danger" meaning that Socrates sees an infant of inapposite heritage—a mere babe—as posing a threat to the city's very foundations. So essential to order are state reproductive proscriptions that if a man, even while in his prime, inseminates a woman without the approval of the rulers, then "bastard and unaccredited and unsanctified we shall call that child which he dumps upon the city."[18]

Meanwhile, in Socrates' hypothetical enclave, children fathered by "worthless" citizens will be programmatically exposed. Socrates accepts that this will have to be done through subterfuge, to avoid outright rebellion by the lower classes. One way of getting around any outcry concerning such a policy, he suggests, would be to institute a lottery. Thus, parents robbed of infants will "blame [their] bad luck on any conjunction" of stars or weather, instead of on their rulers. These children "of the inferior sort, and any one of the others who may be born defective," will be handed over to nurses, who "will put [them] away as is proper in some mysterious, unknown place"—those trackless mountains, say, or pathless hills. All other infants the nurses will raise in a rustic setting, which Socrates deems far preferable to the city for meeting the needs of children. The infants will be attended by professional wet nurses, as well as by mothers who have just given birth, although officials will have to take "every precaution that no mother shall recognise her own [baby]," presumably so that mothers do not become emotionally attached to, or attempt to reclaim, their children.[19]

In a slave-holding society like Athens, such notions hardly would have struck anyone as radical. Theories advocating state control of reproduction appear to have been accepted without question and promulgated without much debate. Like chattel, women and children throughout most of the classical world lacked political standing; they owed the limited privileges they had to the largesse of the patriarchy.[20] Greek dramatists, prosodists, and poets frequently portrayed women as liabilities, both for men singly and for the state. Hesiod, Aeschylus, Menander, and many others bemoaned women's laziness, uselessness, duplicity, sexual wantonness, and cowardice; their spendthrift ways and gossipy, refractory, and tempestuous nature. In *Medea*, the Athenian playwright Euripides has Jason deliver the roundly condemnatory line, "Men ought to beget children somewhere else, and there should be no female race." His Hippolytus, in the play of the same name, upbraids Zeus, saying that women were not necessary either for the creation of the "mortal race" or for domestic satisfaction. Zeus should instead have arranged it so that men could simply purchase "the seed of children" by depositing "a sum of bronze or iron or gold" in his temples, the amount of the sum pegged to the wealth of the man.

Such misogyny saw expression in the household and community. Women generally had few property rights and little disposable worth. Law bound Athenian women to their homes, where their primary function, apart from attending to domestic matters, was to produce heirs. Since heirs were ipso facto male, female infants were expendable. According to classicist Sarah Pomeroy, the rule of husbands over wives was supplemented by state control for the purposes of keeping bloodlines clean. Pomeroy wrote, "The sexuality of female citizens was regulated by law, lest a child who was not the offspring of a citizen be insinuated in the citizen body."[21] Greeks aimed mainly to restrict access to privilege; however, Athenians in particular considered it important to keep the enfranchised population free of the taint of slave or barbarian.[22]

As both student and teacher at the Academy, Aristotle steeped himself in Platonic precepts for twenty years beginning around 368 B.C., and his notions concerning childbearing, laid out in Book VII, Chapter 14, of his *Politics*, closely resemble those of his elder colleague. His principal departure lies in his rejection of Plato's proposition that men and women, apart from their gender differences, possess equal capabilities.[23] Aristotle firmly placed women below men on the scale of creation, arguing that men are endowed with "deliberative faculties" and "moral virtues," whereas these characteristics are present in women to a far lesser degree. The inequality between the sexes, he declared, is permanent.

However, like Plato, Aristotle held that the needs and concerns of the state must always take precedence over those of the family or individual. The state, he averred in *Politics*, has an obligation to foster in its citizens (i.e., males in the ruling class) the development of reason and intelligence, which derive directly, he said, from "our engendering."[24] Furthermore, for the sake of a stable economy, it must limit and monitor the quality of population. To this end, he asserted,

> let there be a law that no deformed child shall be reared; but on the ground of number of children, if the regular customs hinder any of those born being exposed, there must be a limit fixed to the procreation of offspring, and if any people have a child as a result of intercourse in contravention of these regulations, abortion must be practised on it before it has developed sensation and life; for the line between lawful and unlawful abortion will be marked by the fact of having sensation and being alive.[25]

By following guidelines similar to those suggested by Plato, which would allow couples to reproduce only between certain ages, the state will help ensure that offspring are healthy. But Aristotle saw other justifications, financial and psychological, for imposing restrictions. Fathers should not be too old, he said, "since elderly fathers get no good from their children's return of their favours, nor do the children from the help they get from the fathers."[26]

A desire to garner the greatest possible recompense from their investment of effort and money, and to guarantee offspring the benefits of their own social standing and connections, should impel men to sire children earlier rather than later in life. Neither, however, should parents be too young, for then their children will view them more as peers than masters, and "also the nearness of age leads to friction in household affairs."[27]

Aristotle, ever the ardent natural historian, further expanded the Platonic argument by offering biological reasons for state oversight. Children of "too elderly parents, as those of too young ones, are born imperfect both in body and mind, and the children of those that have arrived at old age are weaklings."[28] In addition, young humans, like young animals, he said, are likely to produce tinier offspring, and also a greater proportion of females, neither desirable from his point of view.[29] Clearly, caring for such enfeebled infants would run counter to the state's best interest.

Like an ancient surgeon general, Aristotle advised that proper prenatal care must be given if women are to deliver healthy babies. Pregnant women "must take care of their bodies, not avoiding exercise nor adopting a low diet." This is because "children before birth are evidently affected by the mother just as growing plants are by the earth."[30] Here Aristotle made explicit the analogy between womb and soil. Women, tilth, crops, and livestock exist on a continuum, all subject to natural laws governing growth and efflorescence.

Elsewhere around the Mediterranean during antiquity, people hoping to propitiate angered deities or call down favor in times of crisis still engaged in the sacrifice of children, following rituals which had emerged thousands of years earlier in Babylonia, Phoenicia, Canaan, and Egypt. This form of worship had been extirpated in some spheres. Early on, for example, the Hebrews had engaged in child sacrifice along with everyone else in their immediate vicinity, but by the fourth century B.C. had long since succeeded in banishing this practice from their repertoire of religious devotions, although it had taken mighty thunderings from the prophets to root out the worship of the ravening gods Moloch and Baal.[31] Moloch, celebrated originally by the Ammonites and other Semitic tribes, required a steady stream of infants to placate him. Presiding over his temples in the form of a huge bronze effigy, the god sat or stood with his larger-than-life arms perennially outstretched to receive their due. Devotees mounted a flight of steps and dropped between his limbs the sacrificial victims, who fell into a fiery chamber below and were consumed. Baal, too, exercised an appetite for children, although scholars cannot agree whether young victims were typically forced only to walk through flames or were actually consigned to, and perished in, a pyre.

Whichever was the case, the Hebraic Scriptures repeatedly inveigh against both gods, admonishing the Jews to avoid imitating either their new neighbors, or their old captors, the Egyptians.[32]

Those same Egyptians, of course, had, in a fit of anti-immigrant fury during the time of the Hebrew captivity, undertaken to programmatically slaughter baby boys born to the descendants of Jacob, whom Pharaoh feared because of their high birthrates. " 'Look,' [the pharaoh] said to his people, 'the Israelites have become much too numerous for us. Come, we must deal shrewdly with them or they will become even more numerous and, if war breaks out, will join our enemies, fight against us and leave the country.' "[33] Under fiat from Pharaoh, midwives were supposed to systematically eliminate boys born to Hebrew women. When that stratagem failed, the ruler commanded that all male infants be drowned in the Nile. (In an attempt to get around this dictate, Moses' mother set him in his basket among the reeds, where Pharaoh's daughter found him and, ironically, returned him unknowingly to his own mother to be wet-nursed.) After lashing Egypt with plagues of bloodied rivers, teeming frogs, gnats, flies, cattle disease, festering boils, hailstorms, locusts, and darkness, the Lord of the Israelites finally evened the score by killing off the firstborn son of every Egyptian, an event commemorated in the Jewish festival of Passover.

As common as such large-scale excesses against children were according to the Hebrews, classical authors and modern historians of dynastic Egypt claim that infanticide for the purpose of controlling family size or eliminating girls was notably absent among Egyptians, as indeed it was among the Jews. This had changed somewhat by the third century B.C., due to the influence of the Greeks. Under the Ptolemies, a Hellenized Egypt adopted bad habits. Documents attest to the fact that women sometimes practiced infanticide, notably a Ptolemaic law specifying the number of days of ritual purification that women must undergo after menstruation, abortion, childbirth, or the exposure of an infant. This law not only implies that the act was tacitly accepted, but also firmly locates exposure within the natural cycle of reproduction. In a letter to his wife from Alexandria written in the first century B.C., a man named Hilarion casually tells his wife, "If—good luck to you!— you bear offspring, if it is a male, let it live, if it is a female, expose it."[34] Posidippus, a first century B.C. poet, claimed, "Everyone, even a poor man, raises a son; everyone, even a rich man, exposes a daughter."[35] Here Posidippus has overlooked a few exceptions to the rule, to wit, households of the Ptolemies and some very wealthy families who did raise daughters, sometimes three or four at a time.[36]

To the east of Egypt, Arabian desert–dwellers seem to have routinely killed off girls, or culled a few when their numbers outstripped those of boys, up until the time of Muhammad, who enacted a firm proscription against infanticide. To the west, the Carthaginians reportedly continued the bloody custom of appeasing Baal until the practice was quashed once and for all with the razing of their city in 146 B.C., during the Third Punic War.

From its bluff overlooking the sea, Carthage, established in 814 B.C., ran repeated sorties against the Greeks, and ultimately controlled the entire western Mediterranean basin. Carthaginians gained infamy among ancient historians for their enthusiastic compliance with Baal's demands for fresh offerings. According to Greek and later Roman critics, the city motto could have been "Sacrificed their sons to Baal"—as well as to the gods Cronus and Saturn, who were identified with him.[37] Plutarch claimed that the Carthaginians conventionally "offered up their own children, and those who had no children would buy little ones from poor people and cut their throats as if they were so many lambs or young birds."[38] Sir James Frazer, in *The Golden Bough*, tells the story of one mass sacrifice in the late fourth or early third century B.C., when, under siege by forces of the Syracusan general Agathocles, the Carthaginians immolated five hundred children in hopes of placating Baal.

However, recent archaeological evidence muddies the picture offered by such accounts. Indeed, since the 1900s, large numbers of graves containing sacrifices to the affiliated sun gods Baal, Cronus, and Saturn, as well as to the earth deity Tanit (identified with the Phoenician goddess of love and fertility, Astarte), have been unearthed at the site of ancient Carthage, many of the burials marked by inscribed stones. However, the American archaeologist Jeffrey Schwartz, who emptied hundreds of waterlogged urns dug from one large cemetery, discovered something intriguing about the burned bones contained within. Although a few burial urns held toddlers and adolescents, Schwartz estimated that anywhere from 51 to 84 percent of the bones—with the uncertainty owed to the difficulty of pinpointing precise ages from fragmentary remains—had belonged to late-third-trimester fetuses. Most of the rest of the bones came from newborns. All of which led Schwartz to suspect that in many cases Baal's victims had come to him already dead, that parents frequently gave over to the god stillbirths or babies who had been born weak or prematurely and soon perished. Perhaps Plutarch and others had excessively demonized Greece's old nemesis, and the Carthaginians were guilty of nothing more than cremating these small unfortunate ones.[39]

Despite many a faded glory, present-day Rome, that erstwhile seat of a far-flung empire, still boasts an embarrassment of statuary, some of it exemplary, much of it mass-produced and arrayed in endless parade upon the tables of street-side vendors. Of all the trinkets on show, whether *en plein air* or in more exclusive shops, there is one image which seems to have been executed in every possible medium—in clay, wax, glass, plastic, copper, brass, bronze, iron, steel, nickel, zinc, alloyed silver, pure silver, and gold of assorted karats. Fittingly, it is the image of the city's fabled founders, the twins Romulus and Remus, their faces upturned, their mouths straining for the dangling teats of the she-wolf who suckled them after their parents abandoned them to the wilds outside the town of Alba Longa. (An especially fine version of she-wolf and kids stands atop a column in a corner of the cobbled piazza of Michelangelo's Campidoglio, at the head of a path leading along one side of the Senator's Palace down into the Roman Forum.)

In the lingo of pop psychology, the founders of Rome might be dubbed exposure survivors. Their personal histories are rather tangled. One legend has it that their parents were Mars, the war god, and his consort Rhea Silvia. Another holds that their father was unknown, their mother a handmaiden to the daughter of Tarchetius, king of Alba Longa.

There is also the story attributing maternity to a daughter of Aumulius, a descendant of the brave Aeneas (father, again, unidentified). According to Plutarch, Aumulius had bound over his daughter, who may have been named Ilia, Rhea, or Silvia, to service as a vestal virgin, but, unfortunately, she had already violated a major prerequisite for the job: When she gave birth to two boys "of more than human size or beauty," Plutarch wrote, an angry Aumulius bade his servant Faustulus to expose them. Faustulus tucked the babes into a small wooden trough on a riverbank outside of town, but the river flooded and the crude bark floated away. It landed downstream by a wild fig-tree, whereupon a she-wolf with a highly developed sense of trans-species altruism happened along. The she-wolf nursed the infants, and a woodpecker also fed and watched over them. Once grown, the boys took the names Acca and Larentia, or, alternatively, Romulus and Remus. Traditionally, their birth-date is put at 753 B.C. Later, after they had established the city of Rome, supposedly at a location on the southern flank of the Palatine hill (you can see the site today, marked by an unprepossessing stone), the brothers had a falling out, and Romulus killed Remus.

A font of Roman law, Romulus took steps intended to spare at least some infants the trauma he had experienced. According to a modern reconstruction of his ancient proclamations,

> Romulus compelled the citizens to rear every male child and the first-born of the females, and he forbade them to put to death any child under three years of age, unless it was a cripple or a monster from birth. He did not prevent the parents from exposing such children, provided that they had displayed them first to the five nearest neighbours and had secured their approval.[40]

However, the Romulan dictates appear to have been ignored. Romans likely committed infanticide with regularity, sometimes to control family size,

although scholars have not succeeded in quantifying its occurrence. According to British historian William Lecky, Romans did not consider exposure
the same as infanticide (despite the fact that exposed children might well
die), and while publicly condemning infant abandonment, they did nothing
actively to stamp out the practice. In fact, Lecky contended, "[exposure] was
practised on a gigantic scale and with absolute impunity, noticed by writers
with the most frigid indifference, and, at least in the case of destitute parents,
considered a very venial offence."[41] Illegitimacy also provided a reason for
abandonment. Even political grief or paranoia prompted by current events
could lead Romans to divest themselves of infants. After the popular general
Germanicus died abroad under mysterious circumstances in the first century
B.C., Suetonius recounted, people ran wild in the streets: "The temples were
stoned and the altars of the gods thrown down, while some flung their household gods into the street and cast out their newly born children."[42] When
Nero murdered Agrippina, someone placed a child in the Forum along with a
sign declaring, "I will not raise you, lest you cut your mother's throat."[43]

   References to exposure abound in Latin literature from Republican times
on. They pepper the bawdy dramas that were the ancient equivalent of
French bedroom farces. For instance, the exposure of infants serves as a plot
device in several works by Terence, the successful second century B.C. playwright. In *The Lady of Andros,* a typical Roman comedy involving assumed
identities, overturned class divisions, and sexual peccadilloes coming home to
roost, a character named Glycerium has a child, and as part of the business of
the play it is taken to be exposed, although by play's end the infant is
retrieved and taken into the family bosom. In *The Mother-in-Law,* a young
woman, Philumena, is raped by a well-to-do fellow, Pamphilus. A few
months later, she unknowingly marries her rapist, who immediately sails off
on business. Philumena, to her chagrin, discovers that she is pregnant—the
result of the rape—and goes home to her mother. When Pamphilus returns,
he does his math and, concluding that the baby can't be his, disavows his
wife. His mother-in-law goes to him and pleads with him not to shun her
daughter, saying,

> I am doing my best to keep the birth secret from her father and
> from everybody. If they can't be prevented from becoming aware of
> it, I shall say there has been a miscarriage. I am sure no one will
> have any suspicion, since it looks so like it, but that the child is
> yours. It shall be at once exposed: it will cause you no inconve-
> nience, and you will have concealed the shameful wrong done to my
> unhappy child.[44]

The daughter's out-of-wedlock pregnancy, not the exposure of the baby, provokes dismay. With similar sangfroid, the husband in another Terence play,

The Self-Tormentor, instructs his pregnant wife that if she delivers the baby while he is away on a trip, she should kill it if it is female. His wife does give birth to a baby girl, but, unable to bring herself to kill it, has it exposed. Upon his return, her husband scolds her for her disobedience and irrationality: Now, he says, she has simply guaranteed her daughter a life of prostitution.

Terence was alluding to the well-known fact that exposed infants regularly wound up on the sale block as slaves, or indentured to pimps. Certain unscrupulous traffickers maimed or mutilated infants and then used them for beggars. Romans confronted the reality of exposure daily as they traversed the Forum, for people traditionally abandoned infants at the Columna Lactaria. One historian has estimated that 20 to 40 percent of Roman newborns would have been deposited here by the third century A.D.[45] Slave traders might find there newborns, still bloody from birth, whom they undertook to raise—in what was considered a speculative enterprise. Oddly, infants from wealthier families stood a lesser chance of surviving exposure than those from poorer ones. One Roman commentator, pseudo-Quintilian, reported that the rich generally made very certain that the infants they were exposing would die, by having them carried out into the wastelands beyond the city limits, whereas the poor, who he said abandoned their children reluctantly and only on account of poverty, "did all they could to ensure that the infant might someday be reclaimed," or at least succeed eventually in buying its freedom from the slavery into which its hapless parents had thrust it.[46]

Infants displaying obvious deformities wound up in the Tiber, as it was considered gravely unfair to the child, the parents, and society to keep such creatures alive. Fathers in Rome, who in Latin were said to "take" children rather than "have" them, had the absolute right to reject babies after birth. Midwives lay newborns on the ground, and fathers ceremonially took them—up into their arms—or left them on the ground, thus signaling that they should be exposed. The essayist Seneca offered a rationalization for this process of culling children, which reflected popular opinion. Seneca wrote, "What is good must be set apart from what is good for nothing."[47] The excision of the weak or deformed from the body of society is not motivated by hatred or anger:

> Does a man hate the members of his own body when he uses the knife upon them? There is no anger there, but the pitiful desire to heal. Mad dogs we knock on the head; the fierce and savage ox we slay; sickly sheep we put to the knife to keep them from infecting the flock; unnatural progeny we destroy; we drown even children who at birth are weakly and abnormal. Yet it is not anger, but reason that separates the harmful from the sound.[48]

Reason, perhaps, but also fear: For the ever superstitious Romans, multiple and other unusual births represented bad political omens. Hermaphrodites

terrified people to such a degree that midwives were expected to box up such
infants and have them taken to the coast and cast into the sea. Only occasionally, owing to a perverse fascination with anomalies, did children with
milder disabilities win reprieve. Suetonius remarked that "usually Romans
took great interest in such creatures as dwarfs, giants, congenital idiots," who
were viewed in the manner of sideshow freaks.[49]

D uring the time of the late Republic, the government began what was to
be a several centuries long pronatalist campaign. From early times,
Romans had considered three the ideal number of children, but few families
ever actually approached that size, especially among the ruling class, where
women frequently raised no children at all. Over time, the low fertility rates
had led to a population decline, which especially worried military strategists,
for without a constant supply of young men entering the ranks of the legion,
Rome's foreign policy was severely compromised. For the sake of state survival, several rulers, starting with Julius Caesar in 59 B.C., passed decrees
encouraging people to have children and discouraging them from committing infanticide. The emperor Augustus urged the Senate to penalize those
who did not do their reproductive duties. Laws passed during his reign limited a man's ability to pass on his estate if he had not produced that proper
trio of children. Later, the government tried the carrot instead of the stick,
offering tax breaks and extra grain allotments—so-called alimentary payments—to fathers for each additional child they raised. Nothing, though,
convinced Romans to reproduce with more enthusiasm. Nor did laws regarding exposure trim the numbers of abandoned infants, who continued to serve
an essential economic function. Children, after all, were the moral equivalent
of slaves: Both were referred to by the same Latin word, *liberi*.[50] In fact, children might have even rated less consideration than slaves. The death of a
good slave was mourned mightily, whereas there was a cultural tendency to
discourage any show of grief at a child's death. Plutarch felt that this attitude
only made sense, remarking that "excessive grief was unnecessary for one who
had barely belonged to society." Plutarch also believed that women must have
been endowed with an innate maternal instinct, "because [a woman is] the
only one willing to handle such a repulsive, distasteful object," as an infant.[51]

The protracted Roman pronatalist campaign involved only exhortations
regarding quantity, not quality. No Roman appears to have arrived at a program
for producing well-bred children that was as comprehensive as those devised by
Plato and Aristotle. This may have been because Romans, unlike the Greeks,
did not place great emphasis on direct inheritance. For Roman men, far more
important than maintaining bloodlines was perpetuating the family name. To
their minds, this could be accomplished as easily by choosing one's heirs as by

fathering them. In fact, adoption was seen to offer as many benefits, if not more, than actual parenthood. Parents, subject in some measure to the luck of the natural draw, might find the quality of their biological offspring lacking. What's worse, this might not become obvious until one had invested large sums of money and effort in raising and tutoring a child. A parsimonious and practical paterfamilias, possessed of a sizable fortune and good standing in the world, often found it in his best interest to adopt a full-grown, unrelated individual, or, say, a grown son of a sibling, and designate him as heir. This path, Roman pundits averred, could prove all-around more cost effective and satisfying than having one's wife bear children.

This is not to say that Romans entirely devalued the old concept of *eugeneia*, or noble birth. They displayed admiration and nostalgia for Sparta, for instance, and when they ruled the city in the second century B.C. reinstituted the *agōgē* in modified form for both boys and girls. However, the quality of being wellborn had more to do with one's own person than with one's heritage. Explained historian Peter Brown,

> A specific body image, formed from a conglomeration of notions inherited from the long past of Greek medicine and moral philosophy, was the physiological anchor of the moral codes of the wellborn. In this model personal health and public deportment converged with the utmost ease. . . . [The body] was regarded as the most sensitive and visible gauge of correct deportment; and the harmonious control of the body, through traditional Greek methods of exercise, diet, and bathing, was the most intimate guarantor of the maintenance of correct deportment.[52]

Romans understood in principle certain truths of heredity. Pliny the Elder's *Natural History* contains a passage that encapsulates a fairly sophisticated understanding of the ins and outs of inheritance. Pliny explained that "sound parents may have deformed children and deformed parents sound children or children with the same deformity," and adds "that some marks and moles and even scars reappear in the offspring, in some cases a birthmark on the arm reappearing in the fourth generation."[53] But unlike Greek thinkers, who remained rather close to their agricultural roots, the urbanized Romans acknowledged the tenets of husbandry, yet exhibited a strong preference for explanations drawn from the mind-over-matter school. Their theories concerning *eugeneia* tended to focus not on the mechanics of breeding "the best" with "the best," but on events occurring during conception, gestation, or nursing that would shape or mar the individual's body—and thereby influence his or her character.

In the first place, to assure that children were well conceived, members of the upper class endeavored to conduct their lovemaking sessions with decorum. During the late Empire, the historian Peter Brown claimed, it was widely believed that those forms of intercourse which "were in some manner continuous with public codes of deportment would produce better children." The flouting of such norms, "whether by oral foreplay, by inappropriate postures, or by access to the woman in menstruation," would manifest itself in poorer-quality children.[54]

Couples seriously desiring to conceive a child would have thought twice before deviating from the norm, for fear of harming any potential offspring. They might also have been especially concerned that no sudden noises interrupt their lovemaking, since these were also thought to have an impact on a conceptus. Pliny contended that a child's likeness to one or the other parent was produced by "accidental circumstances," that is, "sights and sounds and actual sense-impressions received at the time of conception." Not only could the infant be molded by the setting in which its parents copulated, its form could be owed to their mental states during the act:

> A thought suddenly flitting across the mind of either parent is supposed to produce likeness or to cause a combination of features [in the resulting offspring], and the reason why there are more differences [of appearance] in man than in all the other animals is that his swiftness of thought and quickness of mind and variety of mental character impress a great diversity of patterns, whereas the minds of the other animals are sluggish, and are alike for all and sundry, each in their own kind.[55]

Due to the readiness with which thought might impress itself upon flesh, pregnant Roman women guarded themselves against all manner of mental impulses which might ruin the child. For example, they believed that if they craved a certain food and touched themselves, the baby would be born with a birthmark in the shape of the desired food. Since the birthmark was supposed to appear wherever their fingers landed, women endeavored to touch themselves only on the buttocks when a craving descended, because at least then the stain they caused would not be readily apparent. (Straightforward dietary strictures also were followed. To avoid bearing babies lacking fingernails, for instance, women were advised not to eat food that was too salty, no doubt a hard rule to follow, given the Roman predilection for that condiment.)[56] If a flaw in a child could not be pinned upon the mother, Romans happily blamed the wet nurse, who was thought to have a profound effect on her charges through the agency of her milk. According to popular belief, wet nurses imparted qualities to babies, ranging from a proclivity for lying to a facility for language.

The emphasis on the particular circumstances of a person's birth as opposed to his or her biological parentage or membership in a group defined by biological relationship reflected, or was reflected in, the mobility built into Roman society. Despite institutionalized slavery, individuals could and did move up in the world, whether through hard work, loyalty to a powerful paterfamilias, political connivance, or lucky investments. Ill fortune, for a Roman, arrived on the wings of a two-headed lamb or a hailstorm, demanding stepped-up obeisances to the gods. Collective actions could bring about or alleviate collective bad times; Romans rarely echoed the Greek notion that national decay followed inevitably from the decline of the human stock. Pliny the Elder indulged in some bleak musings in *Natural History* on this score, noting that "few men are taller than their fathers." The age was rushing toward a "conflagration" and a "crisis" which was "exhausting the fertility of the semen" and causing the "entire human race" to shrink.[57]

But the Romans, who had conquered the world with faith in the power of the human will and of labor, generally preferred not to locate the source of any problem within themselves. The coming of the Christian Era, though, changed that, as polemicists waged a war to regulate human sexuality and reproduction for spiritual, and ultimately political, purposes.

4
Sanctity, Body Heat, Bad Cess, and Blood

During the Middle Ages in Europe, babies served as emblems of their parents' sanctity or sinfulness. As Christian theologians moved into the cultural foreground and classical traditions waned, as the Church of Rome inexorably extended its reach northward into regions once under sway of pagan Celtic and Germanic traditions, sex became a political battleground, and reproduction a matter of religious concern.

During antiquity, women's sexual behavior had already been highly subject to regulation. By Greek and Roman lights, the good woman was virginal before marriage and faithful to her husband afterward. Few restraints were placed on men's sexuality, though, and the tightening of medieval standards functioned primarily to rein in male behavior. Whole spheres of activity that had been tolerated during antiquity became taboo, especially homosexuality, and even the long-standing prerogative to rape women without consequence was curtailed. By the sixth century A.D., simple nudity, untinged by eroticism, gained the power to shock. The naked Christ had to be swathed and veiled, lest the sight of his body provoke barely converted worshipers—women especially—to relapse into phallic worship.[1]

Eventually, almost the entire medieval world would agree that monogamous marriage constituted the only forum in which sex was acceptable. Prostitution, concubinage, adultery, pederasty, and other sexual arrangements not deemed acceptable to the church did of course persist; however, if caught by authorities, people engaging in prohibited activities were subject to heavy fines, public humiliation, physical punishment, or even death.

An infant born outside of marriage thus represented clear evidence of perfidy: It stood as the literal embodiment of its parents' sin. To be reminded of this, medieval congregations had only to direct their gaze to a convenient altarpiece portraying the infant Jesus in the arms of his mother Mary, who had been cast by the early church in that most unusual and exemplary role: a woman who conceived while still a virgin.

The Christian emphasis on sex within marriage stemmed in large measure from a deep-seated antipathy for the body, which was expressed forcefully by the apostle Paul, who deplored sexual urges because they deflected a person's attention from God. Indulging bodily desires meant reveling in the fallen world, accepting its terms, as it were, and turning away from the spiritual realm. Paul held up celibacy as the ideal state, viewing marriage as a tactic to which a person should resort only if the "temptation to immorality" was so great that he or she could not resist it.[2]

Paul's convictions engendered a tug-of-war in the early church, as exhortations to chastity ran counter to its attempts to sanctify marriage (and thereby exert control over both society and relations within the household). Saint Jerome in the fourth century resolved this conflict to his satisfaction by reasoning that "it is not disparaging wedlock to prefer virginity," and allowing that marriage was praiseworthy because it produced more virgins.[3] Virgins, for Jerome, existed in an earthly approximation of an Edenic state. Having eschewed the things of the flesh, they stood as close to heaven as it was possible to get in life. Jerome admitted that those who followed the biblical charge to "go forth and multiply" did not technically violate any law, but they did commit the fundamental error of focusing on the wrong sphere of action, since their true "company" was in heaven, not on earth. Men, he said, were better off not marrying.

The propagation of the human race was in this respect of no concern at all: Better that humans should fade away from the earth altogether, Jerome thought, than that they surrender to lust.[4] Such notions filtered across Europe, with support for them reaching its zenith during the twelfth and thirteenth centuries, when monasteries proliferated and many upper-class men and women staged a retreat from secular life. At its most extreme, the position that chastity was a person's highest calling spawned acidulous tracts which chronicled the endless indignities and sorrows of marriage and motherhood, all in an attempt to persuade young women to get themselves to nunneries.[5]

Saint Augustine, also writing in the fourth century, profoundly influenced Catholic doctrine regarding sex and childbearing. Augustine's pronounce-

ments against sexual relations, like those of Jerome, were made in the context
of a larger argument urging Christians to embrace chastity. But realizing that
not everyone was cut out for this path, Augustine attempted to establish categories of sexual behavior, to distinguish between intercourse that was tolera-
ble and that which was beyond the pale. He effectively invented an
intermediary state of chastity to which married couples might aspire, the
state of "conjugal chastity."

To qualify, couples could copulate for one purpose and one purpose alone:
to produce children. Augustine took pains to point out that the worth of
these children lay in the fact that they might be "born again and become 'sons
of God,' " which is to say, be initiated into Christian belief. (He further stipulated that only believers were eligible to participate in the sex-with-virtue
program, declaring that even if unbelieving parents partook of sex solely for
reproduction's sake, they were censurable because they did not perform the
critical follow-through motion of assigning their offspring over to God.)
Under no circumstances were believers to indulge in sex for "bestial" reasons,
that is, merely for the sake of physical pleasure.[6] Here Augustine, casting
about for evidence of ulterior motive, settled upon a simple test: Proof that a
couple was engaging in "carnal concupiscence" came in their use of contraceptives of any sort. This showed that they had copulated out of debased
physical desire and that their minds and hearts were not concentrated upon
that shining, elevated goal of chastity.

Augustine's related concept of ensoulment, that is, the notion that the soul
enters an embryo at conception, served as a disincentive to abortion, for it
meant that ridding oneself of a fetus was then classed as murder. However,
although the distinction is a subtle one, there is a difference between arguing
against contraception and abortion, and for reproduction, and the bishop of
Hippo clearly placed little value in reproduction per se. Pronatalism did not
motivate his ban of birth control, rather hatred of fleshly appetites. Ideally, he
had rather people forgo sex entirely.[7]

The medieval push to institutionalize virginity and chastity had not only
ecclesiastical but also social backing. The secular concern had to do not with
maintaining spiritual virtue but with ensuring that lineages remained invio-
late. If a man married a virgin, he guaranteed that any children she bore
would be his and that his family bloodlines would remain "pure" and uncon-
taminated by outsiders. Settling into the marriage bed for the first time, a
bride had to pass a crucial test, the stain of her hymenal blood paradoxically
signaling to all involved her unstained, unsullied nature. If she failed the test,
showing herself to have been previously corrupted, she was reviled and
rejected—even if she had lost her virginity as a result of rape. Later in marriage, if a wife violated her vows and slept with another man, she might in
some locales be put to death by being drowned or buried alive. The taint of an
adulteress seeped down through the generations. "More than rape or abduc-

tion," wrote French historian Michel Rouche, " . . . adultery resulted in pollution of the woman and her offspring, poisoning the future." An adulterous wife "subverted her children's claim to authenticity and destroyed the charisma in her blood."[8] (This idea of "polluted" blood persisted into the twentieth century, subverting the logic of genetics and governing the twisted racist arguments of the Nazis.)

Throughout much of the medieval period, which spanned roughly a millennium from the fifth to the fifteenth century, understanding of human reproduction advanced very little or not at all. By and large, physicians did not strike out in new directions; rather, they relied upon Greek precepts, gleaned from texts that had found an audience among Arab scholars, and from venerable Latin works that had been preserved in the wake of the Roman Empire's disintegration. Beginning in the early Middle Ages, Arab scholars writing in Latin, including Avicenna and Averroës, disseminated many ancient Greek works throughout Europe in thirdhand form.

Medieval doctors also dipped into the ample folklore that had been shared and embroidered upon within the community of women for millennia. Midwives probably had as much or more understanding of women's bodies and the details of pregnancy and childbirth than physicians—at least male physicians, who for reasons of propriety (and fear of women's sexual powers) refrained from examining women directly, either visually or by touch. Thus, male physicians often worked in partnership with midwives, who relayed information to them from their female patients. (However, as the Middle Ages wore on, midwives also suffered increasing persecution from the church, which tended to identify them with witchery, and the medical community itself eventually acted to repress midwivery as it expanded its own purview.)[9]

When it came to the subject of childbearing, physicians got by with assorted versions of ancient explanations. Depending on when and where he (and sometimes she) lived in Europe, a doctor might have access to such texts as Hippocrates' *Diseases of Women*. He might have read, in one form or another, the Roman author Pliny's pronouncements regarding menstruation; the Hellenistic physician Galen's discussions of the female "testes" (the ovaries); the Christian polemicist Tertullian's arguments concerning sperm; Avicenna's pointers on labor and delivery; or the Roman philosopher Favorinus' advice about breast-feeding. Late in the Middle Ages, homegrown medical writers appeared, including Bartholomeus Metlinger of Augsberg, whose 1495 book on childbearing won a wide readership, and Trotula of Salerno, a female practitioner who at some point between the eleventh and thirteenth centuries penned a lengthy manual also titled *The Diseases of Women*.

Medieval physicians conceptualized sickness and well-being within the framework of humoral theory, which had been developed by the Greeks. The basic premise was that the body contained four fluids (blood, phlegm, yellow bile or choler, and black bile or melancholy) which flowed and eddied around inside it. Like a car with circulating brake fluid, transmission fluid, oil, and antifreeze, a person whose level of blood, phlegm, or yellow or black bile sank too low or bubbled over required servicing. In addition to his or her salubrity, a person's moods and overall disposition were owed to these fluids, hence the so-called melancholic or phlegmatic character, about which we sometimes hear still.

Within this overall scheme fit the notion, also inherited from the Greeks, that physical, mental, and temperamental differences between males and females stemmed from the excess or lack of body heat.[10] Males, it was said, were hot and dry, women cold and wet. This, explained Trotula of Salerno, was

> so that the excess of each other's embrace might be restrained by the mutual opposition of contrary qualities. The man's constitution being hot and dry might assuage the woman's coldness and wetness and on the contrary her nature being cold and wet might soothe his hot and dry embrace.[11]

Sexual passion was thus reduced to a circular exercise, the innate need of wet bodies occasionally to get dried, and of dry ones to be doused. The gender of an infant was also determined by the degree of moisture the fetus possessed. Females had been exposed to less heat during gestation: Males were males because they had been well "cooked" in their mothers' wombs. Men owed their strength to their inner fires, which benefited them by creating an arid, and therefore healthful, interior landscape. Men were like the dazzling Sahara, women like some miasmic Hibernian bog always overhung by drizzle. In the cool bodies of women, harmful "moistures" pooled and caused sickness unless they were expelled in the menses.[12]

Several explanations of how conception occurred jostled with one another for acceptance in the Middle Ages, but perhaps most popular were those, derived from Aristotle, which gave all credit to men as the font of movement, of life. The "spermatic secretion," Aristotle had proposed, provides either the material from which a new being is fashioned or the first cause for its creation. He tended to prefer the latter option, outlining in several passages in his writings how semen, which is motivated by a rational soul, triggers the formation of beings from menstrual blood, the female "semen," which otherwise would lie fallow, like a plot of ground unsown, since it possesses only a passive soul, of the type Aristotle dubbed "nutritive" or "vegetative."[13] The male semen, composed of particles gathered throughout the body, is not incorporated into the embryo, merely endows it with soul and incites the menstrual blood to gain form over time, with the organs appearing in serial fashion, heart first. Pliny, in the first century A.D., also contended that semen's effect was to stimulate changes in the menstrual blood, causing it to coagulate and assume human form which would "in due time" acquire life.[14]

Galen, who held a high-profile job as official physician to the emperor Marcus Aurelius in second-century Rome and left a legacy of medical investigations which profoundly influenced medieval thought, carried out first-hand observations of fetal development, examining chicken eggs and aborted fetuses of animals and humans. Galen opined that the male reproductive organs processed blood into semen, and that this process was matched in females, whose ovaries produced an actual seminal substance—not menstrual blood—which flowed down the oviduct to the womb. Upon mingling together, the two semens, male and female, were transmuted into bodily parts, passing through four distinct stages of development in the womb. This type of scheme, in which structures emerge from an undifferentiated embryo, has been dubbed epigenetic by historians of science.

Epigenesis claimed many adherents, but medical philosophers in antiquity also embraced a class of explanation that has come to be known as preformationist.[15] Preformation, broadly, is the notion that bodily parts exist in complete but extremely minute form before conception, and that gestation involves a kind of ballooning process, whereby the tiny organs and limbs inflate. Tertullian, the Christian theologian who resided in Carthage probably in the third century, fully accepted the idea that humans were found in embryonic form in semen, a premise which led him to the bizarre yet logical conclusion that fellatio was cannibalism.[16]

Whatever theory they might ascribe to regarding reproduction, many physicians, whose advice could run counter to the pronouncements of the church, continued to dispense Roman wisdom when it came to telling patients how best to influence the quality of their future offspring. They counseled that a good deal of vigor in lovemaking was necessary for conception (sterility was linked to frigidity on the part of women) and for the well-being of any child conceived. As long as a couple did not engage in "unnatural" practices such as oral or anal sex, or intercourse during menstruation, their mutual pleasure redounded to the good of future offspring. It was thought by thirteenth-century Tuscans that only a woman who had been passionately aroused would produce a handsome child.

Such ideas continued to spark discussion throughout the Middle Ages and beyond, and were accompanied by sometimes acrimonious debates over another, terribly odd type of reproduction known as spontaneous generation. Although Aristotle had suggested that lower forms of life, such as worms, insects, and fish, could reproduce without the need for a germinal factor like semen, some medieval thinkers decided that even humans might arise in such fashion. By the early 1500s, Philippus Aureolus Theophrastus Bombastus von Hohenheim, better known by his Latin moniker Paracelsus, was contending that "men can be generated without natural father and mother," through an elaborate process which he detailed in his writings:

> But neither must we by any means forget the generation of homun-culi. For there is some truth in this thing, although for a long time it was held in a most occult manner and with secrecy. . . . In order to accomplish it, you must proceed thus. Let the semen of a man putrefy by itself in a cucurbite [a vessel used for alchemical experi-ments] . . . for forty days, or until it begins at last to live, move, and be agitated, which can easily be seen. After this time it will be in some degree like a human being, but, nevertheless, transparent and without body. If now, after this, it be every day nourished and fed cautiously and prudently with the arcanum of human blood, and kept for forty weeks in the perpetual and equal heat of a *venter equi-nus,* it becomes, thenceforth a true and living infant, having all the members of a child that is born from a woman, but much smaller. This we call a homunculus; and it should be afterwards educated with the greatest care and zeal, until it grows up and begins to dis-play intelligence. Now, this is one of the greatest secrets which God has revealed to mortal and fallible man.[7]

So ardent a proponent of spontaneous generation was Paracelsus that one day he arrived for a lecture he was giving before doctors in the city of Basel armed with a supply of excrement, from which he imagined creatures such as serpents, toads, and spiders arose. When his audience marched out of the odoriferous hall in protest, Paracelsus shouted, "If you will not hear the mys-teries of putrefactive fermentation, you are unworthy of the name of physi-cians."[8] (According to the *Oxford English Dictionary,* at least one use of the word "bombastic" derives from a reference to Paracelsus, an attribution which, on the basis of this and many other tales, seems quite sound.)

Disputes within the medical community affected the thinking of the nobility and aristocracy, but hardly would have trickled down to the lower classes. Especially among the uneducated peasantry, superstition, not acade-mic discourse, exerted the major influence. In the thousands of tiny, insular villages dotting the European countryside, where official church doctrine contended with homegrown heresies and soothsaying, sex and pregnancy were seen as fraught with spiritual danger, always subject to the meddlings of malign otherworldly forces. Folk wisdom preached that Sundays and certain holy days were bad days upon which to conceive, and cautious people avoided sex accordingly. The gaze of a hunchbacked woman or a jealous neighbor; the mere presence of a corpse; the nefarious doings of spirits, who were well known to victimize women daring to venture abroad after dark—all had the power to damage a child in utero. The French historian LeRoy Ladurie provided an example of such common beliefs in the heretical statements made before a team of fourteenth-century inquisitors by Bernard Bélibaste, a cattle breeder living in the Pyrenees. Quizzed by his interrogators, Bélibaste said,

> When the spirits come out of a fleshy tunic, that is a dead body, they run very fast, for they are fearful. They run so fast that if a spirit came out of a dead body in Valencia and had to go into another living body in the Comté de Foix [now southern France], if it was raining hard, scarcely three drops of rain would touch it! Running like this, the terrified spirit hurls itself into the first hole it finds free! In other words into the womb of some animal which has just conceived an embryo not yet supplied with a soul; whether a bitch, a female rabbit or a mare. *Or even in the womb of a woman.*[19] [Emphasis added.]

The spirits of dead animals might enter humans, or those of humans enter animals, thereby conveying to the new host qualities both good and evil.

So, too, the stars and planets shifting in their distant spheres might cast a pall over pregnant women, transmogrifying fetuses into monsters. Horrific rumors of metaphysical maleficence routinely passed for fact, and medieval listeners readily accepted the truth of tales attributing disastrous events to the heavens (much as many Americans today give credence to accounts of alien landings and abductions by spaceship). Few would have thought twice, for instance, about the news from Sicily that after a solar eclipse pregnant women bled from their mouths and delivered two-headed babies.[20]

As in classical times, scapegoating the mother whenever a child was born with flaws was a popular pastime. Suffused by the supernatural, women with child might themselves cause harm to their fetuses through missteps—quite literally so in the case of the old Irish belief that a woman who twisted her ankle while walking in a graveyard would cause her child to be born with a clubfoot. A pregnant woman's path was strewn with obstacles, and even her moods could ruin the baby, sentencing it to a life of depression, insanity, or idiocy.

People in the Middle Ages believed firmly that it was possible to forecast a child's future, and they developed elaborate systems for doing so, which connected circumstance, character, and fate. The day of week or season an infant entered the world, or its physical features, determined what it would do and who it would be. For example, in Ireland it was thought that children born on Whitsunday, or Pentecost, the seventh Sunday after Easter, were destined to become murderers, or to die a violent death. Those born with crooked fingers would be liars.

Parents took such signs seriously indeed, as two separate incidents involving the same odd birth anomaly reveal. Occasionally, babies emerge from the birth canal capped by a small piece of the amnion, or innermost layer of the amniotic sac. To medieval observers, this bit of extraneous membrane, known as a caul, brought to mind nothing so much as a monk's hood. For one French father, the implication of his newborn's caul was clear: Obviously, his wife had slept with the local friar. So convinced was he of her guilt based on the evidence provided by the infant, that he proceeded to beat the poor, just-delivered woman senseless. For another parent, the same irregularity was an incitement to a more benign yet equally extreme act: A new mother read her son's caul as a message from God, who she assumed had dressed her child in the womb for his future profession. Obediently, she bound the baby over to a Franciscan order—and to a lifetime of poverty, prayer, and celibacy that he might not have chosen for himself.[21]

It is a truth universally acknowledged that patriarchies prefer sons, and the patriarchies of the Middle Ages proved no exception. No matter their economic standing, peasant to prince, husbands wished their wives to produce male offspring. As a consequence of this bias, boys almost always received far more favorable treatment than girls. In many countries, they were nursed for twice as long, usually two years as opposed to one. As toddlers, they were given preference at the table and fed diets richer in protein and fat, which enabled them to grow sturdier and more able to resist disease. Those classes able to afford medical care called in doctors to treat boys with far greater frequency than they did to treat girls, and almost always at an earlier stage in the course of an illness. Too, because girls in moneyed households tended to be confined to the domestic sphere after puberty, especially in southern Europe, boys received better educations.

The dictates of patrilineal inheritance provided both the motivation and the continuing justification for the medieval preference for sons. In terms of the all-important lineage, that *idée fixe* among everyone from the Norse to the Spanish, women were all but invisible. Individuals were tied to one another through birth and intermarriage, and tracing the lines of connection was a primary occupation, one which secured people within the web of the local community and within a greater, regional hierarchy.[22] Those without a rightful place in a lineage, or those whose more tenuous relations to a family put them far from the mainstream of inheritance, suffered economically and might themselves find it difficult to establish themselves in anything but a menial position.

Obsessed with the consolidation of property, the highest level of feudal society was perpetually wracked by inter- and intrafamilial struggles over marriage, divorce, widowhood, and remarriage, which all bore upon the legitimacy of male heirs. These fiercely fought disputes often pitched secular interests against sacred ones, and the holy side did not always win—as the later, sixteenth-century wranglings of Henry VIII of England so acutely reveal.[23] Daughters, in this game, represented a weakness. Not only did a family stand to lose property through dowry payments, they also might sully the purity of their lineage by being forced to accede to a marriage pairing a daughter to a man of lesser rank. (Of course, those with sons preferred such matches because then the "blood of [their] lineage could be irrigated by that of kings, princes, and counts.")[24]

The quest for sons led people to seek out medical and mystical advice. Because it was widely thought that the womb was segregated by gender—males grew on the warmer right side, females on the cooler left—couples ardently desiring sons would copulate only with the woman lying on her right side in the hopes of bringing about a conception there.[25] A well-configured vagina was also deemed necessary if a woman was to turn out sons: Semen needed to be able to proceed in a straight shot to the womb if it were to retain momentum sufficient to engender a male. Other prescriptions said to guarantee boys involved sympathetic magic: Some people associated hares with male potency and before intercourse ate with relish the internal organs of these beasts on the assumption that doing so would ensure the birth of a son. More elaborate proceedings described by the thirteenth-century physician Gilbertus Anglicus required young men to dig up two comfrey plants before 3:00 A.M. on the eve of Saint John the Baptist, then, while pacing back and forth reciting the Lord's Prayer three times, to milk the juice from the plants. Using the juice for ink, the man was to write the same prayer three times on a card, which he should wear around his neck while having intercourse.[26]

Although it appears that people did not as programmatically rid themselves of female babies during the Middle Ages as during antiquity, parents could and did dispense with unwanted girls. In several instances, where historians have been presented with detailed records of local populations, they have found clear evidence of skewed sex ratios. The historian Emily Coleman suggested that among peasants in France extremely poor care for female infants probably caused a high rate of attrition and led to such imbalances.[27] Parents throughout Europe, of all classes, appear to have neglected young girls in this manner. The prevalence of this type of "passive" infanticide may account for the general lack of legal documentation for "active" infanticide. Or it may be the case that child murder was far more common than any records indicate, and that drowning, overlaying (in which parents rolled over upon and "accidentally" smothered infants sleeping in the same bed), and other methods for getting rid of unwanted children were so common and accepted that only the most egregious cases ever prompted courts to prosecute.[28]

This in spite of the fact that in many communities, across all religions, bans on infanticide had begun to be promulgated. Among Spanish Visigoths in the seventh century, for example, infanticide was made punishable by

blinding or death. Hebraic laws codified by the Jewish philosopher Maimonides in the twelfth century tolerated infanticide only in the case of
deformed infants or those delivered prematurely; otherwise, taking the life of
a full-term infant after its head or the greater part of its body had emerged
from the birth canal was deemed a capital crime.[29] Meanwhile, in the larger,
Christian world, laws against infanticide grew increasingly harsh as the Middle Ages proceeded. By the reign of Charles V in the sixteenth century, it
became punishable by extreme penalties—for example, to have one's flesh
ripped by hot tongs and then be buried alive.

Universally, leniency seems to have been granted when the infant showed
severe deformities. In fact, as in antiquity, societal pressures against perpetuating the lives of such infants remained quite strong. A strange story
recounted by the historian Michel Rouche attests to the confusion which
might result when Christian dogma came in conflict with ancient precepts:

> A woman of Berry gave birth to a crippled, blind, and mute son, more
> monster than human being. She tearfully confessed that he had
> been conceived on a Sunday night and that she did not dare kill him,
> as mothers often do in such cases. She gave him to some beggars, who
> put him in a cart and dragged him around for people to see.[30]

Her shame, palpable five hundred years later, seems to have risen out of a
sense that she had violated at least three moral dicta, the first banning intercourse on the Sabbath, the second mandating infanticide for the sake of the
child and society, and the third demanding respect for life.

The woman of Berry's solution—to hand the child over to someone else—
was effectively a variant on the Roman habit of exposing infants in the
Forum. This was a route that became increasingly popular on the Continent
as the Middle Ages progressed. Parents without the will either to dispatch or
to raise their infants abandoned them. Annually across Europe, thousands of
parents laid their unwanted babies upon the church's doorstep.

The Catholic Church encouraged this trend, feeling that abandonment
was far preferable to infanticide. Possibly as early as the fourth century,
parishes began taking in and endeavoring to raise abandoned children. Some
sources report that a foundling hospital was established in Trier in southwest-
ern Germany in the sixth century and in Angers in western France by the seventh, and during this period, the infants would have been raised as serfs
attached to the church that had saved them. Certainly, foundling hospitals existed
in Milan by the end of the eighth century, in Rouen in northwestern France by
the ninth century, and in Rome (where, as one explanation has it, the number of
dead babies netted by Tiber fishermen had finally come to the attention of the
Pope) by the twelfth century, and the church's network of institutions continued
to expand, especially in France and Italy, into the nineteenth century.

Early on, the system was seen by the church as a method for saving souls. The poorest among the peasantry along with women bearing children out of wedlock had few incentives to keep their offspring, given the economic hardships or opprobrium they faced, and the church calculated, probably correctly, that such parents more often than not dispensed with their unwanted babies through infanticide. This raised the specter of untold numbers of unbaptized souls condemned to Limbo for eternity, which sufficiently disquieted Catholic policymakers that they decided to offer less affluent couples and unmarried women an escape valve.

Literally and figuratively, the church provided a revolving door for unwanted infants. Without an assurance of anonymity, mothers would not dare to leave a child, consequently many churches installed devices called wheels. These small, rotating chambers with sliding doors enabled a person standing outside the church to deliver a child unseen to a person within. A cathedral in a large town might receive dozens of children each night through the wheel, their parents depositing them and hurrying away under cover of darkness. But of the thousands of infants yearly taken in by the church, only a very small percentage survived. The turnover could be in the 90 percent range, due to poor hygiene, malnutrition, and the ministrations of the wet nurses hired by parishes to tend their tiny charges.

Indeed, consignment to a wet nurse could be tantamount to a death sentence, and people throughout Europe acknowledged this fact by the epithets they assigned to these women: in England "angel maker," in France *faiseuses d'anges,*" in Germany *"Engelmacherin"*—baby killer by any other name. In the hovels of the generally impoverished wet nurses, several babies at a time might be subjected to slapdash feeding, the hardships of extreme heat and cold in dank, dirty lodgings, and rough treatment at the hands of a woman who in many cases had smothered or drowned her own baby in order to have enough milk to sell. Wet nurses frequently also dabbled as prostitutes, and after syphilis arrived on the Continent from the New World, wet nurses who became infected could and did pass the disease on through their milk. (The notion also persisted that character traits could be transmitted through breast milk. The Italian sculptor Michelangelo, son of a government official in Caprese, an isolated hamlet north of Arezzo, was nursed by a stonemason's wife and used to say that he "drank in with the milk the hammer and chisel.")[31]

Trotula of Salerno had advised parents on the selection of wet nurses, borrowing from classical authors:

> The nurse ought to be young and have a pink and white complexion. Let her be not too near to prospective parturition nor too far removed from preceding parturition. Let her not be dirty. She should have neither weak nor too heavy teats, but breasts full and generous, and she should be moderately fat. Let her not eat salt,

sharp, acid or styptic things—leeks, onions, nor the other kinds of things that are mixed with foodstuffs such as pepper, garlic and colewort. Especially have her avoid garlic. Let her beware of anxiety and guard herself during menstruation.[32]

But by the Renaissance, when virtually every well-to-do family across Europe farmed infants out to wet nurses, it may have been the case that parents could only afford to be just so choosy. In Italy, for example, it was widely known that poor women doubled up, whoring and wet-nursing as need be to make ends meet. Fathers apparently groused about the low quality of wet nurses, but their dissatisfaction did not translate into a push for reforms. Despite a growing sentimentalization of children and childhood by upper-class parents, infants and toddlers were still assigned only marginal humanity. In fifteenth-century Florence, reported historians David Herlihy and Christiane Klapisch-Zuber,

> adults tended to ignore, neglect, and forget their offspring: infants in early life possessed a kind of transparency. . . . The custom of placing babies soon after birth with often distant wetnurses increased the chances that the father or guardian would fail to remember their existence and report their names [to fiscal surveyors gathering information for the city]. And girls were more likely to be sent for nursing outside the home than were their brothers.[33]

Withal, the trend during the Middle Ages was toward greater valuation of the young, but as the popular historian Barbara Tuchman has pointed out, this did not mean that adults held a firm place in their hearts for children. There was, she wrote, "a comparative absence of interest in children. Emotion in relation to them rarely appears in art or literature or documentary evidence."[34] High infant mortality rates—perhaps reaching 45 percent in places—must have acted as a check on parental emotions, a warning not to become too attached to offspring until they had successfully traversed the dangerous years of infancy and early childhood. But surely the process must have been circular, with an already flimsy attachment to what Plutarch had called "repulsive, distasteful" objects eroded even further by the constant depredations of disease. Upon reaching age six or seven, poorer children plunged immediately into the adult world of rural labor or servitude, while sons of noblemen often left home to become pages and complete their education in the arts of war and rulership. Tuchman conjectured that the brutality for which the Middle Ages are renowned stemmed from the indifference children faced. She wrote, "Possibly the relative emotional blankness of a medieval infancy may account for the casual attitude toward life and suffering of the medieval man."[35]

The pestilence arrived in Europe in 1347, ferried to the Sicilian port city of Messina by a boatload of sailors who had set out from the Black Sea.[36] Within months, contagion had entered the Continent through several other points and was spreading in all directions at a fantastic clip. By early the next year the Black Death had felled millions, in Florence alone killing 80,000 to 100,000 people, or approximately two-thirds of the residents.[37] Taking its highest toll in cities during late spring and summer, the harrow of this disease, the plague, passed again and again over Europe from the fourteenth to the sixteenth century, reducing the total population by about one-third its former level.

The bubonic plague terrified everyone, spread, as it seemed to be, by the mere glance of an infected person. Physicians, unaware that the disease was caused by the bacterium *Yersinia pestis,* which is transmitted between rats and humans by fleas, and between humans through fleas and, in its almost always fatal pneumonic form, by coughs and sneezes, concluded that the epidemics that decimated Europe were owed to manure and rubbish in towns and cities, standing water in bogs and marshes, and strange celestial conditions, all of which produced a "corruption and infection of the air." Rank odors were a sure signal that such corruption had taken place, and that people were in danger of viscous "miasmas" that would precipitate, like malign dews, upon a village or town. Afterward, anyone touching an object tainted with miasmatic atoms would take ill.

Children, of course, died in droves, and along with the elderly accounted for a substantial fraction of the dead, particularly in later plague epidemics when surviving adults had gained some immunity to infection. But in many places, the wholesale elimination of a generation of infants by plague was followed immediately by baby boomlets. These surges in births show up in parish baptismal records, in city tax surveys, and censuses taken by various government officials and monarchs who for purposes of taxation and military planning had begun attempting to keep track of their subjects.

Indeed, as feudal states began to consolidate power and monarchies rose, people's reproductive habits became of great interest to rulers. Whereas the church had justified its incursions into the bedroom theologically, states did so on military grounds. Living in times roiled by wars, internal revolts, and local skirmishes, leaders could not fail to miss the connection between a large population (and therefore a ready supply of young men able to bear arms) and political survival. Rulers readily grasped the equation that "the strength and riches of kings consist in the numbers and opulence of their subjects."[38] Periodic depopulations from plague and, starting about 1500, from typhus, influenza, and malaria brought the lesson home.

By the mid-sixteenth century, political theorists had homed in on population as a key element in the fate of any country and were attempting to understand the dynamics of population growth and control. An Italian author, Giovanni Botero, pondered the raw rate of increase of which humans are capable, and calculated that in just three thousand years, the descendants of Adam and Eve should have overrun the globe. That humans had not multiplied geometrically Botero attributed, as the Greeks had before, to the awesome toll taken by disease, wars, and natural disasters. But the sharpest scythe swinging across the centuries, he decided, was starvation. Population ceases to grow, despite a steady production of babies, when people bump up against the ecological limits of the land, when, wrote Botero, "the fruits of the earth do not suffice to feed a greater number."[39]

Botero surveyed a Europe in which population had stagnated for several centuries, but he lived on the cusp of a massive explosion in people. In coming centuries, the challenge to accurately monitor population would increasingly occupy the minds of scientific investigators and political theorists. John Graunt, a London mathematician who would in 1662 make primitive attempts to project the city's growth using data tallied from birth and death records kept by municipal authorities, offered this rationale for monitoring population:

> There seems to be good reason, why the *Magistrate* should himself take notice of the numbers of *Burials* and *Christnings, viz.* to see whether the City increase or decrease in People; whether it increase proportionably with the rest of the Nation; whether it be grown big enough, or too big, *&c.*[40]

During the eighteenth and nineteenth centuries, the study of populations would become intricately interwoven with ideas concerning the supposed superiority of certain individuals and groups, and with the possibility that any group possessed of exalted qualities might "degenerate" over time. Meanwhile, plant and animal breeders would gain insights into hybridization and stock improvement which would carry over into thinking about the human species, and medical researchers, newly armed with microscopes and a fervor for empirical investigation, would probe further into the mysteries of reproduction, claiming for the emerging discipline of science what had, for a thousand years or more, belonged unequivocally to theology.

5

Of Eggs
and Sperm

By the time the master frescoist Giotto began working on the elegant yet emotionally charged cycle of paintings whose brilliant hues cover every square inch of the barrel-vaulted Madonna dell' Arena Chapel in Padua, then an independent republic neighboring Venice, the university there was already a century old, having opened its doors in 1222. Just as Giotto would in Padua revolutionize painting, rejecting the flat Byzantine portrayals of the human body and bringing a classical naturalism to his figures, so, too, the teachers and students of medicine at the nearby university would repudiate the medieval tradition and embrace a philosophy that looked to the material workings of the body for the sources of disease.

From the fifteenth to the seventeenth century, the university in Padua graduated a host of men (it granted no degrees to women until the mid-1600s, when Elena Cornaro Piscopia took the first) who gained fame as Europe's finest medical practitioners. It proved a training ground for dozens of pioneering anatomists, whose meticulous investigations of tissues, organs, and bodily systems radically altered the art of medicine. The school counts among its alumni dozens of illustrious figures, including, for example, Thomas Linacre, court physician to Henry VIII and founder of England's College of Physicians. Among his other accomplishments, Linacre stunned the medical world with fresh translations of Galen from the original Greek into Latin which corrected numerous errors that had crept in over the centuries.

Padua also boasted a stellar faculty. While teaching there, Andreas Vesalius all but single-handedly restored the practice of human dissection, which had fallen into disrepute during the Middle Ages. Remarkably observant and talented as a draftsman, Vesalius challenged conventional wisdom by claiming that Galen's supposed insights into human anatomy, which had the force of law among medieval practitioners, had been gained through dissections not of humans but of other animals. From his own studies, Vesalius could assert with confidence that Galen had, as a consequence of analogizing from animals, gotten many things wrong. In 1543, in an ambitious bid to correct understanding, Vesalius published *De humani corporis fabrica libri septum (The Seven Books on the Structure of the Human Body)*, complete with illustrations commissioned from the studio of Titian. This masterpiece of research, commonly known as the *Fabrica*, is recognized as a foundation text of modern anatomy. On the strength of its publication, Vesalius won a position as physician to Charles V, the Holy Roman Emperor. For the remainder of his career, he garnered wealth and acclaim, all the while pushing his progressive medical views with the goal of overthrowing the humoral theory of the body.

Arriving at the university on the heels of Vesalius and teaching there from 1551 to 1562 was Gabriele Fallopio, a former canon of the cathedral of Modena who had turned physician and come to Padua after stints at the universities in Ferrara and Pisa.[1] Simply hearing the terms that have entered the lexicon— "fallopian aqueduct," "fallopian arch," "fallopian canal"—one might assume Fallopio was an architect. In fact, like Vesalius, he was a student of human architecture, a supremely skilled explorer of bodily structures. Peering into the abdomen of a female corpse, Fallopio was the first postclassical anatomist to describe the pair of ducts linking the outlying ovaries to the uterus, which thenceforward bore the name fallopian tubes.[2] He also was the first to use the terms "vagina," which he borrowed from the Latin for sheath or scabbard, and "clitoris," from the Greek *kleitoris,* or mound. Stressing the importance of the hands-on approach to learning, Fallopio fought along with Vesalius and other physicians to discourage the scholastic exercise of endlessly explicating and amending ancient texts, advocating instead the importance of direct observation for the advancement of medical understanding.

Padua's heyday coincided with a huge expansion in scientific investigation across Europe. During the Middle Ages, the word "science" had been used interchangeably with the word "art" and signified simply a branch of study, as in the "sciences" of grammar, divinity, and law, or the "medicinal science." Although the word "scientist" would not be coined until 1840, the fifteenth century was full of thinkers who questioned the old methods of inquiry and asked how it was that they might know something as true.[3] Rather than viewing experience merely as confirmation of received general principles (e.g., men are hot and dry, women are cold and wet), and attributing exceptions to the intervention of mysterious powers, these revolutionaries proceeded from the assumption that knowledge must be tested. If this process challenged accepted paradigms, so be it.

Lines between disciplines were loose, and no clear career paths existed for those who would pursue scientific studies. Any educated person might engage in legitimate research. Thus, for example, Leonardo da Vinci could paint and write as well as carry out pioneering work in anatomy, studying live models and dissecting corpses for his intricate pen-and-ink studies of the musculoskeletal system. Even those who dedicated themselves solely to scientific investigations recognized few hard-and-fast boundaries between fields, and most moved without constraint from pursuit to pursuit, now mulling over mathematics, now studying the heavens, now pondering plants and animals.

Many Renaissance thinkers adopted the precept that knowledge must be based first and foremost upon observation. Some practitioners of the attempt to acquire understanding through material means displayed greater rigor than others, and certainly, the approach was not adhered to universally by those engaged in scientific pursuits, but those investigators dedicated to observation drove the field forward in the days before philosophers began codifying the so-called scientific method.

By 1620, the Englishman Francis Bacon was championing the materialist view, proclaiming the bankruptcy of reasoning on the basis of abstractions in his widely distributed treatise *Novum Organum*. In pursuing knowledge, Bacon argued, it was necessary to amass facts first with no preconceptions, and only afterward draw conclusions. But Frenchman René Descartes would counter Bacon in 1637 with his *Discourse on Method* (originally titled "The Project of a Universal Science which can elevate our Nature to its highest degree of Perfection"), in which he contended that reason operates in the absence of sense experience. According to Descartes, logic and certain fundamental concepts exist in a realm outside the material world yet accessible to the human mind. The mind, straddling the metaphysical and physical, is capable of working upon abstractions in the absence of experience to gain knowledge deductively.

Bacon's preferred method of inquiry, known as induction, served as the cornerstone for the school of philosophy known as empiricism, which was propounded most ardently by the Englishmen John Locke and David Hume. Descartes, in turn, spawned the rationalist school, and provoked a battle which raged for several centuries, concerning the merits of so-called a priori knowledge, derived through pure reason, and a posteriori knowledge, attained through the accumulation of mundane detail. (Although the debate continues in philosophical forums today, modern experimentalists in practice employ both approaches in a fluctuating mix, with hypotheses—the a priori contribution—constantly being tested and refined through experiment and observation—the a posteriori contribution.)

Men like Fallopio, involved in the effort to unlock the secrets of reproduction, were empiricist at heart, and for them, close inspection and dissection were the key to understanding. Partially incubated bird eggs were a favorite object of study, as were aborted fetuses of any sort. Not only human bodies went under the knife, but also hens, mares, ducks, does, cats, sows, ewes, bitches, rabbits—any female, pregnant or otherwise, the would-be anatomist could obtain. One Italian physician cut open the body of a hanged woman and claimed to have discovered in her uterus semen from an intercourse several days before her death.[4] Anatomists attempting to comprehend reproduction also carried out nonsurgical examinations of gravid animals and pregnant women. These researchers communicated their results to one another in personal correspondence and, by the mid-1600s, in formal letters to organizations like the Royal Society of London for the Improvement of Natural Knowledge and, still later in the century, in papers appearing in fledgling scientific journals.

It is generally agreed that Fallopio's student and successor Hieronymus Fabricius ab Aquapendente, also known as Girolamo Fabrici, instituted the field of embryology. Fabricius carried out studies of embryos from a range of species, presenting his results in an illustrated volume titled *De formato foetu (On the Formation of the Fetus)*, which came out sometime between 1600 and 1604. Published posthumously, his more focused *De formatione ovi et pulli (On the Formation of the Egg and of the Chick)* tracked the step-by-step development of embryonic chicks. The book contained a suite of generally accurate illustrations portraying emerging physical features, noted by the naked eye periodically during incubation.

For all his visual acuity, Fabricius remained confused, as did his peers, about the fine points of conception. He, along with many others, believed that the process differed fundamentally between egg-laying (oviparous) animals like birds and non-egg-laying (viviparous) creatures, which bear live young.

Fabricius knew that male birds—for example, roosters—emit semen. Common folk held that in the case of chickens, the semen could be seen inside the egg in the guise of the chalazae, the twisted white strands within the albumen which hold the yolk in place. But Fabricius rejected this barnyard lore, finding no evidence that the semen of male birds either made its way into the female or penetrated the eggshell. He concluded that

> the semen of the cock thrown into the commencement of the uterus, produces an influence on the whole of the uterus, and at the same time renders fruitful the whole of the yelks [yolks], and finally of the perfect eggs which fall into it; and this the semen effects by its peculiar property or irradiative spirituous substance, in the same manner as we see other animals rendered fruitful by the testicles and semen.[5]

In suggesting that semen might work such broad biological effects at a distance, fecundating the entire female reproductive system through its "peculiar property," Fabricius was not laying claims that would have been considered bold or outlandish. Aristotle, in fact, believed that females might be impregnated by the wind, and that a stiff northerly breeze would cause women to give birth to sons. Fabricius considered that semen provided the all-important motive force which fashioned the material within the shell—specifically the chalazae—into a chick.

The viviparous animals posed a far greater puzzle. Fabricius had opened up the bodies of numerous mammals and never discerned a shell of any sort inside them. At a loss to explain this absence, he fell back on Galenic notions, and assumed that among viviparous animals reproduction did not involve eggs. Rather, male and female semens must mingle in the womb to create the fetus. This alternative interpretation had the added benefit of accounting for the gender of offspring, which resulted simply from the proportions in which males and females contributed semen. If the male semen were abundant, the offspring would be male; if vice versa, female. Having divided up the zoological kingdom into two unique types, Fabricius also left open a third possibility, namely, that some creatures might enter the world through spontaneous generation.

Although Fabricius performed the service of dispelling many inaccuracies regarding embryological growth, his work eventually drew fire on methodological and theoretical grounds from one of his former students, William Harvey, who had listened to many a lecture by the master anatomist in the candlelit precincts of Padua's theater. Harvey had upon receiving his medical degree from Padua in 1602 returned to his native England and almost immediately ascended to the highest ranks of the profession. As a lecturer to the College of Physicians in London, he vigorously pressed the case for empiricism, contending that medicine should be taught "not from books, but from dissections, not from the positions of the philosophers, but from the fabric of nature."[6] Most widely known for his inquiries into the heart and circulation of the blood, Harvey also made major contributions to the field of embryology. In 1628, he became personal physician to King Charles I, and from that time onward intensified his study of reproduction, having been granted the right to dissect any game killed on royal preserves. Harvey thus gained access to hundreds of deer and other animals. He finally summed up his observations in *Exercitationes de generatione animalium (Exercises Concerning the Generation of Animals)*, released in 1651, when he was seventy-three.

Hardly a dry tome on matters scientific, the work offers a chatty, digressive discussion of the sexual anatomy, mating habits, and embryology of birds (including the common fowl, Harvey's wife's pet parrot, and a rare European specimen of emu which was presented to King James by Maurice, Prince of Orange), and of a mixed bag of other animals, from snakes to deer to humans. Harvey haunted the henhouse to produce a painstaking account of the daily changes that take place in chicken eggs during incubation, an account differing in critical details from that earlier presented by Fabricius. He also carried out systematic examinations of aborted fetuses, and gathered eyewitness accounts of the reproductive behaviors of wild and domesticated animals. As a result, he felt justified in leveling critical remarks not only at Fabricius, but also at The Philosopher himself, Aristotle. Harvey veers between cold logic and the occasional scathing outburst in dealing with the two thinkers he had chosen as his "especial guides and sources of information on the subject of animal generation," although he is notably more gingerly in addressing the flaws of the venerable Greek than of his own, less exalted, instructor.[7]

Harvey's most penetrating charge against his former professor was that Fabricius had betrayed the cause of medical science by choosing, at a certain point, to follow "probabilities according to previous notions rather than inspection" and by abandoning daily observations of the egg "up to the period when the foetus is perfected" for "mere opinions or musty conjectures."[8]

As regards Aristotle, Harvey begged to differ with a number of points concerning the growth of embryos, most importantly Aristotle's declaration that the first item to form in the course of development was the heart. Harvey insisted instead that blood had primacy. Blood, he argued, is "the generative part, the fountain of life, the first to live, the last to die, and the primary seat of the soul," and is concocted from the same substance that composes all the other parts of the chick.[9] Not only that, but the quality of any individual of any species, he said, lies in the blood. In this, Harvey concurred with Aristotle, who had considered wisdom and nobility, timidity, courage, passion, and furiousness to be qualities inhering in the blood. "To the blood, therefore," Harvey wrote, "we may refer as the cause not only of life in general . . . but also of longer or shorter life, of sleep and watching, of genius or aptitude, strength, etc."[10]

Harvey had no doubt that both physical and psychological characteristics could be passed on from parents to offspring in certain instances, and happily drew parallels between animals and humans:

> And this too is a remarkable fact, that virtues and vices, marks and moles, and even particular dispositions to disease are transmitted from parents to their offspring; and that while some inherit in this way, all do not. Among our poultry some are courageous, and pugnaciously inclined, and will sooner die than yield and flee from an adversary; their descendants, once or twice removed, however, unless they have come of equally well-bred parents, gradually lose this quality; according to the adage, "the brave are begotten by the brave." In various other species of animals, and particularly in the human family, a certain nobility of race is observed; numerous qualities, in fact, both of mind and body, are derived by hereditary descent.[11]

In essence, this is Greco-Roman eugenicism given new expression. Here, blood is a river that flows from parents to children, transmitting and permuting life down the generations. Soul and individual character course through the same channel. Numerous thinkers advanced similar arguments during this period, some suggesting as well that the parts of the body could impart "virtues," meaning positive attributes, to the blood. A strong heart, for example, would somehow affect the blood, so that when that blood was conveyed to offspring, they, too, would have strong hearts.[12] This meant that just as blood can be lost through a wound, so, too, soul and character might vanish from a line, should insufficient care be taken in breeding.

While Harvey normally exercised enormous care to reason only from the facts at hand, on the matter of the blood he betrayed a metaphysical bent typical of his times, abandoning mechanical explanation to assert that blood is "the instrument of the Great Workman," of God.[13] "Inasmuch," he opined, "as it seems to partake of the nature of another more divine body, and is transfused by divine animal heat, it obtains remarkable and most excellent powers, and is analogous to the essence of the stars."[14] This tendency to mystify the blood and inheritance would manifest itself again and again in Europe over the ensuing centuries, attaining its most perverse expression in the Nazi ideology of *Blut und Boden*, or "Blood and Soil."

Harvey's principal theoretical dispute with Aristotle and his followers stemmed from their assertion that semen alone conveys form and vitality to the eggs of oviparous animals and to the menstrual blood of viviparous ones. Harvey contended that an egg contains "matter, organ, efficient cause, place, and everything else requisite to the generation of the chick."[15] Elsewhere, with greater biological specificity, he added, "An egg is a body, the fluids of which serve both for the matter and the nourishment of the parts of the foetus."[16] Clearly, a rooster contributes something which is "transfused into the hen, and from here is given to the uterus, to the ovary, to the egg," but this generative force is matched by a similar force within the egg.[17] Harvey argued that, like all eggs, a chicken egg

> is a true generative seed, analogous to the seed of a plant; the original conception arising between the two parents, and being the mixed fruit or product of both. For as the egg is not formed without the hen, so is it not made fruitful without the concurrence of the cock.[18]

In equating the eggs of oviparous creatures with the seeds or "sperma" of plants and in providing for a contribution of "plastic power" from the female side, Harvey shifted males out of the central role in reproduction which they had held unassailably since classical times. Neither males nor females alone are "genetic," he argued, "but become so united *in coitu.*"[19] Distinctively, Harvey placed eggs in a biological continuum, seeing them as "the mid-passage or transition stage between parents and offspring, between those who are, or were, and those who are about to be."[20]

Like Fabricius, Harvey had failed repeatedly in dissecting viviparous animals to locate an egglike entity. Neither had his investigations revealed any trace of semen in the wombs of animals after copulation, nor even the presence of an embryo in the wombs of animals which, from external signs, appeared to have conceived. So he could hardly credit the physicians who maintained that the semen, or "spermatic fluid," constitutes "the prime matter of the conception."[21] Moreover, women he had surveyed did not universally claim to emit semen during intercourse, as the physicians said they should; if anything, they might discharge "a fluid in the sexual embrace," a fluid similar, Harvey said, to urine, and unlikely to qualify as a "fruitful," fecundating substance.[22] Having dismissed the possibility that fetuses were shaped from semen, he also rejected the idea that they arose from coagulated menstrual blood. In fact, in his inspections of dogs, rabbits, and other animals, Harvey found "that there is not necessarily even a trace of the conception to be seen immediately after a fruitful union of the sexes," nor even several days afterward.[23]

How, then, did conceptions occur in viviparous creatures? From what substance were fetuses made? Harvey knew, he had witnessed, that women miscarrying after about a month of pregnancy sloughed off amid the blood a smooth, shiny mass "the size of a pheasant's or pigeon's egg." Clad in a membrane of "white mucor" and containing only "limpid and sluggish water," such abortuses resembled eggs minus the shell.[24] This must mean, Harvey decided, that even if he could not identify where eggs were manufactured, mammals must indeed produce them (perhaps in the uterus).[25] The logical conclusion, then, was that the ovum, or egg, constitutes the universal germ, and that fetal development proceeds basically in the same fashion for all species, by "epigenesis, or the superaddition of parts," not metamorphically or instantaneously, but through a gradual process by which tissues, organs, and bones "emerge in their due succession and order."[26]

Harvey's pronouncements by no means persuaded the entire community of those endeavoring to comprehend reproduction that such a thing as a mammalian egg existed. In fact, for the next two hundred years, the field continued to be riven by messy disputes, inaccurate observations passing as fact, and intellectual balkanization. There were indeed broad divisions: ovists (the egg men) versus spermatogeneticists (those believing in the exalted powers of semen); and epigeneticists versus preformationists. But these categories were nonexclusionary and allowed for a kind of multiple-choice position taking, so that, for instance, an ovist might also be either an epigeneticist or a preformationist, while a preformationist might also be an ovist or a spermatogeneticist. In general, ovism and preformation were more popular than spermatogenesis and epigenesis, but even individuals flip-flopped, so keeping historical score is an intensely frustrating exercise. An attempt to track people's positions reveals, however, how thoroughly theological concepts influenced scientific theories—and impeded scientific progress. Most investigators were at pains to fit their explanations of how embryos arise and grow into the larger scheme of God's creation of the world. Had He brought everything into being all at once, or was He bringing matter constantly into being? (Usually stripped of reference to divine agency, similar questions plagued cosmologists in the 1940s and 1950s, who struggled to determine mathematically whether the universe came into existence instantaneously, in a big bang, or had always and would always exist, accreting slowly, with matter added in an ongoing fashion to maintain a steady state.)

To the modern observer, the weirdest of the ideas circulating during this period has to be preformation, the premise that the embryo does not develop by stages, but that bodily structures or entire creatures preexist in minute form and are scaled up in size during incubation or gestation. Preformationism (also, rather confusingly for post-Darwinian readers, known as evolution) was ascendant during the eighteenth century, especially among German and Swiss philosophers, and represented an accommodation to religious views, in that it presumed a single act of creation in which God had provisioned the planet with all the elements life would need to keep running until the end of time.[27]

Many preformationists espoused a variant of the theory known as encasement (referred to as *Einschatelung* in German and *emboîtement* in French). Imagining that every creature on the planet must have existed in perfect completion from the beginning, encasement theorists envisioned a near-infinite series of beings stacked one within the other like Chinese boxes. For Gottfried Wilhelm Leibniz, the great mathematician and theologian, these telescoped beings had no substance, rather were immaterial souls. But others, like the respected Swiss natural historian Charles Bonnet, who was among the first to suggest that fossils represented species that had been destroyed in earlier geologic catastrophes, took the model literally, and proposed that the great hordes of humanity had, from the moment God crafted man and woman, been nested neatly inside Adam's sperm or Eve's ovaries. In a strange iteration of the medieval quest to estimate how many angels might fit upon the head of a pin, a few proponents of encasement absurdly attempted to calculate the number of submicroscopic humans that might have been packed into Eve, in hopes of thereby being able to forecast when the last "box" would empty, and the human race die out.

In addition, many, including Harvey himself, continued to allow for that exceptional form of reproduction, spontaneous generation, whereby bees, mice, lice, worms, and other (usually verminous) creatures were supposedly produced. Throughout the seventeenth and eighteenth centuries, supporters of spontaneous generation disregarded several clear demonstrations that the phenomenon was bogus and some diehard advocates persisted in making a case for it until the 1870s.[28]

The lack of adequate magnifying equipment hampered Harvey and other embryologists, who, when their eyesight proved inadequate, sometimes fell back on intuition. This soon changed. The year before Harvey's death, the man who would boldly lead embryology into the future took up a position teaching medicine at the University of Pisa.

Although his forty-year career was plagued by controversy and personal misfortunes which forced him to move from institution to institution, Marcello Malpighi managed to produce a prodigious body of work and stands as one of the great medical researchers of history. Where Galileo peered upward at the moon and planets using the primitive telescope, Malpighi gazed into the depths of living matter through handheld lenses and the eyepieces of both simple single-lensed and compound microscopes. The compound microscope had been invented earlier in the century, although it would not be fully perfected until the 1800s, and Malpighi may not have always had access to this superior instrument, but he ardently advocated use of the new tools.[29] Having with the aid of magnification identified the capillary system in 1661, Malpighi went on to elucidate cellular structures of assorted organs, including the brain and eye; of tissues, like fat and deeper layers of skin; and of the blood. He distinguished red blood cells in 1666. For this and other findings, he gained international recognition, not to mention the enmity of envious Italian colleagues.

Malpighi made his primary contributions to embryology while at the University of Bologna, where he studied the development of insects as well as chickens. Reiterating Fabricius and Harvey's studies of incubating eggs, he spied upon the early development of embryos (prior to the fifth day), making out features his predecessors had never been able to see with the naked eye. Malpighi was such an acute observer, he often perceived what even other embryologists working with the microscope around the same period missed.[30] He himself attributed his success in discerning patterns within the "obscure and chaotic" hen's egg to a technique he had developed for cutting away a bit of the yolk and thereby removing the blastoderm, or disk-shaped mass of dividing cells that forms the early embryo, so that it could be placed upon a glass slide and studied over and over again.[31]

Operating with few presuppositions, Malpighi endeavored to portray in his pencil drawings precisely what he saw upon his slides, and to describe embryological development in unfashionably direct narratives that left out the customary colorful asides and references to earlier philosophers. His *De formatione pulli in ovo (On the Formation of the Chick in the Egg)*, delivered in 1673 as a letter to London's Royal Society, is a model of restraint that prefigures the modern scientific paper in its no-fuss, no-bother approach. In it, Malpighi elucidated the changes to be seen during the incubation period with far greater specificity than anyone previously had achieved. Some years later, providing a thumbnail description of the process of development, Malpighi wrote that

> it is permissible to conclude that from fluid material Nature forms cavities surrounded by solid substance, that these cavities admit appropriate matter, which is consumed in the construction of the parts, and that when this solid texture of the parts has been laid down as the basis of the body, canals become visible, which with their own fluids perform motion, sensation, nutrition, and the remaining functions.[32]

Despite the accuracy of his microscopy, Malpighi persisted in believing that some anatomical parts preexist, although never prior to fertilization. Rather, he thought, the act of fertilization forms certain structures, like the umbilical vessels, and incubation renders them visible. Some historians have lamented that Malpighi, whose findings so greatly clarified the pursuit of embryology, wound up perpetuating the erroneous postulates of preformationism rather than smashing them with empiricist vigor. In fact, Malpighi's rather subtle views bore only tangential relation to mainstream preformationism and he took no pains to defend the theory. However, his statements on the subject were subsequently distorted, mainly by the French priest and philosopher Nicolas Malebranche, a devotee of Descartes and proponent of *emboîtement*.[33]

Malpighi was hardly alone among microscopists in failing to hack a path through the theoretical thicket. Jan Swammerdam, a brilliant Dutchman, carried out nonpareil investigations of insects and frogs, and described, in 1672, the ovarian follicle of mammals, yet he nominally supported preformationism, and, like Malpighi, was often cited by Malebranche.[34] His countryman, Antonie van Leeuwenhoek, the first to make scientific report of the existence of sperm, accepted the reasonableness of preformation as well.

Notwithstanding Leeuwenhoek's attitude on this matter, his identification of human sperm would, along with the discovery of the mammalian egg, ultimately contribute to the demise of preformationism.

In November 1677, Leeuwenhoek with a certain degree of reluctance conveyed a piece of intelligence to the Royal Society of London. A draper by training, the Delft resident had already become well known for his outstanding microscopic observations, performed with the aid of lenses he ground himself by methods that were a closely guarded secret. His friend and fan, the

Delft physician Regnier de Graaf, had put him in touch with the Royal Society of London in 1673, and Leeuwenhoek afterward dispatched regular bulletins across the North Sea. His reputation spread locally as well, and accordingly he was visited at home one day early in the fall of 1677 by a Mr. Ham, who arrived bearing a sample of semen in a glass vial. The semen had issued from a man infected with gonorrhea, and Mr. Ham had been startled to see swimming in the fluid animalcules. Leeuwenhoek reported in a letter to the Royal Society that Mr. Ham believed the creatures "to have arisen by some sort of putrefaction. He judged these animalcules to possess tails, and not to remain alive above twenty-four hours."[35]

Often appearing in its original Latin form *animalculum* (meaning "little animal"), animalcule had before this time been used to refer to spiders, beetles, worms, and other tiny creatures. The *Oxford English Dictionary* credits Leeuwenhoek as the first to have applied the word in the sense of an animal so small as to be visible only by microscope. Indeed, when Leeuwenhoek peered down through his instrument at the semen Mr. Ham had provided, he, too, "saw some living creatures in it." After disclosing this astonishing information, Leeuwenhoek in an entirely offhand way confessed that he had

> divers times examined the same matter (human semen) from a healthy man (not from a sick man, nor spoiled by keeping for a long time, and not liquified after the lapse of some minutes; but immediately after ejaculation, before six beats of the pulse had intervened): and I have seen so great a number of living creatures in it, that sometimes more than a thousand were moving about in an amount of material the size of a grain of sand.[36]

His English audience greeted this revelation with a certain amount of doubt and shock. As everyone of that day knew, animalcules frequently were the direct product of decay, found in fermenting fluids or rotting flesh. To aver that semen swarmed with animalcules thus implied that a process of rot must be occurring in men's bodies, which was needless to say an extremely disturbing thought to that most august of male bodies, the Royal Society. Secretary Nehemiah Grew responded to the Dutchman's communiqué, suggesting that he might dispel the skepticism of the society's members by perusing "the semen of animals, as of dogs, horses, and others," to which Leeuwenhoek responded testily, "I well know there are whole Universities that won't believe there are living creatures in the male seed: but such things don't worry me, I know I'm in the right."[37] Leeuwenhoek went ahead and expanded his research anyway, and sketched the additional spermatic animalcules he saw, publishing these results in the *Philosophical Transactions,* the Royal Society's house organ.

As it happened, another Low Country researcher, Nicolaus Hartsoeker, independently recognized spermatozoa in human semen in 1678, before news of Leeuwenhoek's findings had circulated. Hartsoeker claimed not only to have spotted spermatozoa, but to have seen inside their teardrop-shaped heads fully formed human beings, or homunculi. In a 1694 work he included a drawing of a prototypical homunculus, huddled within a long-tailed sperm. The little man had his knees drawn up but was faceless: Oddly, a large, pale sphere blazoned with a dark diamond obliterated his head.

Hartsoeker's disclosures sowed greater confusion in the field, causing a schism among preformationists, a fair number of whom took up the case for animalculism, also known as spermism. This splinter group contended that God had installed the world's preexistent creatures in Adam's sperm rather than Eve's eggs. The homunculus gained such fame that the English comic-novelist Laurence Sterne referred freely to him in his wildly successful novel *Tristram Shandy,* serialized from 1759 to 1767, for which he was lionized in London and abroad. The homunculus, Sterne wrote, "may be benefited, he may be injured,————he may obtain redress;————in a word, he has all the claims and rights of humanity which Tully, Puffendorff, or the best ethic writers allow to arise out of that state and relation."[38]

The sheer waste involved in the spermist scheme did give some thinkers pause, for hordes of neatly crafted homunculi must perish for every one that took hold in a womb. But such reservations did not worry true supporters of the theory overmuch, because they assumed there must be a reason behind this divine design, even if they couldn't immediately perceive it.

Battle lines were drawn, and throughout the 1700s researchers periodically announced victory, claiming to have categorically proved or disproved spermism. In 1748, the Jesuit father John Turberville Needham, a confederate of the influential French naturalist Georges-Louis Leclerc comte de Buffon, asserted that putrefaction yields not only sperm but also other animalcules (today we identify them as bacteria, yeasts, and other microorganisms), which might spontaneously generate into new beings. Needham argued that there is "a vegetative Force in every microscopical Point of Matter, and every visible Filament of which the whole animal or vegetable Texture consists."[39] Thus, sperm were not alive, merely potentially alive, falling into the class of "vital atoms" which continuously combined and recombined to form new beings.

Needham's views so distressed the Italian polymath Lazzaro Spallanzani that he undertook extensive experiments with sperm and other substances to disprove them. A man of diverse interests, Spallanzani also translated the *Iliad* and penned a disquisition on the physics of stone skimming. He appears to have been among the first to perform artificial insemination, attempting the procedure with frogs and, notably, dogs. He described the result of his manual insemination of a female spaniel:

> Sixty-two days after the injection of the seed, the bitch brought
> forth three lively whelps, two male and one female, resembling in colour

and shape not the bitch only, but the dog also from which the seed had
been taken. Thus did I succeed in fecundating this quadruped; and I
can truly say, that I never received greater pleasure upon any occasion,
since I first cultivated experimental philosophy.[40]

A confirmed ovist, Spallanzani objected to assertions by Buffon and Needham that sperm and other animalcules were not alive. In one set of experiments, he demonstrated that animalcules vanished after liquids were boiled
for a sufficient length of time and then covered tightly so that no air could
touch them. This showed, he said, that animalcules were parasites. Heat also
killed sperm, so they, too, must be parasites. Following this line of reasoning,
Spallanzani experimented with toad semen and eggs, and with filtered semen
to see what powers it might retain, in each case coming to the conclusion that
while direct contact with the male ejaculate was essential to conception, the
viscous medium in which sperm swam was the fecundating substance, not
the sperm themselves.

The debate over spermism raged on until the 1800s, but by 1827, it was
clear that the spermists had lost. The demise had been long in coming, and
was owed, finally, to the discovery of the mammalian ovum, that item which
Harvey had proclaimed so essential but had never quite located. After Harvey, the hunt had centered on the ovaries, then known as the *testes muliebre,* or
"woman's testes." Leeuwenhoek's colleague, the physician Regnier de Graaf,
had argued that the *testes muliebre* were not akin to the male testes, as it had
been thought, rather to the ovaries of birds. In 1664, de Graaf peered at a rabbit ovary and saw a distinctive nodular mass upon it, and concluded that he
had found the mammalian egg. As it turned out, he had only located the
source of the egg, but he put the searchers on the right track. Malpighi got
wind of de Graaf's finding, and agreed, in a June 7, 1672, letter to the secretary
of the Royal Society, that "eggs are to be found in female testes, even in the
young of animals shortly after birth," and that they entered the uterus
through the fallopian tubes.[41] Jan Swammerdam, who made similar investigations to de Graaf, may have been the one to suggest that the *testes muliebre* be
renamed the *ovarium.*

But it took another century and a half for egg men to deliver the crowning
blow to spermism. It came in 1827 when German embryologist Karl Ernst
von Baer cut open a pregnant bitch and by chance noticed a *Köperchen,* a
small yellow speck, within each of the follicles inside the dog's ovaries.

In his autobiography, von Baer described his reaction:

> "Odd!" I thought, "What can this be?" I opened a follicle and
> removed the fleck with a knife into a water-filled watchglass which
> I took to the microscope. When I had inspected it I fell back as
> though struck by lightning because I saw a very small, clearly

defined, yellow yolk sphere. I had to recover from this before I had the courage to look again because I feared that a ghost had deluded me. It seems strange that a sight that one has anticipated and hoped for can startle one when it materialises.[42]

Von Baer went on to show that humans start life in the same fashion as dogs and other mammals: as ova. Although not immediately recognized, von Baer's dictum that the "foetus derives its origin from an ovum already formed in the ovary before fecundation" was within ten years accepted as a profound contribution to the study of reproduction.

By that time, anyone interested in the subject was busy in the laboratory carrying out their own artificial inseminations of everything from worms to frogs to rabbits to dogs. Such investigations revealed over the coming decades that semen did not have "spirituous" powers or ability to act at a distance; that semen itself was not a fecundating fluid, but instead a medium for sperm; that sperm and egg together initiated conception; that eggs and sperm were specialized cells, each containing half the normal store of hereditary material in a central pronuclei; and that to initiate embryonic growth, the pronuclei of sperm and egg must fuse.

Not only eggs but sperm, not only sperm but eggs: As Harvey had anticipated, both parents make a contribution to offspring in sexually reproducing species. More than that, German embryologist August Weismann contended, the mechanism whereby parents passed on characteristics to their offspring was contained in the egg and sperm, or "germ" cells, in the form of "germ plasm." A person's traits were owed to certain "determiners" in the germ plasm, and this germ plasm was inviolable. So that it would not be the case, as French biologist Jean-Baptiste Lamarck had suggested, that adaptations made by a parent in response to environmental pressures would be conveyed to children. While their bodily, or somatic cells, might be altered, said Weismann, their germ plasm would remain just as they had inherited it.

Even as Weismann's ideas were winning converts throughout the biological world, an obscure monk named Gregor Mendel was cultivating various strains of peas in his monastery garden in what is now Brno, Czechoslovakia. Mendel's experiments would reveal the patterns whereby traits appeared, disappeared, and reappeared in ensuing generations, and thereby open the way for systematic explorations of the intricacies of heredity.

6
Quantity
Is the Problem

It is a melancholy object to those who walk through this great town [Dublin], or travel in the country, when they see the streets, the roads, and cabin-doors crowded with beggars of the female sex, followed by three, four, or six children, all in rags, and importuning every passenger for an alms. . . .

I think it is agreed by all parties, that this prodigious number of children in the arms, or on the backs, or at the heels of their mothers, and frequently of their fathers, is, in the present deplorable state of the kingdom, a very great additional grievance; and, therefore, whoever could find out a fair, cheap, and easy method of making these children sound and useful members of the commonwealth, would deserve so well of the public, as to have his statue set up for a preserver of the nation.[1]

So begins Jonathan Swift's "A Modest Proposal," which delivers among other things the most darkly ironic commentary on the relationship between children and politics ever penned. In Swift's Ireland of 1729, English landholders, often absentee, controlled vast stretches of the countryside, and English ships bellied into the docks loaded with cheap imports which destroyed the market for domestic goods. Forbidden to cut its own trade deals abroad, Ireland steadily lost jobs and wealth. The Crown's imperialistic policies drove an already impoverished populace further into the hole.

Aiming for maximal political impact, Swift offered a sinister solution to the Irish difficulties: The excess babies should be fattened and sold for food. They would be "most delicious and nourishing . . . whether stewed, roasted, baked or boiled," and might serve equally well "in a fricassee or a ragout." Swift calculated "that of the hundred and twenty thousand children [of poor parents annually born], twenty thousand may be reserved for breed, whereof only one-fourth part to be males; which is more than we allow to sheep, black cattle, or swine." The remaining "hundred thousand may, at a year old, be offered in sale to the persons of quality and fortune through the kingdom." One child should be enough to make two dishes at a dinner party; "the fore or hind quarter will make a reasonable dish" when the family dines by itself, and, "seasoned with a little pepper or salt, will be very good boiled on the fourth day, especially in winter," wrote Swift.[2]

Through cannibalism, Ireland could launch a homegrown (literally) industry, which would fatten its coffers. This macabre expedient would have many advantages, which Swift enumerated, among them that, owing to their new-found fiduciary interest in their offspring, "men would become as fond of their wives during the time of their pregnancy, as they are now of their mares in foal, their cows in calf, or sows when they are ready to farrow."[3]

For all its obvious satire, Swift's proposal adumbrated a number of very real circumstances and cultural assumptions of the age. Dublin, like London, Paris, Milan, and other cities of the period, overflowed with those who had migrated from rural areas, hoping to leave the iffiness and penury of the agricultural life behind. For despite the fact that overall food production climbed steadily throughout the 1800s, agriculture was still a hit-or-miss enterprise, and food shortages and outright famines flared throughout the century.

In England, the uncertainties of farming had been exacerbated by shifts in land use, which removed the safety net for many in the lowest rural classes. Villagers had by long tradition been allowed by the gentry to hunt, forage, graze animals, and garden upon portions of their holdings. Access to these so-called commons enabled many a family to supplement their diets in good years and scrape by in years when harvests were not up to par. The process of denying the use of this acreage had begun in the late 1500s, but accelerated late in the following century.

During the early 1700s, aristocrats moved decisively to consolidate their holdings in order to expand their sheep-raising enterprises as well as their farming efforts. Among the rich, this movement proved both faddish and lucrative and led to a series of parliamentary acts known as the Enclosures. The Enclosures, combined with other economic factors, drove hundreds of thousands of rural dwellers, particularly those whose existence was already marginal, off the land. The English writer Oliver Goldsmith, in his 1770 poem, "The Deserted Village," lamented the effects not only of the Enclosures but also of the new rage among the wealthy for landscape gardens, with their artfully arranged sight lines and carefully placed ornamental outbuildings. He dolefully asked, "Where then, ah where, shall poverty reside," when "those fenceless fields the sons of wealth divide, / And even the bare-worn common is denied."[4] The answer was that a certain fraction of those displaced became more-or-less permanent vagabonds, always on the move, in and out of towns. Others became city dwellers, lured to urban areas by the promises of industrialization and a burgeoning mercantilism.

In France, too, mass migrations took place. Land ownership was concentrated in the hands of the aristocrats, who owned about two-thirds of all acreage but made up about 3 percent of the population. As many as three-quarters of all Frenchmen lived in the countryside, most of them peasants who were tenuously tied to the land and willing to uproot during hard times.[5] Historian Simon Schama described the reality of prerevolutionary France:

> armies of emaciated beggars dying on the roads; Paris streets slopping with ordure and butcher's offal; relentless *feudistes* screwing the last sou out of peasants barely subsisting on chestnut gruel; prisoners rotting in the hulks for stealing a loaf of sugar or smuggling a box of salt; horse and hound laying waste to standing crops in the name of the lord's *droit de chasse;* filthy bundles of rags deposited every morning on the steps of Paris churches containing newborn babies with pathetic notes claiming baptism; four to a bed in the Hôtel-Dieu, expiring in companionable dysentery.[6]

The new urban dwellers in most countries suffered overcrowding in ramshackle tenements. Until the mid-nineteenth century, these buildings usually lacked sanitation. Thus, the legions of the emerging working class lived amid filth many degrees worse than in the towns and villages whence they hailed, where there might be cesspits or rudimentary sewers, or at least the piles of stinking manure from animals quartered within the village might periodically be carted off to the fields for fertilizer. Despite the fact that certain advances had been made in hygiene, for example, in the preservation of milk (the spoilage of which had previously contributed greatly to childhood mortality), public health worsened during the Industrial Revolution because the dense packing of people in the cities promoted the transmission of disease. Deadly infections like cholera, typhus, smallpox, and measles ripped through urban Europe, their readiest victims being those with the worst nutrition and meanest living conditions. Well-to-do inhabitants generally survived such purges, and in fact life expectancy climbed steadily among the European upper classes throughout the eighteenth and nineteenth centuries.

It was, many demographers agree, this rise in longevity in the 1700s, rather than an increase in birthrates, which contributed to the marked upswing in population that commenced at mid-century. In a mere eighty years, global population, which stood at about 750 million in 1750, jumped to 1 billion. The rapidity of the gain was staggering. The British almost doubled during the century, from 5.5 to 9.9 million. Even as its birthrate fell, France's population surged to 26 million, after contracting to around 19 million during the early part of the century.

People at the time possessed some quantitative awareness of these demographic upheavals because governments had, starting in the seventeenth century, gradually put into place mechanisms for gathering what were termed initially "state numbers," and later, "statistics." During the Renaissance some cities like Florence had made counts of households in their immediate vicinity, and by the 1700s, municipalities and states throughout Europe and colonial America took intermittent censuses, for the sake of collecting and levying taxes and making projections of military strength (by predicting the numbers of young men who would come of age over time). In some places, clergymen might tabulate their parishioners, but in most instances censuses were administered by government officials, who became increasingly adept at manipulating their data.

The mathematical tools for analyzing such information had been developed by individuals like the Londoner John Graunt, who in the late 1600s took a stab at compiling vital statistics, making crude estimates of birthrates and death rates from the city's bills of mortality, which included information on both christenings and burials. Graunt also attempted to forecast the city's growth, and suggested that its population might double in sixty-four years, discounting inflows of immigrants, a prediction that seemed to astonish him and to prompt others to engage in further population studies.[7] This interest in population, argued French philosopher Michel Foucault in *The History of Sexuality,* sparked a near obsession by states with the sexual activities of their citizens:

> [I]t was necessary to analyze the birthrate, the age of marriage, the legitimate and illegitimate births, the precocity and frequency of sexual relations, the ways of making them fertile or sterile, the effects of unmarried life or the prohibitions, the impact of contraceptive practices. . . .[8]

In part what governmental scrutiny revealed was a trend, starting in the eighteenth century, for people to marry later in life and bear fewer children. This was, in the main, troubling: Falling birthrates were perceived as a national security threat. A shibboleth derived from the Greeks held that the health of a nation was tied directly to its maintenance of a stable population through the balancing of birthrates and death rates. Every country required a certain number of people to function smoothly and maintain a going economy—each according to its natural endowments, such as the fertility of its land, its access to the sea, and the favorableness of its climate. Thus, political analysts strove to understand how the control of wages, emigration, immigration, taxes, inheritance laws, and employment might lead them to the optimal population—or keep them at that critical balance point.[9]

It was generally perceived that the forces of death were mightier than those of life, and that keeping birthrates high was always the essential task. Depopulation was to be guarded against with vigilance. The fewer children born, the fewer citizens, and thus the less a country's "strength and richness." (Early students of population did not comprehend that populations might also grow if death rates fell.) A country whose population dwindled was bound to watch its power shrink. In France in some circles, fears of depopulation rose almost to hysteria. Baron Montesquieu had alerted Frenchmen to this phenomenon, lamenting the depopulation of Rome, Sicily, Greece, Spain, northern Europe, Poland, Turkey, Asia Minor, Egypt, and northern Africa.[10] Thus, leaders exhibited a reflexive desire to keep fertility up even as populations began to grow after mid-century.

That the state had every right to involve itself in reproductive matters was tacitly accepted. On this score, Graunt expressed an opinion which can be taken as emblematic of views held by upper-class Europeans well into the twentieth century:

> Now forasmuch as Princes are not only Powerful, but Rich, according to the number of their People (Hands being the Father, as Lands are the Mother and Womb of Wealth) it is no wonder why States, by encouraging Marriage, and hindering Licentiousness, advance their own Interest as well as preserve the Laws of God from contempt and violation.[11]

Licentiousness, libertinism, prostitution, premarital sex, celibacy (even among the clergy)—all weakened the state by undermining procreation, and needed to be suppressed. Because states were seen to be in perpetual competition with one another, it was vitally important for a state's citizens to breed with as much vigor—or more—than those of rival powers. Many countries enacted populationist or pronatalist programs, which encouraged marriage and childbearing; like ancient Rome, few ever met with discernible success in exhorting birthrates higher.

Amid this climate in which high birthrates were associated with wealth, a few dissidents began to express the disquieting notion that humans were inherently capable of reproducing at a far faster clip than the environment could support, a possibility which the Italian Giovanni Botero had previously advanced (see page 84). The Anglican minister William Derham made this suggestion regarding animals in his 1713 work, *Physico-theology*. Derham wrote, "The whole surface of our globe can afford room and support only to such a number of all sorts of creatures. And if . . . they should increase to double or treble that number, they must starve, or devour one another."[12] Derham believed that to prevent this from happening God acted to balance populations. Other analysts would soon claim that human populations might also be subject to similar checks, a contention that would send out tremors whose aftershocks continue to be felt today.

S uch ruminations hardly interested the hordes of people streaming into the burgeoning cities, who found themselves earning cash wages for perhaps the first time in their lives yet still saddled with the eternal problem of making ends meet. Children represented more mouths to feed, and the abandonment of infants by financially strapped urban couples served as a means of regulating family size. Indeed, they began jettisoning infants almost as commonly as unwed mothers.

Throughout Europe, illegitimacy was on the rise despite the best efforts of societal watchdogs. In the dislocations brought about by industrialization, people left their traditional homes, and at the same time escaped the oversight of local magistrates or parish priests, who had often acted as the paternalistic moral police and forced men to marry women they had impregnated. Still subject to opprobrium and punishment, single mothers rarely opted to raise illegitimate children.

Everywhere, abandonment rates climbed. In Rome in 1720, about a quarter of newborns wound up in foundling homes, and at times national rates in Italy and France may have soared significantly higher. Between 1784 and 1822, the number of foundlings on the rolls of French homes tripled, rising from 42,000 to 138,000. During the nine years from 1824 to 1833, a staggering 336,297 babies rode the revolving boxes into the arms of the church in France.[13] From 1754 to 1760, 14,934 children were consigned to the London Foundling Hospital, which had been swamped upon opening its doors in 1741. According to a nineteenth-century history of the institution, a "disgraceful scene" occurred on its first day of operation, with "women scrambling and fighting to get to the door, that they might be of the fortunate few to reap the benefit of the Asylum."[14] Couples seemed equally eager to consign unwanted offspring to St. Petersburg's well-staffed and amply funded foundling home, run in somewhat grand style out of the former palaces of two counts. It accepted anywhere from 5,000 to 9,000 infants a year.[15]

Across Europe into Russia then, parents abandoned millions of babies during the eighteenth and nineteenth centuries. By the mid-1800s, when private charities increasingly assumed the task of running such institutions, foundling homes took in an average of 37,000 infants annually in Italy; 34,160 in France, 17,769 in Spain, 15,119 in Portugal, 7,674 in Belgium, and 15,475 in Moscow alone.[16] As during the late Middle Ages, children who survived a sojourn in these humanitarian institutions had to be either extraordinarily hardy or extremely lucky. Historian David Kertzer reported that in one Parisian home prior to 1850, 25 percent of the infants died within the first few days of arriving, and up to 75 percent of those remaining died before they reached age one.[17]

Some historians have suggested that the high abandonment rates may be a sign that disapprobation of infanticide had finally begun to have an effect, and they point to an apparent decline in the incidence of the act during the eighteenth and nineteenth centuries. Whether infanticide rates fell is open to question, given the difficulty of establishing the prevalence of the act in the first place.

Peter Hoffer and N. E. H. Hull, legal scholars who made a close study of infanticide in England and New England from 1558 to 1803, noted that social attitudes toward it shifted several times during this period, and that prosecutions of baby murder, one of the few solid indicators of this "hidden crime," fluctuated accordingly.

Hoffer and Hull pointed out that the filing of charges and outcome of prosecutions depended in large measure on race and class. In the colonies during the 1600s, the preponderance of those tried for infanticide were black women. And it was generally believed by judges that women who had consorted with men not of their race were guilty of infanticide if they were brought before the bench to explain an unwitnessed infant death. In 1705, Hoffer and Hull recounted, Justice Samuel Sewall (of the Salem witchcraft trials) "warned his colleagues in council against a law on miscegenation that might lead to 'murders and other abominations against the infant products of such mixed unions.' " Unwed mothers and servants also received harsh treatment. Married women fared best, especially if they could prove "benefit-of-linen," that is, if they could show that they had equipped themselves with diapers in preparation for the birth.[18]

During the eighteenth century in England and the colonies, Hoffer and Hull contended, "growing toleration for illicit sexuality and improving material conditions joined with the rise of romantic sentimentality to alter the views of judges and jurors," and laws against infanticide loosened.[19] In the previous century, extramarital sex had been anathema, and women were frequently reviled as Jezebels and temptresses. It seemed obvious to almost everyone that an unwed mother had ample motive for doing away with her bastard child; this was, in effect, a recognition of the fact that society had so successfully demonized and punished "illicit" sex that single mothers would do anything to get rid of the evidence of their guilt, as it were, and avoid public disclosure.

Superstitious fears that deformed infants were "changelings" with devilish powers also accounted for the killings. There would be "a sudden fatal 'accident' to the child, followed by a hurried burial."[20] Everywhere throughout Europe, the belief persisted that evil spirits posed perpetual threats to pregnant women. In the German poet and novelist Johann Wolfgang von Goethe's epic *Faust*, witches are portrayed as having a constant demand for the fat of unborn babies, which fuels their preferred vehicle, the flying broomstick. Witness to a frenzied Walpurgis Night bash hosted by Beelzebub on the Brocken in the Harz Mountains, long considered a mystical site by Germanic peoples, Faust hears a chorus of witches intone,

> The road is wide, the road is long,
> Ah, what an antic, frantic throng!
> The pitchfork pokes, the broomstick thrusts,
> The infant chokes, the mother busts.[21]

Faust, Part I, published in 1808, tells a desperate tale of seduction and betrayal, presenting a sympathetic portrait of the unfortunate Gretchen, who gives in to Faust's overtures and then, wracked with guilt, kills the child produced by their unrighteous union. Goethe, in fact, based Gretchen's story on a real case, and deplored the severity of laws against infanticide then extant in Germany.[22]

If actual rates of baby killing did fall in some places during this period, it may have been partly due to changing attitudes toward motherhood. Increasingly throughout Europe, literature and paintings extolled the virtues of motherhood, sometimes eroticizing it, while publishers scored successes with pamphlets and books on child care, and a popular movement urging women to breast-feed their own babies won high-profile adherents. Popular travelogues published in France beginning in the early 1700s contributed to this trend. According to historian Joseph Spengler,

> In these fictitious narratives, as well as in actual accounts of the new world, inequality and luxury were condemned, the right of each to live a happy life affirmed, youthful marriage advocated, the need of preserving the species asserted, celibacy (for whatever reason) rejected, eugenic selection and the importance of health stressed, breastfeeding of children by mothers advocated, and mercenary wet nursing attacked.[23]

Feelings toward children softened and expanded: No longer merely miniature adults, they now were perceived as inhabitants of another imaginative universe, and newly indulgent parents accordingly stocked nurseries with toys, games, and entertaining books. By Victorian times, there was a veritable cult of childhood among the well-to-do British. Yet for all that, baby killing once again loomed as a matter of public concern in the mid-1800s, especially in England, where news of the almost outright obliteration of female infants by the Rajputs and other high castes of northern India scandalized the reading public—and triggered twinges of guilt on account of the tiny carcasses daily found stuffed into London culverts, tossed onto waste heaps, and bobbing down the Thames.

The British were alerted to the female infanticide in northern India by colonial chaplains, military men, and imperial officials from the early 1800s on. Benjamin Disraeli was aware enough of the issue to compare England's infanticide rate unfavorably to that of India in his 1846 novel *Sybil*. A comprehensive treatment of the subject was published in London in 1857 by John Cave-Browne, an Anglican chaplain who had been posted to Bengal. *Indian Infanticide: Its Origin, Progress, and Suppression* described the techniques for female infanticide employed by the Indians and attempted a quantitative assessment of its prevalence. Cave-Browne concluded that the practice prevailed in "Ajmere, Jyepore, Odeypore, Malwa, and Rajasthan," as well as in Punjab, among the highest castes (apart from the Brahmans) and that fathers killed their daughters mainly to avoid the costs of marrying them off and to prevent the possibility of sullying their family line by making a match with a lesser clan, which might inflict a fatal wound to their "pride of caste."

For European women of the eighteenth and nineteenth centuries, childbearing was fraught with terror. Mortality rates during childbearing were high, and on the grounds that female midwives, with their herbal recipes and tricks for easing delivery, caused more harm than good, the rising medical establishment attempted to render them obsolete. William Smellie of Scotland waged a campaign to bring greater rigor to the fields of obstetrics and midwivery, encouraging the presence of a medical doctor at all deliveries and the entrance of men into midwivery.

The new obstetrical practices, however successful in broader terms, served to exacerbate women's fears (especially before the advent of truly effective anesthesia in the mid-1800s). Obstetrical tools from the period, on display in medical museums, could pass for instruments of war. Royal physician to the wives of James I and Charles I of England, Peter Chamberlen added forceps to the armamentarium in 1630.[24] Irish physician Fielding Ould developed episiotomy in 1741, whereby an incision is made to the vulva to spare tearing during delivery (again, something that women might appreciate after the fact, but hope not to experience without being suitably braced by a stiff belt of alcohol). Smellie stocked the physician's bag additionally with a pair of lance-pointed sheers, known as Smellie's scissors, which he used to perform craniotomies, a procedure in which a physician perforated the skull and removed the brains of a fetus that proved undeliverable.

Laurence Sterne, the English novelist who seems to have been a veritable font of gynecological knowledge, alluded to all these developments in his best-seller *Tristram Shandy*. Tristram begins his fictional autobiography with the events surrounding his conception and birth. "My father," he reports, "was for having [a] man midwife by all means,————my mother by no means." His father "begged and intreated"; his mother "insisted upon her privilege in this matter to choose for herself." The compromise is struck: "In a word, my mother was to have the old woman,————and the operator [male midwife] was to have licence to drink a bottle of wine with my father and my uncle Toby Shandy in the back parlour————for which he was to be paid five guineas."[25]

The "operator" in question, Dr. Slop, prefers to be known by the French term, *accoucheur*, which is a sign of his pretentiousness. (The first *accoucheur*, or male midwife, had been the physician Boucher, who attended Louise de La Vallière, Louis XIV's mistress; soon, male midwives had become faddish among both French and English aristocrats.) Squat and full of himself, Dr. Slop makes his entrance into the Shandy household on the day of the delivery covered in muck, having taken a tumble off his horse. While the labor proceeds upstairs, the doctor expounds upon advances in his field. "Sir," he says at one point, "it would astonish you to know what Improvements we have made of late years in all branches of obstetrical knowledge, but particularly in that one single point of the safe and expeditious extraction of the *foetus*. . . . "[26] When the female midwife fails in her attempts to deliver Tristram, Dr. Slop succeeds in pulling baby Tristram out with forceps—in the process irreparably smashing the child's nose.

Tristram's father also, before the birth, toys with the idea that his wife should be delivered by caesarean section. His readings have led him to believe that the "470 pounds avoirdupois" of pressure which the child's head is subjected to during normal labor must wreak "havoc and destruction . . . in the infinitely fine and tender texture of the cerebellum!"[27] For the sake of the intelligence of his future child, he broaches the subject with his wife, "but seeing her turn as pale as ashes at the very mention of it . . . he thought it as well to say no more of it." While poking fun at the medical establishment, Sterne manages deftly to convey the disquiet women felt at the incursions men were making into the previously female preserves of pregnancy and birth.

I n the eighteenth century, attempts to understand the dynamics of populations invariably edged over into discussions of social economy, and talk of social economy into dissertations upon political utopias. Were they pipe dreams or could mankind attain a state of perfection? The French philosopher Marie-Jean-Antoine-Nicolas de Caritat, marquis de Condorcet, for one, proclaimed that man had already risen from barbarism to civilization, and would ascend further to a glorious state of material and intellectual perfection. An empiricist, a technological and scientific enthusiast, a banner-waving revolutionist (who ultimately fell out of favor with the Jacobins and may have committed suicide to avoid the guillotine), Condorcet believed humanity capable of indefinite improvement, of eternal progress. With his friends the encyclopedist Denis Diderot and mathematician Jean Le Rond d'Alembert (himself an abandoned child raised by a glazier), he contended that ignorance and prejudice are the main impediments on that upward trending course: Individuals are not born with biological limitations, rather exposed from infancy to certain ideas and emotions which ring them in.[28] Social inheritance, far more than familial, shapes people. The shortcomings of the cultural legacy can be overcome through education, Condorcet and the Encyclopedists said, which will free people's minds and open the way for collective advancement.[29]

Condorcet's ideas spread rapidly to England, embraced by the likes of William Godwin (whose daughter Mary Wollstonecraft Shelley would write that classic admonitory tale on the dangers of technological overreaching, *Frankenstein*). Godwin carried the argument further by denying any role for innate tendencies. "Alter men's opinions and they will act differently," he wrote. "Make their opinions comfortable to justice and benevolence, and you will have a benevolent society."[30]

While these thinkers elaborated optimistic philosophies, others viewed humanity's possible futures from a more crabbed perspective. In England, for example, many expressed agreement with the outlook of Robert Wallace, a prominent Edinburgh minister whose 1761 work, *The Various Prospects of Mankind, Nature, and Providence*, considered which policies should be pursued with regard to poverty.[31] Wallace forecast a gloomy outcome should governments enact social welfare programs. A government which ensured that the material needs of all its citizens were met would create a situation in which

> mankind would encrease so prodigiously, that the earth would at last be overstocked, and become unable to support its numerous inhabitants . . . [unless] some wise adept in the occult sciences, whould invent a method of supporting mankind quite different from any thing known at present.[32]

Here, the operation of a supposed natural law of population increase is seen as swamping human efforts; only a supernatural solution might counter that overwhelming force.

Other commentators distinguished a flaw in human nature which invariably led people, once their basic wants were met, to indulge in sloth and self-indulgence. Along these lines, the Reverend Joseph Townsend, in his 1786 work, *A Dissertation on the Poor Laws, by a well-wisher to Mankind*, argued that the eighteenth-century equivalent of modern social welfare did not ease poverty but exacerbated it. Why should the poor be industrious and frugal, Townsend asked, if they know that their children will be taken care of and they will be "abundantly supplied, not only with food and raiment, but with their accustomed luxuries, at the expense of others." (Americans may be reminded of President Ronald Reagan's stories of Cadillac-driving welfare moms.) The lowest classes, Townsend believed, know no "pride, honour, and ambition." Their only spur to work and strive is hunger, and if the state enables them to put food in their bellies, then it undermines its own socioeconomic goals.[33]

Within the household of Daniel Malthus, a peripatetic British intellectual living on inherited wealth, the musings of Godwin and Condorcet, of Jean-Jacques Rousseau and David Hume, provided constant fare for discussion. Growing up listening to heady contemplations of human perfectibility, young Thomas Robert Malthus, known always as Robert, gradually constructed an opposing viewpoint. He honed his opinions at Jesus College, Cambridge (where he graduated as Ninth Wrangler, or ninth best mathematician at the university).[34] They were still clearly in flux in 1796, when, at age thirty, just prior to taking Anglican orders, he wrote a brief essay in support of workhouses, the system of workfare set up under the poor laws. But the next year, he engaged his father in a heated debate in which he debunked optimism and utopian philosophies. When Daniel responded by advising his son to put goose quill to page, Robert sat down and quickly turned out *An Essay on the Principle of Population as it affects the Future Improvement of Society, with Remarks on the Speculations of Mr. Godwin, M. Condorcet, and other Writers.*[35]

The *Essay*, published anonymously in 1798, gained an ardent following and went through six editions during Malthus's lifetime. His authorship soon discovered, Malthus garnered first a fellowship at his alma mater, then a rectorship in Lincolnshire, from which he derived income, and finally a professorship of political economy—reputedly the first such position created—at a college for civil servants run by the East India Company. Until his death in 1834, he ranked as a major player in an ongoing policy debate regarding the poor and working classes. His *Essay* was highly controversial, serving to ignite debate for decades and spawning any number of supporting and dissenting volumes by other authors. Godwin himself initially applauded Malthus's work and even met with him a few times, but countered eventually, in 1820, with *Of Population: An Enquiry Concerning the Power of Increase in the Numbers of Mankind, Being an Answer to Mr. Malthus's Essay on that Subject*. Godwin explained his reversal by saying that he had assumed that "the Essay on Population, like other erroneous and exaggerated representations of things, would soon find its own level."

However, he added, "in this I have been hitherto disappointed," in part because "the theory of this writer [Malthus] flattered the vices and corruption of the rich and great," who were all too happy to be excused of any responsibility for poverty and the poor.[36]

Malthus, who read widely and eclectically, had absorbed the wisdom of Plato and Aristotle on the dangers of population growth; had harkened to Wallace, Townsend, Benjamin Franklin, Adam Smith, and others who drew connections between poverty, population dynamics, and the wealth of nations. He had considered the functions and effects of the poor laws. Prior to rewriting the *Essay* for the 1803 edition, he corresponded with numerous statisticians and demographers throughout Europe and toured the Continent gathering more data, even visiting the St. Petersburg foundling hospital.[37]

From his experience and knowledge Malthus attempted to determine what had "impeded the progress of mankind toward happiness" and whether the impediments might be removed. In considering these questions, he began with two key *"postulata"*: "first, that food is necessary to the existence of man: secondly, that the passion between the sexes is necessary, and will remain nearly in its present state."[38] Remembering well Wallace's warnings about unchecked population growth bumping up against food supplies, Malthus further postulated that all animals and plants, including humans, would reproduce exponentially in an ideal world in which there were sufficient means of subsistence. What would the rate of increase be for humans? Casting around for an answer, Malthus found a ready example in North America, where, he said, from the time of first settlement by Europeans, population had doubled every twenty-five years. In some isolated settlements where "vicious customs and unwholesome occupations were unknown" it had supposedly doubled every fifteen years. Malthus thought that "even this extraordinary rate of increase is probably short of the utmost power of population."[39]

Comparing the "slowest" of the rates of population increase—the doubling every quarter century—to the food production, Malthus discovered the central impediment to human perfectibility: While population sped along in geometric high gear, agriculture plodded behind the dray horse of arithmetical increase. Population grew by leaps and bounds, while food yields, even with extraordinary efforts, could only inch up gradually: This is the "Malthusian law" which became so famous.

To Malthus, the disparity between the two ratios could mean just one thing: Humanity would exert a "constant effort towards population," while food output lagged behind, and would ultimately cease expanding because the amount of land which could be cultivated was finite. Earlier writers, including the German political scientist J. F. de Bielfeld, who published his works primarily in France, had argued that agricultural yields "could be pushed excessively far, if there were more hands to work on the land."[40] Some philosophers even contended that the earth was inexhaustible, always able to produce more if there were sufficient people to till the earth—but not Malthus.

He saw the history of every country as a repeating cycle in which population would rise past the supportable point and "depress the whole body of the people in want and misery," and vice.[41] By vice he meant "promiscuous intercourse, unnatural passions, violations of the marriage bed, and improper arts to conceal the consequences of irregular connexions," i.e., contraception, among other things; by misery, the whole catalog of woes, from infanticide to war to epidemic illnesses, which cut short life and fell heaviest upon the "lower classes of people."[42] All these terrible privations would spur the cultivation of more land and employment of more labor, which would boost yields. But as food supplies again became adequate, people would begin having more children, food would again become less available, vice and misery would worsen, and so on, in perpetuity.[43] This dynamic, Malthus argued, was the reason why all efforts of a nation to better its general conditions, including technological improvements, had been thwarted and would continue to be.

Malthus dismissed any suggestion that poverty was owed to a concentration of wealth and land in the hands of a few. No reform "could remove the pressure of [the law I have stated] even for a single century," and therefore dreams of equality and of societies in which everyone "should live in ease, happiness, and comparative leisure; and feel not anxiety about providing the means of subsistence for themselves and families," were an impossibility. If his premises were correct, Malthus wrote, "the argument is conclusive against the perfectibility of the mass of mankind."[44]

Nature would always militate against perfectibility through the mechanisms of vice and misery, the checks imposed upon population, which tended to cull those at the lowest reaches of society. And Malthus, at least in the initial version of his *Essay*, found this entirely as it should be:

> [T]here is one right which man has generally been thought to possess, which I am confident he neither does nor can possess—a right to subsistence when his labour will not fairly purchase it. . . . A man who is born into a world already possessed, if he cannot get subsistence from his parents, and if the society do not want his labour, has no claim of *right* to the smallest portion of food, and in fact has no business to be where he is. At nature's mighty feast there is no vacant cover for him. She tells him to be gone, and will quickly execute her own orders, if he do not work upon the compassion of some of her guests.[45]

Castigated for the inhumanity of such passages, he toned down his rhetoric in later versions, yet remained committed to the belief that God built struggle into existence so that humans would not rest in a state of barbarism, but instead, driven to labor by hunger, would rise to civilization, the model for which was western Europe. Poor laws and other relief, by removing this spur and encouraging increased fertility, doomed the lower classes to suffering. Malthus advocated the gradual removal of state support for the poor and their children, especially those who were illegitimate.[46] (Today in the United States, we hear Malthusian echoes in conservative cries to curtail government benefits to unemployed or marginally employed single mothers.) Only those who work hard and "are suffering in spite of the best-directed endeavours to avoid it, and from causes which they could not be expected to foresee, are the genuine objects of charity." Those who are "idle and improvident," though, "are deservedly at the bottom scale of society."[47]

It must be said that Malthus did somewhat amend his initial views. In each successive revision of the *Essay*, he attempted to shift emphasis further away from his cold-hearted justifications of nature's plan and instead concentrate on the need for the poor to take responsibility for their lives by exercising prudential, or moral, restraint. Though impelled by passion, humans, unlike animals, could choose whether or not to reproduce. It was, he said, everyone's duty to defer marriage until they had the financial means to feed their children and in the meantime not to indulge in "vicious gratifications"—by which they might beget illegitimate children. The poor, lacking "personal respectability," did not generally meet these obligations. Thus, society needed to cultivate in them "a spirit of independence, a decent pride, and a taste for cleanliness and comfort."[48] Better representation in government and greater equality in the eyes of the law would help endow the poor with self-esteem, as would "a system of general education" which he felt government was obliged to supply, mainly because educated people seemed to breed with less abandon.[49]

But in the end, Malthus could never quite rid himself of the belief that there was a divine plan in human want. In a fallen world, occupied by sinners, suffering served a purpose. He wrote:

> I have always considered the principle of population as a law peculiarly suited to a state of discipline and trial. Indeed I believe that, in the whole range of the laws of nature with which we are acquainted, not one can be pointed out which in so remarkable a manner tends to strengthen and confirm this scriptural view of the state of man on earth.
> ... It must be allowed that the ways of God to man with regard to this great law of nature are completely vindicated.[50]

The ideas of Malthus, repeatedly attacked, corrected, defended, and elaborated upon, shaped discussion across Europe well into the twentieth century. Criticized by various and sundry for his inhumanity, by other political and economic observers for faulty hypothesizing, and by religious factions for his putative support of birth control—which he in fact reprobated; dismissed by Marx and his followers as a "miserable parson" and tool of the property-owning class; embraced by a range of economists, demographers, and social reformers dubbed neo-Malthusians, including economist John Stuart Mill and British birth control activists Francis Place, Charles Bradlaugh, and Annie Besant, who distributed literature which advocated, and detailed methods of, birth control; Malthus became many things to many people.[51] Charles Darwin would read him "for amusement" in October 1838 and in the wake of doing so hit upon the theory of natural selection, and Darwin's cousin, Francis Galton, would also absorb the Malthusian message to major effect. When the first issue of the *Annals of Eugenics* came out in 1925, a portrait of Malthus graced its frontispiece, which was supplied with a caption assigning to him an enormous honor. The caption read: "Strewer of the Seed which reached its Harvest in the Ideas of Charles Darwin and Francis Galton."[52]

7

Quality
Is the Problem

Ironically, even as Malthus was proclaiming the inelasticity of food production, agriculture throughout his own country and most of northern Europe was undergoing a tremendous upheaval and an overall surge in output. In England alone, the total amount of wheat harvested annually appears to have jumped 225 percent between 1750 and 1850, and that of other grains to have risen substantially, while the number of animals brought to slaughter at major markets may have doubled.[1] The groundwork for this upswing had been laid in the 1600s, and perhaps earlier in certain locales, and not all regions experienced outright gains, yet the trend was on the whole positive and was brought about in some measure by advances in technology. Agriculture, no longer the sole purview of the yeoman farmer or the serf, was coming under the scrutiny of science. Agricultural societies and clubs sprang up, especially in England and Holland, their members fired both by the mission of applying scientific principles to farming and animal husbandry, and by a patriotic zeal to feed their nations' expanding populations. With the sponsorship and participation of the wealthy gentry, the societies hosted fairs and exhibitions at which they spread technical news, held competitions for well-bred animals, proselytized, and, not incidentally, tacitly reinforced class distinctions, their very actions reminding "ordinary farmers that the men who could afford to raise prize animals were their natural leaders."[2]

By the end of the eighteenth century, intensive husbandry had pushed the average weights of cattle and sheep up. Specialists reengineered these animals by breeding for desired characteristics such as tastiness; the ratio of flesh to bone, and of fat to muscle; and speed of maturation. Through crossbreeding and inbreeding they developed assorted varieties of long- and short-coated sheep, and long- and short-horned cattle. Meanwhile, "improvers" experimented with methods of tilling, planting, and draining wetlands, and instituted the practices of rotating crops and sowing nitrogen-fixing legumes, like clover, to help restore soil fertility (in addition, manure for the same purpose became more generally available thanks to larger herd sizes). New crops came on the scene, notably root vegetables like turnips, and farm machinery improved. By the time Victoria ascended the throne in 1837, agricultural science had been thoroughly legitimized, a department of agriculture having been established at the realm's preeminent universities in Oxford and Edinburgh. In 1843, the government-run Rothamstead Experimental Station opened for the purpose of conducting controlled field trials. English landholders and innovators spearheaded the movement that brought greater systematization and profitability to agriculture, but the Germans, the French, and eventually the Americans also participated in the movement, setting up research facilities within their own institutions of higher learning and bringing experimental methods to the study of plant and animal hybridization.

In one way or another, breeding was on everyone's mind in the eighteenth and nineteenth centuries: breeding in the sense of reproduction (as in, "the leghorn pigeon is a good breeder," or as in Swift's calculation that "there may be about two hundred thousand couple, whose wives are breeders"); breeding in the sense of extraction or parentage; or in the sense of bringing up or educating the young; or in the sense of good manners and behavior. From the "blood" to politesse from physical inheritance to public comportment, a continuous line was drawn, tying interior, inborn quality to exterior manifestations of that quality.

Because people perceived no great divide separating the realms of animals and humans, comparisons like those made by Swift in "A Modest Proposal," yoking women with "sheep, black cattle, or swine," and with "mares in foal, . . . cows in calf, or sows when they are ready to farrow," were so commonplace as to be unremarkable. There were human tribes, races, species, breeds, strains, or types, just as there were animal ones, all vaguely coincidental. Writers might refer to the "tribe of canines" or the "human tribe," to a "race of cattle" or the "infant race" (meaning babies). Those of any political stripe felt comfortable describing groups of people and animals in the same language. Godwin, for example, surveys the world and finds it "a wilderness, a wide and desolate place, where men crawl about in little herds."[3]

Indeed, so blurred was the line between humans and animals that in some instances additions to a man's family were considered in the same light as the multiplication of his farm animals, as when, in an 1822 issue of the *Annals of Sporting and Fancy Gazette*, congratulations were extended to "a Yorkshire gentleman, who, on a single August morning, was presented with twin baby daughters, eight piglets, and seven kittens."[4] Under the tongue-in-cheek humor lurks a serious subtext, for aristocrats, particularly in England, became so enamored of scientific husbandry that some wondered aloud why men did not exercise as much care in choosing their spouses as they did in choosing mates for their prize cattle.[5]

Malthus himself, although preoccupied by population, mused upon the possibilities of applying to humans the principles of the wonderful new science of breeding. Mankind was probably not able to attain physical perfection any more than political perfection, he thought. Yet, he wrote,

> It does not, however, by any means seem impossible that, by an attention to breed, a certain degree of improvement, similar to that among animals, might take place among men. Whether intellect could be communicated may be a matter of doubt: but size, strength, beauty, complexion, and perhaps even longevity, are in a degree transmissible. . . . As the human race, however, could not be improved in this way without condemning all the bad specimens to celibacy, it is not probable that an attention to breed should ever become general; indeed I know of no well-directed attempts of the kind, except in the ancient family of the Bickerstaffs, who are said to have been very successful in whitening the skins, and increasing the height of their race by prudent marriages, particularly by that very judicious cross with Maud the milk-maid, by which some capital defects in the constitutions of the family were corrected.[6]

That last bit of the passage represents a rare Malthusian attempt at wit: The Bickerstaffs were an imaginary dynasty whose antics had been reported in the magazine *The Tatler* during its short-lived yet ballyhooed run from 1709 to 1711. Brainchildren of Swift, the Bickerstaffs had, like the sitcom Cleavers or cartoon Simpsons, gained an independent life in the popular imagination. The short, swarthy progenitor of the family, Sir Isaac, a knight of the Round Table, had undertaken his own experiment in breeding, having, with "Design of Lengthening and Whitening his Posterity," married his eldest son off to a lady who "had little else to recommend her, but that she was very tall and very fair." Later, in the 1400s, according to a genealogy provided by the latest Isaac in the lineage, a bad marriage by a Bickerstaff heiress to a courtier "gave us Spindle-Shanks, and Cramps in our Bones, insomuch that we did not recover our Health and Legs till Sir Walter Bickerstaff Married Maud the Milk-Maid, of whom the then Garter King at Arms (a facetious Person) said pleasantly enough, That she had spoil'd our Blood, but mended our Constitutions."[7]

Talk of ailments that seemed to run in families dovetailed in certain circles with the larger discussion of degeneration, a term which had entered the English language with reference to breeding as early as the fifteenth century. A degenerated stock or race had lost the proper qualities, or "declined from a higher to a lower type"; it could therefore be considered "debased" or "degraded," and was to be guarded against as well as reviled, because it posed a threat to the very enterprise of civilization. Degraded animal stocks—cows that gave little milk, hens that did not lay well, sheep prone to sickness—cost the farmer money and in extreme manifestations might bring him to fiscal ruination. So, too, according to common belief, degraded human stocks could spell the demise of nations.

The fear of degeneration preyed upon all eighteenth- and nineteenth-century European elites, even as they expanded their hegemony over vast segments of the globe. It fell especially hard upon the minds of French intellectuals, whose writings seem almost always to have the theme of degeneration lurking somewhere just under the surface. Take, for example, the 1756 work *L'ami des hommes, ou Traité de la population (The Friend of Man, or Treatise on Population)*, which went through twenty editions in three years. In it, the author, Victor Riqueti, marquis de Mirabeau (flamboyant father of the revolutionary leader Honoré-Gabriel, comte de Mirabeau), railed against luxury, which he contended was responsible for the sorry economic state of the country. But the indulgent, self-aggrandizing, and nonproductive activities of the idle rich had also damaged France more profoundly: They had discouraged childbearing among rural laborers and aristocrats alike. Rural workers, Mirabeau said, limited their families because they perceived their children would be denied upward mobility. Meanwhile, wealthy women refused to bear children so as to preserve their figures, or, having deigned to give birth, would not suckle the infants. In this fashion, Mirabeau declared, too much luxury had caused the outright degeneration of the upper classes.[8]

Writing ten years later, J. Faiguet de Villeneuve, a military paymaster who produced two important works on population stimulation, declared that the quality of the French people as a whole had deteriorated due to centuries of bad breeding, and that currently women were bearing excessive numbers of unfit, feeble children. To remedy the situation, Villeneuve proposed a series of reforms centering on marriage laws, taxation of families, education, and employment, which he believed would restore the French to robustness.[9]

Another influential French commentator, the Abbé Pluquet, campaigned in the 1780s against luxury, which he held responsible for creating a situation whereby the "lowest classes of citizens contribute most to the population." This issue of the disparity in birthrates between rich and poor came to be known as the problem of the "differential birth rate," and fanned fears of degeneration wherever it was raised. (The differential birthrate continues to provoke discussion throughout the industrialized West, and is often used

with veiled—and not-so-veiled—racist intent.) Pluquet foresaw only negative consequences emerging from the high birthrate among the poor. He
wrote, "If the unhappy [members] of this class desire to perpetuate themselves, it is only in moments of drunkenness, caused by strong liquors: fathers
thus constituted produce only feeble, debilitated infants, whom misery and
indigence enfeeble even more than their fathers."[10] Pluquet, as did so many of
his fellow social analysts, considered enfeeblement not only a physiological
state, but also a moral and intellectual one. Although he offered no hard data
in support of his statements, they bore the force of received wisdom. Later,
medical researchers would elaborate upon the notion that physical ailments
such as dementia and venereal disease corrupted the bloodlines, led to infertility, or, worst of all, doomed offspring to become sexual deviants.[11]

Many political and economic theorists readily accepted another tenet
implicit in Pluquet's work: that the quality of a population—which is to say
of a nation—is simply a function of the quality of the individuals composing
it. They equally acceded to the corollary statement that the poor must invariably stand as inferior to the ruling classes. Even before the founders of
anthropology had agreed upon what, by their ostensibly scientific standards,
quality was and how best to measure it, Europeans busily slotted humanity
into hierarchical pigeonholes. Malthus thus ranked "savages" by degree,
declaring that the "wretched inhabitants of Tierra del Fuego have been
placed by the general consent of voyagers at the bottom of the scale of human
beings." Next up came the "barbarous" natives of Van Diemen's Land,
although Malthus supposed the Andaman Islanders might possibly share the
dubious honor of being the second-worst specimens of humanity. According
to Malthus's not untypical opinion, the natives' inferiority manifested itself
not only in their subsistence economy, but also in their physical appearance.
"Their stature seldom exceeds five feet; their bellies are protuberant, with
high shoulders, large heads, and limbs disproportionately slender."[12] Malthus
clearly found contemplating such physiques distasteful, his aesthetic aversion
nakedly signaling his moral contempt, his formal judgment mirroring his
religious one. As the slave trade expanded in the eighteenth century, and
imperialism rose in the nineteenth century, rankings like those made by
Malthus were increasingly put forward as justifications for the subjugation of
vast segments of humanity.

Just as parents in the West for millennia systematically refused to grant
infants human status until they had passed a certain age, Europeans, from
their first forays into the Americas, Africa, and Asia in the fifteenth century,
had reduced the natives they contacted into mere bodies, soulless and mind-
less brutes who did not deserve consideration. Philosophers might argue

among themselves about whether savages were "noble," or irredeemably hea-
then and violent; whether they represented arrested progression along the
species' path to civilization, or were effectively children in adult bodies. But
whatever the particular slant of the debaters, one assumption went unchallenged: Aboriginal peoples were incontestably less valuable, less advanced,
less worthy members in the Great Chain of Being, a belief which was incorporated into the very foundations of the emerging science of anthropology.

From at least the turn of the sixteenth century, *race* had been used in the
French language to allude to "the extraction of a man, of a dog, of a horse; as
one says 'of good or bad race.' "[13] By the eighteenth century, Europeans generally had begun to speak of human "races" and to stereotype them, claiming
superiority for one or another, as when, in 1727, the Earl of Boulanvilliers
contended that the French peasantry descended from inferior Celtic stock,
whereas noblemen belonged to a more elevated race of "long-headed"
Franks.[14] The first apparent use of the term in a scientific context appears to
have been by the academician Georges-Louis Leclerc, comte de Buffon,
whose *Histoire naturelle, générale et particulière,* issued between 1749 and 1789,
attempted to chronicle all of nature.[15] As manager of the king's botanical garden and museum, Buffon won a large readership for his ambitious work, but
his scientific peers often disputed and dismissed his efforts as lacking rigor.
Indeed, although he functioned as a popularizer of science, few of his ideas
except those regarding geology and paleontology stimulated contemporary
research. However, Buffon did employ the term "race" to refer to divisions
within the human species based on physical characteristics, arguing that the
white-skinned northern European races stood at the apex of humanity, and
this appears to have been his principal, if highly equivocal, legacy.

To the lasting detriment of science, attempts to quantify physiological differences between putative races became one of the obsessions of anthropology.
When at age twenty-three German physiologist Johann Friedrich Blumenbach,
who is generally considered the founder of physical anthropology, published *On
the Natural Variety of Mankind,* he set the stage for what became an ongoing
attempt—continuing today despite the best efforts to expose the bankruptcy of
its premises and methods—to establish a scientific basis for the view that race
is "the prime determiner of all the important traits of body and soul, of character and personality, of human beings and nations."[16] Based on his measurements of skulls from around the world, Blumenbach systematized what had
theretofore been merely free-floating concepts, laying out a fivefold division of
Caucasian, Mongolian, Malayan, Ethiopian, and American races. The last
four were merely degenerated versions of the Caucasian race, which in this
scheme constituted the original human type. Classification of humans
through skull measurements, or craniometry, went on to have a long and
inglorious run, drawing many adherents from the French brain scientist Paul
Broca to, in the 1970s, American psychology and education professor Arthur

Jensen. (Although thoroughly discredited, craniometric data played a role in the recent arguments advanced by Charles Murray and Richard Herrnstein in their controversial tome *The Bell Curve*.)

Blumenbach engendered a number of contemporary emulators who put forward their own preferred racial categories. Some discerned dozens of distinct races, while others opted for the German's reductive approach and limited themselves to a small number of overarching groups, which might contain several subcategories. Perhaps most widely attended to upon this subject was Joseph-Arthur, comte de Gobineau, a French royalist whose disquisitions profoundly influenced the shape of modern European history.

Although almost unknown today, Gobineau was a leading intellectual of the late nineteenth century, celebrated both at home and abroad. As a young man he frequented conservative Parisian literary salons and for a time held a post as secretary to Alexis de Tocqueville. Known initially for his fiction, Gobineau soon gained international attention for his *Essai sur l'inégalité des races humaines (Essay on the Inequality of Human Races)*, which came out in four volumes from 1853 to 1855. In the dedication to George V, king of Hanover, the author explained that in undertaking to write the book he had intended mainly to identify the reasons for "the bloody wars, the revolutions, and the breaking up of laws—which have been rife for so many years in the States of Europe."[7] Presuming that he lived in a fallen age, Gobineau looked to other great powers which had risen to glory only to come to naught, hoping that somewhere in their histories might lie clues to the puzzle of the modern dilemma. As he surveyed the glories and wreckage of Greece, Rome, and the Germanic peoples, he was

> gradually penetrated by the conviction that the racial question overshadows all other problems of history, that it holds the key to them all, and that *the inequality of the races from whose fusion a people is formed is enough to explain the whole course of its destiny.*[8]

All great civilizations, he concluded, fell not because of the defects conventionally cited—"luxury, effeminacy, misgovernment, fanaticism, and corruption of morals"—but because of race degeneration. With quasi-scientific exactitude, Gobineau specified what he meant by this:

> The word *degenerate,* when applied to a people means (as it ought to mean) that the people has no longer the same intrinsic value as it had before, because it has no longer the same blood in its veins, continual adulterations having gradually affected the quality of that blood. . . . [A civilization dies when its] primordial race-unit is so broken up and swamped by the influx of foreign elements that its effective qualities have no longer a sufficient freedom of action.[9]

While Gobineau declaimed authoritatively, his statements were grounded not in any understanding of the true mechanisms of heredity, which were then unknown, but in the same naive beliefs about the blood which pervaded medieval thought. Gobineau quivered with a "secret repulsion" at "blood mixing." There is, from his paranoiac point of view, a certain inevitability about degeneration. All conquering peoples must associate with those they have conquered, and therein lies their ruination: They will interbreed with their underlings and doom their own race. This degradation happens with mystical instantaneousness. Gobineau warned, "From the very day when the conquest is accomplished and the fusion begins, there appears a noticeable change of quality in the blood of the masters."[20] Although he allowed that some members of "inferior"races might through miscegenation improve their blood, and that the mixing of the black and white races often produces "artistic genius," Gobineau asserted in a bleak crescendo that interbreeding between the races can in the long run yield only disaster:

> If mixtures of blood are, to a certain extent, beneficial to the mass of mankind, if they raise and ennoble it, this is merely at the expense of mankind itself, which is stunted, abased, enervated, and humiliated in the persons of its noblest sons. Even if we admit that it is better to turn a myriad of degraded beings into mediocre men than to preserve the race of princes whose blood is adulterated and impoverished by being made to suffer this dishonorable change, yet there is still the unfortunate fact that the change does not stop here, for when the mediocre men are once created at the expense of the greater, they combine with other mediocrities, and from such unions, which grow ever more and more degraded, is born a confusion which, like that of Babel, ends in utter impotence, and leads societies down to the abyss of nothingness whence no power on earth can rescue them.[21]

Like Malthus, Gobineau tied external appearance to essential worth. His opinions were purely noxious and his tone could be appallingly snide, as when he wrote that "Oceania has the special privilege of providing the most ugly, degraded, and repulsive specimens of the race, which seem to have been created with the express purpose of forming a link between man and the brute pure and simple."[22] But to many of his contemporaries, Gobineau sounded reasoned and completely justified: Like fulminating radio talk show hosts today, he tapped into popular fears with frightening success.

Almost immediately upon its completion, Gobineau's work found its way into other languages. A stalwart American apologist for slavery published an English version of the four volumes in Philadelphia in 1856. British editions of the work followed. In Germany, the composer Richard Wagner fastened on to the Frenchman's elevation of the so-called Aryan race, whose members Gobineau asserted were the most beautiful and accomplished exemplars of humanity. That no ethnic group in northern Europe actually identified themselves as Aryans, and that this so-called race was as much a construct of imagination as anything else, did not matter. The myth of a pale, blue-eyed race programmed by destiny to stand at the pinnacle of history fed into Wagner's romantic idealization of the Germanic *Volk* and his artistic attempts to glorify German culture and fortify it against foreign "contamination." Wagner also infected the man who was to become his posthumous son-in-law, British-born political philosopher Houston Stewart Chamberlain, with his racist enthusiasms, and Chamberlain incorporated Gobineau's postulations whole hog in his 1899 *Die Grundlagen des neunzehnten Jahrunderts (The Foundations of the Nineteenth Century)*, which Kaiser Wilhelm II read and pronounced his favorite book.[23]

Gobineau's tract is laced with metaphors comparing the works of nations to sexual potency and to the production of children, and equating nations with male bodies. However, it would be left to Francis Galton to take as the explicit charge of nations the literal improvement of their bodies politic.

Explorer, author, statistician, and anthropologist, Galton was more famous in his day than his cousin Charles Darwin, with whom he corresponded throughout his adult life. Son of a prominent Quaker banker whose family fortune derived from the sale of munitions, Francis was born in Birmingham in 1822, the youngest of seven children. He was by most accounts something of a prodigy. Reading by age two and a half, he was by age four penning notes to his eldest sister Adèle, who was bedridden most of her life with scoliosis, a progressive, then untreatable, deformation of the spine. By eight, he was familiar with most classical texts, and was carrying out advanced mathematical computations. His father, intending that his son follow in the footsteps of his renowned maternal grandfather, Erasmus Darwin, sent Francis to Cambridge University to prepare for a career in medicine, but young Galton buckled under the pressure, suffering a nervous breakdown, the first of several, during his third year there.

Galton's own ambitions lay elsewhere. After his father's death in 1844 provided him with a substantial inheritance, he embarked for North Africa with friends. He abandoned the standard tourist routes and set out by camel across the Bishari (now the Nubian) Desert, and proceeded by boat up the Nile beyond the fourth cataract to Khartoum, which was considered a daring excursion in the 1860s, when British explorers like Sir Richard came along. From Khartoum, Galton went on to Beirut and Jerusalem, settling finally near Damascus. Here he set up house, taking as pets two Sudanese monkeys and a mongoose. He learned Arabic, indulged a fascination with Orientalism, and consorted with local prostitutes. Historian Daniel Kevles has suggested that Galton's return to England in 1846 may have been prompted by his having contracted venereal disease. Again in 1850 his wanderlust carried him abroad, this time to southern Africa, little explored by whites. He traveled overland to the Kalahari Desert, where he mapped large regions and claimed to have negotiated peace between warring tribespeople. Back in England, he wrote a how-to book on journeying through remote lands, *The Art of Travel,* which specified, among other things, how to find water and train pack animals. It was snapped up by Britishers and quickly went through five printings.

In 1853, Galton was thirty-one, famous, and well connected through family ties and personal association to some of the leading intellectual families in England, including the Darwins, Wedgwoods, and Huxleys. He was newly elected to the prestigious Royal Geographical Society and recently married to Louisa Butler, daughter of the headmaster of Harrow school. Still wrestling with psychological conflicts, he sought illumination in phrenology, the system developed by the German physician Franz Joseph Gall, which held that a person's character and mental attributes could be read from bumps on the head. Soon rejecting phrenology, Galton began his own investigations into the nature of human behavior and ability.

The debate about whether faculties such as intelligence were inborn or shaped by experience had been raging for some centuries among European philosophers and stemmed from the fundamental dispute between empiricists and nativists, between those who said the mind is a tabula rasa and knowledge must be acquired, and those who said knowledge is built into the mind in the form of certain ideas and truths.

Contemplation of the essential endowments of humans easily segued into an attempt to determine how faculties might be distributed among individuals. Were all men born with equal potential? If not, what factors might contribute to the distribution of various characteristics?

In attempting to address this puzzle, Galton proceeded from a broadly hereditarian perspective as outlined by his cousin Charles, whose *On the Origin of Species by Means of Natural Selection* had issued in 1859. Popular imagination associates Darwin most strongly with the observations he made of exotic animals, especially on the Galápagos Islands, during his epochal round-the-world journey aboard HMS *Beagle.* However, in *Origin of Species,* Darwin was actually almost as concerned to examine domesticated animals and plants as he was wild ones. He saw in the habits of the farmer and plantsman striving to improve stocks through selective breeding and hybridization a direct parallel to the operations of nature. In the wild, weak progeny or hybrids died off, while those with some intrinsic edge survived and proliferated. Over time, this spontaneous winnowing mechanism, which Darwin

dubbed natural selection, produced alterations in species. In the same fashion, but through the consciously directed hand of the breeder or hybridizer,
so did domestication.

Darwin posited that neither natural selection nor domestication proceed
inevitably in a certain direction: Ultimately, either can produce outcomes that
are beneficial or detrimental to the survival of a species. Both can result in
fertile, healthy organisms capable of reproducing themselves prodigiously, or
in sterile or infertile organisms which fail to reproduce effectively, or die off
before reaching maturity. In an age devoted to the myth of progress, however,
Darwin's nondirectional evolution was invariably discarded for a model in
which the process was seen to have trended ever upward, toward that most
perfect end point, the human species.

Darwin was most interested in how variations within species arose, so he
spent a good deal of time contemplating physiological differences, such as
albinism or the thickness of wing bones in wild versus domesticated ducks.
Without knowledge of genetics, he was hard pressed to explain the existence
of "sports," "freaks," and "monstrosities" with any exactitude. However, while
the laws governing inheritance were unknown, Darwin maintained that,
"perhaps, the correct way of viewing the whole subject [of variation among
individuals] would be, to look at the inheritance of every character whatever
as the rule, and non-inheritance as the anomaly."[24] Galton took this message
to heart, and proceeded with the assumption that a person's intellectual,
moral, and physical attributes derive from his ancestors, who pass their traits
on to him in one degree or other down the generations.

Armed with this premise and fixated upon those people who stood head and
shoulders above the common run in terms of their achievements—whether in
the arts, politics, business, or any other arena—Galton wrote a two-part article
which was serialized in the popular magazine *MacMillan's* in June and
August 1865, putting forward for the first time publicly his theory that inborn
ability accounts for worldly success. These articles formed the germ of his book
Hereditary Genius, published four years later to instant acclaim.

The notion that genius, or extraordinary talent and mental acuity, ran in
families was, of course, quite commonplace, and served to bolster class prejudices. But in *Hereditary Genius* Galton gave increased potency to this dogma by
studding his discussions of the subject with numerical calculations which
supposedly provided incontrovertible support for his conclusions. For example,
he considered the case of 286 British judges who sat on the bench between 1660
and 1865 and were listed in the *Lives of the Judges,* a Victorian *Who's Who* of the
judiciary. In tracing the relationships among these men, Galton discovered that
of the total, one out of nine were "either father, son, or brother to another judge,
and the other high legal relationships have been even more numerous."[25]
This, he declared, was proof that judicial ability is inherited. If ascending to the
bench were merely accidental, there should not be such a strong pattern of

familial connection. Clearly, these men were endowed by their heritage with assorted qualities which fitted them to become judges: They were "shrewd, practical, helpful men; glorying in the rough-and-tumble of public life, tough in constitution and strong in digestion, valuing what money brings, aiming at position and influence."[26] In short, they had inherited three key qualities which together constituted natural ability: capacity, zeal, and vigor, without which a man "cannot hope to make a figure in the world."[27]

In his analysis of the judges Galton committed several major errors, among them relying upon a biased sample and failing adequately to control for the effects of wealth, education, class standing, nepotism, and sheer luck upon securing a position in the British high judiciary. In short, he passed off cant and specious logic as scientific validation, a criticism which can readily be applied to his chapters in the same book on statesmen, commanders, men of science, poets, competitive scullers, and assorted others. Just as physical anthropologists assumed the existence of races and then went out and found "proof" that they were right, Galton assumed that ability was inherited, then went out and found evidence for it. He proceeded to push a whole range of human traits—whichever might contribute to a person's rise in any given career—into the same box and call them genius or natural ability. This error can perhaps be counted as Galton's most unfortunate, insofar as his investigations spawned the whole school of behavioral psychology which purports through IQ and other standardized tests to rank human beings by "intelligence," an exercise which has since the early part of this century repeatedly served to buttress racist and classist positions.

A pioneer in the use of statistics, then a relatively new tool for analyzing group phenomena, Galton also applied the law of deviation from the mean to his examinations of natural ability. According to this law, derived by the Astronomer Royal of Belgium Adolphe Quetelet, for any given array of similar values, whether people's heights or magnitudes of stars in the sky, those values will be dispersed in a predictable pattern around an average value (or mean deviation). Invariably, the bulk of events will fall at or near the average, with the remainder distributed in diminishing frequency at either extreme, a verity represented visually in the figure of a bell curve.

When Galton ranked various groups of people in Britain by natural ability, their distribution conformed to Quetelet's law. For Galton the message of this mathematical exercise was clear: English society was configured pyramidally, with a few men of talent at the top holding positions of influence and power and the mass of individuals populating the lower classes, because the immutable laws of nature had made it so. Nature only produced so many men with innate ability—only 1 out of every 4,000 were men of "eminence," only 1 out of every million, men of "genius."

This was not to say that some series of events might not act to shift the average of natural ability higher or lower. Quite familiar with theories of degeneration, and with the work of Cesare Lombroso, Galton had no doubt about which direction the average might most easily slide: toward the undesirable lower end of the scale.

Lombroso, an Italian psychiatrist and physical anthropologist, maintained that criminals and "moral imbeciles" could be identified by a host of giveaway features, such as retreating foreheads, small heads, hirsutism, projecting jaws, patchy beards, and large ears; they composed as a whole "a type resembling the Mongolian and sometimes the Negro."[28] Thieves, forgers, sexual deviants—all were born, not made, their condition owed primarily to the insanity, epilepsy, and alcoholism of their forebears. A product of lineal degeneration, criminality represented an evolutionary throwback to an atavistic bestial state, Lombroso claimed. Even so, he held out the possibility that nurture could overcome nature, and championed the cause of rehabilitation. "Hereditary influences," he wrote, "are not all manifested at any given moment, or once for all. They are latent in the organism and manifest themselves gradually throughout the whole period of development."[29]

Galton, though, tended to hew a harder, deterministic line, asserting that an "evil inheritance" had a greater chance of prevailing than efforts at education or reform. The lower classes *were* the lower classes, Galton believed, precisely because of their innate physical, intellectual, and moral decrepitude. Social Darwinism, which was of course elaborated not by Darwin but primarily by social theorist Herbert Spencer, would employ a similar justification in defense of the status quo, advancing the apothegm that the lower classes were merely losers in a biological game in which the "fittest" thrived, and those who were not fit justly suffered.

If innate inferiority doomed one to failure, the flip side was equally true, Galton reckoned. No impediment, whether lowly birth or inadequate education or accidents of fate, would prevent a man of ability from reaching his deserved station in life. How, then, could one shift the average ability of a population higher? There did seem to be a way to do this, Galton mused, if only the quality of offspring could be improved across the board. . . .

The methods by which Galton concluded such a goal might be accomplished constitute the pseudoscience of eugenics. By "eugenics," Galton meant the scientific "cultivation" or betterment of the human line. Putting forward his plan in *Hereditary Genius* and in an 1883 work, *Inquiries into Human Faculty and Its Development*, Galton launched what quickly became an international movement. On his choice of the term "eugenics," Galton wrote:

> We greatly want a brief word to express the science of improving stock, which is by no means confined to questions of judicious mating, but which, especially in the case of man, takes cognisance of all

influences that tend in however remote a degree to give to the more suitable races or strains of blood a better chance of prevailing speedily over the less suitable than they otherwise would have had. The word *eugenics* would suitably express the idea; it is at least a neater word and a more generalised one than *viriculture,* which I once ventured to use.[30]

Like Gobineau, Galton had no doubt that northern Europeans had greater natural ability or genius on average than "men of lower races." (Galton distinguished dozens of races, within the United Kingdom alone recognizing three—British, Low Dutch, and Norman-French—which is to say he used the term "race" loosely to refer to ethnic groups. By "race" he also sometimes signified economic class.) In fact, he thought, some very smart dogs were more intelligent than the least-intelligent races. However, he asserted, the "noble qualities" which the Europeans possessed, including physical energy, heightened sense perception, moral integrity, sound judgment, and intelligence, were rapidly being bred out of the human population due to higher birthrates among poorer Europeans and nonwhite peoples elsewhere in the world.

The steps Galton proposed by which humanity could pull itself back from the brink of disaster were twofold (later these main lines of attack were classified by Galton's followers as "positive" and "negative" eugenics). First, people of able minds and bodies should only marry others of their kind and class. This self-selected group of "superior" individuals should then endeavor to have as many children as possible, to counterbalance the inflated number of offspring being produced by "inferior" sorts of people. "If the races best fitted to occupy the land are encouraged to marry early," Galton wrote, "they will breed down the others in a very few generations"; that is, the number of their offspring and their offsprings' offspring will far exceed those of the races least "fitted to occupy the land." (It is interesting to note that Galton and his wife never had children, apparently due to infertility.)

At the same time, Galton urged governments to step in and implement measures to discourage or prevent from reproducing themselves members of the "lowest class."[31] That such policies should find support seemed axiomatic to Galton, who remarked that

the striking results of an evil inheritance have already forced themselves so far on the popular mind, that indignation is freely expressed, without any marks of disapproval from others, at the yearly output by unfit parents of weakly children who are constitutionally incapable of growing up into serviceable citizens, and who are a serious encumbrance to the nation.[32]

Thus, a eugenic program could be considered to fall within "the sphere of practical politics," and certainly was feasible biologically:

> There is nothing either in the history of domestic animals or in that of evolution to make us doubt that [through eugenics] a race of sound men may be formed who shall be as much superior mentally and morally to the modern European, as the modern European is to the lowest of the Negro races.[33]

Galton felt that undertaking to improve his race, and humanity in general, was on the order of the highest moral calling, one "which is supposed to be exercised concurrently with, and not in opposition to the old ones upon which the social fabric depends."[34] In fact, he hoped that eugenics would become a new secular religion, supplanting old "sentiments or persuasions" with a creed, "by which the evolution of a higher humanity might be furthered."[35] Eugenics was an enterprise in which everyone in society could participate. Galton suggested that identical twins be studied intensively, since "their history affords means of distinguishing between the effects of tendencies received at birth, and of those that were imposed by the special circumstances of their after lives."[36] He had no doubt that such studies would reveal that intelligence, insanity, and diseases were hereditary. Galton also urged physicians to take more thorough life histories of patients, so that hereditary diseases might in time be identified and eliminated from populations.

Already, "half unconsciously," the species had begun to further its own evolution, but "[man] has not yet risen to the conviction that it is his religious duty to do so deliberately and systematically." A focus on selective breeding in humans could only, Galton thought, profit everyone, engendering a new mental attitude, a "greater sense of moral freedom, responsibility, and opportunity."[37]

In 1884, some nine thousand people flocked to that neo-Gothic temple of science, the British Museum in South Kensington (now the British Natural History Museum), and allowed themselves to be enshrined in the record books of Galton's Anthropocentric Laboratory. Ever fixated upon measurement, Galton, through the agency of his laboratory assistants, took from these willing subjects their height, weight, arm span, breath power, and other characteristics, all with the aim of further refining the art of anthropological taxonomy, or anthropometry. Soon, a Galtonian devotee, Karl Pearson, a statistical wizard himself and developer of the chi-square test, which enables researchers to compare predicted statistical outcomes with results gained from experiments or observations, had opened the Biometric Laboratory on Gower Street, near the University of London, for the purpose of gathering similar physical data. Pearson's lab metamorphosed first into the Eugenics Record Office and then into the Galton Laboratory for National Eugenics under the auspices of the University of London (which received the bulk of Galton's estate after his death in 1911 and installed Pearson as its Galton Eugenics professor).[38]

In 1902, Galton, Pearson, and an Oxford professor of comparative anatomy began publishing *Biometrika*, a journal devoted to attempts to quantify physiology, intelligence, and other human traits. About half of *Biometrika*'s charter subscribers were American, about half European, and many of them eager eugenicists, for Galton's doctrine had spread with astonishing quickness throughout the scientific community.[39] Between 1890 and 1930, some thirty eugenics societies were formed, frequently national in scope.[40] By 1907, the Eugenics Education Society, which operated independently of Pearson's organization in London, boasted six branches throughout Great Britain, and a membership mostly drawn from the ranks of upper-middle-class professionals.[41] To the first international eugenics conference, which was held in London in 1912, flocked some 750 attendees, including luminaries such as Winston Churchill, Alexander Graham Bell, and former Harvard University president Charles Eliot.[42] The following year, many in Britain would trumpet the birth of Eugenette Bolce, declared the world's "first eugenic baby."

In France, Germany, Denmark, Sweden, Norway, Poland, Russia, Switzerland, Italy, Brazil, the United States, and Japan, among other places, scientists, physicians, social theorists, birth control advocates, feminists, and others joined ranks, advocating the Galtonian gospel, which came to be known after 1895 in Germany as *Rassenhygiene*, or "race hygiene," a phrase which spread to other countries as well. Everywhere, practitioners attempted through a mixture of statistical analysis and primitive genetics to provide a quantitative basis for eugenic reforms.

Galton's eugenics can be seen as the culmination of two thousand years of Western attitudes. It reasserted the Platonic doctrine of the state's interest—obligation even—in the regulation of reproduction, and offered a supposedly scientific explanation for the perceived problem of race degeneration. At the same time, Galton did not simply reiterate past views. He furthered the ascendancy of number in the fields of biology, psychology, anthropology, and criminology (fingerprinting is a Galton invention), and promoted state reliance upon statistics as an aid to policy making. In virtually all his work, he sought to reduce individuals to statistics, to relate physical appearance to behavior and to status in life, in a manner that purported to be scientific but was finally not terribly far removed from palmistry—or the phrenology he had rejected as a young man—from what journalist Walter Lippmann would call the "Babu sciences."[43]

Moreover, Galton offered a program for social action which galvanized thousands of people worldwide to engage in eugenic activism. In 1780, a London physician, James Graham, had advertised the overnight use of a "grand celestial bed" in his Temple of Health and Hymen, for fifty guineas selling couples a chance to conceive children of extraordinary beauty. Galton, with less enterprising hucksterism but no less audacity, promised his eugenic acolytes as much and more. The French philosopher Michel Foucault suggested that for the growing middle classes in Europe, emphasis on heredity was a means of co-opting aristocratic pretensions based upon blood and the "antiquity" of one's lineage and alliances. As such,

> included in bourgeois marriages were not only economic imperatives and rules of social homogeneity, not only the promises of inheritance, but the menaces of heredity; families wore and concealed a sort of reversed and somber escutcheon whose defamatory quarters were the diseases or defects of the group of relatives—the grandfather's general paralysis, the mother's neurasthenia, the youngest child's phthisis, the hysterical or erotomanic aunts, the cousins with bad morals.[44]

Eugenics promised individuals a kind of redemption from the embarrassment and taint of socially—and evolutionarily—unworthy ancestry, a seeming way of achieving power through shaping the future.

The commandments of Galtonian eugenics were basically two: Thou shalt encourage fertility among "superior" individuals or classes and thou shalt discourage fertility among "inferior" individuals or classes. Galton had offered a few suggestions as to how these charges might be fulfilled, but eugenics was malleable enough that it could be adapted to serve a variety of needs. As the philosophy took hold throughout Europe and the Americas, its proponents elaborated a wide assortment of educational, public health, immigration, and tax policies that suited their own political beliefs and cultural circumstances. Even within individual countries, eugenics was by no means monolithic, and drew adherents from the entire political spectrum, as appealing to some socialists as it was to some imperialists. Those who argued for increased women's rights, for example, could readily assert, as did the British free-love advocate and sexologist Havelock Ellis, that "the question of Eugenics is to a great extent one with the woman question."[45] Indeed, the eugenics movement as it came to be constructed in England appealed to socially progressive thinkers because it emphasized the right of a woman to practice birth control, a significant advance in women's rights. It also promoted improved prenatal and postnatal care, which benefited women and their children. Meanwhile, those who felt their countries were being overrun by immigrants could, as eugenics supporters in the United States did, use Galtonian arguments to press for the shutting down of the "dysgenic" influx.

By proposing a biomedical fix for what had been and always will be socioeconomic problems—poverty, unemployment, violence, self-destructive behaviors, mental and physical debilities—Galton satisfied a reflexive desire to find simple solutions for complex dilemmas. *His* simple solution was that elusive entity, the baby of quality, upon which the very existence of the human species was said to hinge.

8

The Evils of Eugenics

French amateur anthropologist and confirmed Aryanist Count Georges Vacher de Lapouge in the 1880s proposed a scheme to improve the human stock of his nation. De Lapouge had hearkened to the teachings of Gobineau and Galton, but he had also experienced a stroke of insight concerning the role that artificial insemination might play in achieving eugenic goals. The successful insemination of a spaniel by the Italian anatomist Spallanzani a century earlier had led to a spate of similar experiments with animals; de Lapouge's own countryman, Jean-Louis Armand de Quatrefages, had fecundated dozens of species, from algae to guinea pigs.

Inspired to attempt the same feat with humans, a few doctors had tried the technique to address male infertility. John Hunter, the famed London comparative anatomist and surgeon who taught Edward Jenner, the inventor of the smallpox vaccination, was visited at his offices in the 1770s by a draper with a congenital defect of the penis which impaired his ability to impregnate his wife. Hunter gave the man a syringe and instructed him to go home, warm it slightly, and ejaculate into it. Then Hunter told him how to inject the sperm into his wife's vagina. According to Hunter's nephew, Everett Home—who mentioned the case in a 1799 report he made to the Royal Society of London about his own dissection of a hermaphroditic dog—the method worked, and the draper's wife conceived and bore a child. Reputedly, J. Marion Sims, a pioneering gynecological surgeon who practiced in Montgomery, Alabama, and New York City up until the Civil War, had attempted to impregnate women by "ethereal copulation" an estimated fifty-five times, and succeeded in two or three cases. And we know from civil court records in

Bourdeaux that at least one French physician was performing artificial insemination upon patients as early as 1883: In August that year, a physician in that city sued a patient for failing to remit payment for the procedure.

The ardent Aryanist de Lapouge, however, had more in mind than the occasional clinical application of artificial insemination: To avoid the further deterioration of the French people he proposed a program whereby semen would be gathered from "a very small number of males of absolute perfection" and used "to inseminate all the females worthy of perpetuating the race."[1]

While unrealized, de Lapouge's strategy for banishing the specter of degeneration, that venerable Gallic bugaboo, serves to illustrate several important aspects of eugenics as it came to be practiced from the 1880s through the 1940s: First, and almost invariably, eugenicists proceeded from the conviction that individual liberties mattered less than state prerogatives. The composition of a nation's population was too vital a matter to leave to chance matings. Thus, direct intervention by agents of the state in people's reproductive lives was deemed not only acceptable but entirely desirable, whether through promotion or prohibition of contraception, regulation of abortion, premarital testing for venereal diseases, licensing of marriage, genetic counseling of couples considered to be at risk for producing diseased or congenitally abnormal children, or sterilization.

Second, despite the fact that a few women held leadership positions in the eugenics movement, it was dominated by the paternalistic belief that women's bodies were the purview, if not the property, of men, there to be manipulated at will in the service of the state or for the higher good of civilization. If women could not be persuaded to voluntarily put on the yoke of eugenics, they should, some advocates felt, be forced to do so.

Third, eugenics and medicine were closely allied. Insofar as both aimed to promote the well-being of infants, especially by assuring that women received adequate prenatal nutrition and medical care, they shared laudable goals. However, as de Lapouge demonstrated, the desire to reduce the number of ailing infants cannot always be distinguished from the desire to engineer healthy infants. Physicians at the turn of the century had begun their love affair with technology. Biochemical tests, specialized instruments, and machines were giving them a sense of mastery and power, catapulting them from mere healers to biological engineers. How much better than leaving reproduction to chance was to move it into the medical arena, where its every aspect could be scrutinized and supervised.

Last, de Lapouge, as so many eugenicists, betrayed his classist and racist biases about who should be allowed to procreate. Notwithstanding the progressive thinkers who saw in eugenics the possibility for creating a society in which all children had the opportunity to thrive, the vast majority of eugenicists believed that white, married, middle- or upper-class couples of northern European extraction should be encouraged to bear children and offered tax incentives and government supports for doing so, while all others should be discouraged or banned outright from reproducing.

It has been said that the French granted credence to eugenics less because they were impressed by its weighty scientific justifications than because they were willing to entertain almost any philosophy which suggested an integrated program for overcoming social problems.[2] Indeed, eugenicists made fewer inroads into scientific, medical, and public health institutions in France than they did elsewhere. Always happy to back pronatalism, the French establishment accepted eugenics when it promoted health education for would-be mothers, campaigned in favor of large families, and proposed subsidies that would encourage childbearing, but rejected it when more intrusive plans requiring people to submit to premarital medical exams and involuntary sterilization were at issue.[3]

The French reluctance to push eugenics too far into private realms was not shared by the likes of Charles Davenport, an American ideologue who stumped heavily in France, and throughout Europe, for a full battery of state controls over reproduction. Davenport was not only a diehard eugenicist, but also an adamant hereditarian who believed that virtually all traits, from intelligence to insanity, were passed on in strict Mendelian fashion from parents to children.

Gregor Mendel's horticultural experiments in the mid-1800s had revealed that common garden peas showed considerable variability for a number of traits, such as stem length, pod color, and seed shape. By performing a series of hybridizations and charting the appearance of plants in successive generations, Mendel was able to identify pairs of traits which seemed to be mutually exclusive as well as independently heritable, and appeared with predictable frequencies in a given line. When a round-seeded parent plant and a wrinkle-seeded one were crossed, for instance, all the plants in the next generation bore pods stuffed with round seeds, the wrinkled factor seemingly having disappeared for good. However, when planted, the seeds from the uniform-looking first generation yielded a second generation of plants that produced both round and wrinkled peas, at a ratio of about three to one. Clearly, Mendel concluded, some "factor" with the capacity to determine which trait would be displayed was being conveyed from parent plants to offspring. Moreover, certain factors could override others. Mendel designated these dominant factors, and proceeded to work out the patterns by which they interacted with weaker, or recessive, factors. He presented his results at an 1865 meeting of Brno's modest Natural Science Society, and then published a paper describing his investigations in the society's journal. Although the journal was distributed to over one hundred libraries and institutions across

Europe, Mendel's article was apparently little read and elicited no stir within the scientific community.[4]

Not until 1900 was the piece rediscovered and publicized by the Dutch botanist Hugo de Vries. In the meantime, microscopists, proceeding in ignorance of Mendel's findings, had identified the chromosomes, the threadlike structures residing within the nuclei of all cells, which become highly visible during meiosis, the process of division whereby mature egg and sperm cells are formed from primitive germ cells. Also, Weismann had elucidated his theory regarding the continuity of the germ plasm (see page 99). Unfortunately, many biologists misinterpreted Weismann to mean that the germ plasm entirely determines which traits a person will possess. They further assumed that environmental factors like prenatal nourishment and education would therefore be unable to counteract the effects of a degraded germ plasm.[5]

Although research results militated against such logic, showing that a person's hereditary endowment, or genotype, did not absolutely determine his observable characteristics, or phenotype, and that events during gestation and throughout life could alter the phenotype, some dogged individuals, like Davenport, maintained the primacy of heredity. Also they continued to argue that single genes are responsible for virtually all traits, notwithstanding unassailable proof that many traits are in fact owed to clusters of genes, each of which may be passed on independently of the others. When critics pointed out that arguments made by hard-line eugenicists did not jibe with the real facts as revealed by experimental or population genetics, eugenicists routinely ignored those facts and ginned up additional mathematical defenses of their position, while proclaiming themselves all the more loudly the true defenders of science and the public interest.

What the Harvard-educated Davenport believed, erroneous or not, had substantial impact. For decades he edited *Genetics* and the *Journal of Physical Anthropology*, the leading publications covering those fields, and he was, moreover, the indefatigable point man for eugenics in the United States. Davenport had converted Mary Williamson Averell Harriman, widow of the railroad tycoon E. H. Harriman, to the eugenic faith, and with her financial support set up the Eugenics Record Office on seventy-five hilly acres in Cold Spring Harbor, New York, in 1910. (The Carnegie Institution of Washington later took over administration of Davenport's facility, merging it with an experimental station there. Eventually renamed the Cold Spring Harbor Laboratory and shed of its eugenics arm, this facility has nurtured several Nobelists and remains a preeminent center for genetics research.)

From his bucolic outpost on the shore of Long Island Sound, Davenport waged a campaign to take the hereditary measure of as many Americans as possible, training for this job cohorts of field-workers. Mostly women, they fanned out to virtually every state, going door-to-door to interview families and gather data for Davenport's files back in Cold Spring Harbor. Where possible they recorded family trees, noting instances of pauperism (basically, entrenched poverty), enfeeblement (a catchall for any sort of impaired cognitive or physical ability), alcoholism, tuberculosis (which was thought to be inherited), and other disorders. Generally, Davenport's workers operated on the assumption that every family hid one or more "defective" forebear, and their methods for ferreting out evidence of a flawed ancestry were quite arbitrary.[6] A certain cast of eye or jut of chin sufficed to reveal an unworthy pedigree.

Eugenicists also proselytized heavily, publishing pamphlets and books presenting their case in direct, pared-down fashion, as in *A Eugenics Catechism*, published in 1923 by the American Eugenics Society, which mimed the Catholic catechism's question-and-answer format (and not coincidentally reinforced the notion of eugenics as a secular religion): "*Question. Why are children the most valuable thing in the world?* Because their character determines what kind of world there will be in the future."[7]

Such didactic literature routinely cast infants in the role of fully conscious, fully morally developed actors, their characters a direct reflection of their parents' characters. This inescapable stamp of hereditary, said eugenicists, explained why a century's worth of social welfare efforts to improve children's bodies, minds, characters, and behavior had failed: The only effective angle of attack was upon the germ plasm, by dint of the applied science of eugenics, which one devotee dubbed the fifth "and perhaps greatest" expression of human genius, after tool use, language, fire, and alphabets.[8]

At state and county fairs during the teens and twenties, eugenicists sponsored contests for "fitter" families. The value of the trophy and medal awarded in one such event at the 1924 Kansas Free Fair was, proclaimed a promotional brochure, "worth more than livestock sweepstakes or a Kansas oil well. For health is wealth and a sound mind in a sound body is the most priceless of human possessions."[9] These competitions, along with a variety of exhibitions and public relations campaigns, served to indoctrinate rural as well as urban working-class Americans in the precepts of the pseudoscience.

At the same time, Davenport and his minions lobbied at both the state and national levels for eugenics to become the law of the land, efforts which drew the support of dozens within the power structure. Major backers of eugenics included Paul Popenoe, a California physician who ran an organization known as the Human Betterment Foundation and would repeatedly praise Hitlerian eugenics policies; Hermann Muller, a lifelong eugenicist who would win a Nobel Prize in 1946 for work done twenty years earlier demonstrating that X rays can cause genetic mutations; Adolf Meyer, a psychologist at Johns Hopkins University who would pioneer the treatment of mental illnesses with drugs; Clarence Little, president of the University of Michigan; E. A. Ross, a University of Wisconsin sociologist who was an ardent Lombrosian and foe of immigration; Henry Osborn, head of the American Museum of Natural History who, during World War II, as chairman of the

Joint Army and Navy Committee on Welfare and Recreation and of the
Advisory Committee on Selective Service argued that screening to keep
"defectives" out of military service saved the country money; David Starr Jordan, who was chairman of the eugenics committee of the American Breeder's
Association, out of which emerged the American Eugenics Society, as well as
a vice president of the American Boy Scouts; Gifford Pinchot, conservation-
ist and chief of the U.S. Forest Service under Theodore Roosevelt; and Harry
Laughlin, Davenport's assistant at the Eugenics Record Office and also a
leader in the American Breeder's Association. Laughlin spoke often of the
need to "eliminate" altogether the most disenfranchised portion of the populace, whom he dubbed the "submerged tenth." In universities and colleges
across the nation, many influential scientists, especially geneticists, espoused
eugenics; 42 out of 100 leading American geneticists who sat on the steering
committee of the International Congress of Genetics in 1928 were active in
the movement.[10]

Even Margaret Sanger, whose promotion of birth control exposed her to
repeated legal assaults and won her the enmity of the Catholic Church, supported eugenics for at least a part of her career. Sanger championed birth
control as "a tool for redistributing power fundamentally in the bedroom, the
home, and the larger community," and viewed eugenics as not incompatible
with that goal, especially insofar as it functioned "as a deterrent to poverty
and human waste."[11]

An ardent pronatalist where the educated classes were concerned, Davenport
fought against freer access to contraception, and no doubt applauded the 1905
indictment by President Theodore Roosevelt of American women who put
their selfish desires before their "paramount obligation to the perpetuation of
family, class, and nation."[12] The rhetoric of the day held that the country
would be overrun by the offspring of foreign-born residents, so that it was the
duty of Anglo-Saxon women to procreate with vigor.

By the 1920s, eugenicists had pushed through legislation that they considered instrumental in the battle against further erosion of the population due
to congenital pauperism, insanity, and crime. Installed by dozens of states,
these laws, among the sorriest distortions of American jurisprudence ever,
required the compulsory sterilization of individuals who were deemed to be
unfit. A model law, drafted in 1922 by Laughlin, made a person's inability to
"maintain himself or herself as a useful member of the organized social life of
the state" cause for sterilization. Laughlin enumerated those individuals composing the "socially inadequate classes." They were the

> (1) Feebleminded; (2) Insane (including psychopathic); (3) Crimi-
> nalistic (including delinquent and wayward); (4) Epileptic; (5) Ine-
> briate (including drug habitués); (6) Diseased (including the
> tuberculous, the syphilitic, the leprous, and others with chronic

infectious and legally segregable diseases); (7) Blind (including those with seriously impaired vision); (8) Deaf (including those with seriously impaired hearing); (9) Deformed (including the crippled); and (10) Dependent (including orphans, ne'er-do-wells, the homeless, tramps, and paupers).[13]

The violation of the bodily integrity and rights of those who fell into this taxonomy of the socially inadequate was necessary, Laughlin asserted, because all children of such degenerate types carry in their germ plasm "the genes or genes-complex for one or more 'inferior' or degenerate physical, physiological or psychological qualities."[14] Unless the state took bold steps—Laughlin hoped that some 15 million "inferior" men and women could be rendered sterile by 1980—the ratio of "defectives" would continue to skyrocket (as eugenicists asserted had been happening since World War I), and the nation would surely collapse, able to muster no one capable of functioning at a moderate level, much less of leading.[15] Unchecked, the trend would lead to a whole nation of Jukes or Kallikaks.

The Jukes family provided a cautionary lesson about the hazards of "inferior" blood. Their story received popular treatment in 1877 in a book by Richard Dugdale. Galton then mentioned the family in *Inquiries into Human Faculty*, presenting it as evidence that "the criminal nature tends to be inherited." The Jukes pedigree, Galton wrote, "has been made out, with extraordinary care." Researchers amassed information on "no less than 540 individuals of Jukes blood, of whom a frightful number degraded into criminality, pauperism, or disease."[16]

The Kallikaks, a pseudonymous family living in the New Jersey Pine Barrens, were named and discovered by Henry Goddard, the psychologist who had imported the Binet-Simon scale to the United States and distorted its original purpose. The French psychologist Alfred Binet had developed the test in order to help educators identify children who could be expected to perform poorly in school. These children could then receive special attention. Binet made no pretensions as to the broad applicability of his test; however, Goddard proclaimed the test a means of gauging "intelligence," which he believed was an inborn quality. That intelligence directly correlated to one's standing in society was demonstrated by the Kallikak family, whose common progenitor, a Quaker businessman, had fathered children with two separate women. All those children bred off his "good" wife, a fellow Quaker, had been upright citizens; all those bred off his "bad" wife, a tavern girl, had been miscreants and lowlifes: proof, Goddard said, that degeneracy was inherited and heredity was all.[17] German eugenicists read Goddard with fascination, and would find in the Swiss "Family Zero" the European counterpart to the Jukes and Kallikaks.

By 1935, twenty-eight states boasted forced sterilization laws which provided for the compulsory castration of those incarcerated in state foster homes, mental institutions, or prisons.[18] Although not uniformly applied, the laws enabled California, the state most zealous in prosecuting the war against "defectives," to sterilize 12,941 people, and perhaps as many as 30,000 people altogether went under the knife.[19]

At the same time Americans were being turned against people who were physically and mentally disabled, Gobinesque fearmongers like Madison Grant and Lothrop Stoddard, whose most well known works were, respectively, *Passing of the Great Race* and *Revolt Against Civilization: The Menace of the Under Man*, fanned suspicion and hatred of immigrants. Alarmed by the reputedly high fertility rate among foreign-born women, these eugenicists favored imposing outright bans or quotas on immigration. Especially noisome from their point of view were persons hailing from southern and eastern Europe, who had been declared biologically "inferior" (thanks partly to IQ testing) and more prone to produce "defective" children. These immigrants effectively were lumped with American blacks, already subject to extensive antimiscegenation laws throughout the South. Blacks and immigrants alike—the cant went—would overrun the country as well as interbreed with northern Europeans and thereby sully the purity of the "race" that had made America great.

The putative threat of conquest by race mixing and differential birthrate prompted an enormous outpouring of effort from eugenicists. At every opportunity, they lobbied Congress to shut the nation's doors, testifying at numerous hearings and serving as advisers to the House Committee on Immigration and Naturalization. Their xenophobia prevailed, and legislators tightened immigration restrictions based on eugenic standards, first in 1921 and again in 1924. Eugenicists had succeeded in establishing what they hoped would be an unscalable bulwark against further assaults upon the American germ plasm.

A mid the turmoil of the German Empire's dying days, the old myth of a Nordic people in a titanic struggle for survival against outside forces attained more potency than ever. The convulsions of industry and labor unrest, ongoing urbanization, a rise in crime, and a falling birthrate from the late 1890s on, among other disruptions, led many Germans at the turn of the century to conclude that degeneration had set in and the country would be lost. Among those suffering from this fear was Alfred Ploetz, a young physician whose dreams of finding an earthly paradise had been smashed upon his visit to the French socialist community New Icaria, Iowa, in the late 1800s. Ploetz, to his dismay, had found the town populated by a "low quality of human beings" and had returned to the Fatherland disgusted.[20] The problem, he concluded, was that insufficient care had been taken to preserve the race, and he sat down to outline his recommendations in this regard, producing in 1895 the first of two volumes that would make up *Grundlinien einer Rassenhygiene* (*Basic Details of Race Hygiene*).

Ploetz defined a race as "any interbreeding human population that, over the course of generations, continues to demonstrate similar physical and mental traits."[21] He contended that the best way to ensure that an elevated race like the Teutons would maintain its original character was to wrest from nature control over who would live and who would die, and to exercise selection over the germ plasm. He envisioned a Germany where only the most perfect individuals of any generation would qualify to reproduce; their germ plasm would endow upon the next generation characteristics which would only be improved upon as the breeding program continued. Eventually, the race would be restored to its original glory and perhaps even supersede it. In hopes of hastening this eventuality, Ploetz founded first the *Archiv für Rassen- und Gesellschaftsbiologie* (*Journal of Racial and Social Biology*) in 1904, and then, the following year, the *Gesellschaft für Rassenhygiene*, or Society of Race Hygiene, which by 1930 had over twenty branches.

Ploetz and his associates had a long list of dislikes. They derided feminism, which they believed was sabotaging the family. Women belonged in the home tending children: This was not simply a societally imposed duty but the fulfillment of a biologically programmed need. Ploetz's protégé, the medical geneticist Fritz Lenz, argued that women wish "above all, to be regarded as beautiful and desirable" and have evolved "mainly for the breeding of children and for the allurement of man." A man, on the other hand, "wants to be regarded as a hero and as a person who gets things done."[22]

In many respects, Ploetz and his ilk looked upon themselves as chivalric knights riding to the protection of the weak wherever they found them.[23] They were doing so by ensuring that future generations would be made up of people of strong moral and physical fiber. This was a kind of charity by omission: Get rid of the would-be objects of charity, and you eradicate their need to be aided. It was only perpetuating cruelty, Ploetz said, to allow the weak to procreate. Medicine itself was responsible for much evil, because by enabling sickly and physically impaired people to live and produce offspring, it had added to the ranks of the unfit and propelled the German race downward. In a society where race hygiene mattered, Ploetz envisioned, such "defectives" would be sent to work camps, where they would receive minimal medical care and be prevented from reproducing.

Ploetz in his early writings allowed that a certain degree of race mixing, even with Jews, whom he ranked with Aryans as a "superior" race, could be highly beneficial. This might suggest that Ploetz was not anti-Semitic. But even before World War I, Ploetz belonged to a secret club, the Nordic Ring, which celebrated the Aryan birthright, and as the Nazis rose, Ploetz became a leading proponent of the creation of a Nordic state.[24] He was closely allied with another promoter of race hygiene, the wealthy medical publisher Julius Friedrich Lehmann, himself a comrade of Houston Stewart Chamberlain. Lehmann's magazine *Volk und Rasse* provided an outlet for many anti-Semitic authors. In an article printed in the magazine in 1926, physician Eugen Stähle urged his readers to "think what it might mean, if we could identify non-Aryans in the test tube! Then neither deception, nor baptism, nor name change, nor citizenship, and not even nasal surgery could help [Jews escape detection]. One cannot change one's blood."[25] Dozens of other race hygiene periodicals promoted similar views.

Those who would turn out to be key scientific advocates of race hygiene, including Lenz, the anthropologist Eugen Fischer (with whom Davenport maintained strong ties into the 1930s), and Ernst Rüdin, who elaborated a genetical theory of psychiatry, also loudly championed the cause of the Nordic race.[26] Their rhetoric portrayed the prototypical Nordic or Teutonic man as somewhat remote emotionally but capable of profound contemplation and mental clarity; as objective and sober, yet deeply religious; as capable of self-restraint; as devoted to the value of independence and willing to revolt against unjust authoritarianism; as unsurpassed in courage when he must fight.[27] He was, in sum, the embodiment of all that was best about the species, the epitomic man.

In 1923, three of Germany's most vocal race hygienists published a textbook to rave international reviews. Fritz Lenz, Eugen Fischer, and Erwin Baur, a botanist, each contributed chapters to the weighty volume, published in Germany as *Menschliche Erblichkeitslehre und Rassenhygiene (The Principles of Human Heredity and Race Hygiene)* and issued later in the United States under the title *Human Heredity*. Soon, the book was standard issue for college genetics and heredity courses in Europe and America, and was referred to universally as Baur-Fischer-Lenz. The text covered all the eugenic bases: exceptional aptitude, inbreeding (the ancients had recognized "the bad effects of inbreeding" and "a considerable part of the legislation of the oldest civilised races turns upon this question," claimed the authors), twin studies, degeneration *(Entartung)*, racial characteristics (including a discussion of mental testing of U.S. Army soldiers which purportedly showed that blacks were innately less intelligent than whites).[28]

The third edition of the work, in 1927, gave emphasis to this last topic, which had preoccupied Fischer (whose given name testified to his "well-born" status) for some decades, ever since, as a young professor at the University of Freiburg, he had delved into the issue of the so-called Rehoboth bastards, mulatto children of German soldiers occupying the colony of

South-West Africa (now Namibia). Closer to home, there was another mixed-race group which Fischer believed posed a threat to the Nordic race, the *Rheinlandbastarde* (Rhineland bastards), as they were called, about eight hundred children who had been born to German mothers and French-African occupation troops during World War I.[29] To Fischer, these were not mere children, but rather the very corruption of the German stock. Because the Negro race was "devoid of the power of mental creation," those carrying their blood imperiled the entire nation.[30]

Fischer held that the human was, in effect, a self-domesticating species whose efforts had produced races with all their variability, "in mental respects no less than bodily."[31] Historically, those races lacking innate intelligence had developed neither adequate leadership nor industry and remained stalled in the Stone Age. Baur-Fischer-Lenz stipulated that not only Negroes, but also Mongoloids (who were supposedly dull-witted) and Mediterranean peoples (who were said to exhibit "a childlike cruelty" and "natural cunning but no genuine intelligence"), possessed undesirable racial characteristics.[32] Parroting Gobineau, Fischer argued that nations decline entirely owing to "a reversed selection of the racial constituents of the people concerned." Through a process of negative selection, "the hereditary carriers of the requisite gifts, the racial leaders," vanish, and so, too, does the nation's former prowess.[33] This "anthropological cause" had plunged post–World War I Germany into chaos, Fischer said.[34] Thus, it was imperative that the nation prevent further corruption of the line. Reproduction by destructive elements like the *Rheinlandbastarde* must be checked through strict regulations, and educated Germans must be urged to have as many children as possible. To keep tabs on the condition of the germ plasm, Lenz argued that the state should undertake a "biological registration of the entire population."[35] Within a few years, Lenz would have his wish.

Otmar von Verschuer, head of human heredity at the Kaiser Wilhelm Institute for Anthropology, Human Genetics and Eugenics in Berlin, a professor and twin researcher who taught Josef Mengele as a graduate student, would accord Baur-Fischer-Lenz a central place in the annals of "the Nazi political and philosophical revolution."[36] Indeed, Lenz's son Widukind would report after World War II that his father had been told on good authority that Adolf Hitler had read the textbook, along with the work of Houston Stewart Chamberlain, while serving his term in Landsberg prison after the failed Beer Hall Putsch of 1923.[37] Many passages in Hitler's *Mein Kampf* employ pseudoscientific jargon and present arguments similar to those in the textbook. "For example," said Widukind Lenz, "there is that long passage about syphilis. . . . That was one of my father's obsessions too."[38]

To read *Mein Kampf* is to see long-standing Western anxieties embodied in a single man and become ragingly pathological. Blood can be found, almost literally, upon every page of this diatribe against Marxism, Judaism, labor unions, and syphilis. Hitler's fixation with blood, race, and degeneracy

translated into a blind, perfervid hatred of the press, the government, international bankers (for him synonymous with Jews)—of anyone who allowed
poison "to enter the national bloodstream and infect public life."[39] This "poison" is variously biological, as with syphilis, which Hitler blamed for filling
Germany's insane asylums and hospitals with "adulterated" offspring; economic, as with labor unions, which would destroy the Fatherland fiscally;
hereditary, racial, and political, as with the Jews, Slavs, and *Rheinlandbastarde*,
whose blood itself is supposedly poisonous. Jews especially Hitler robbed of
any human status, describing them repeatedly in microbial terms. "The Jew,"
that reducible entity, was a "moral mildew," an "abscess," a "moral pestilence,"
"worse than the Black Plague," a "bacillus," the "internal cause" of disease, a
"parasite."[40] Hitler also invoked another powerful legend of blood taint by
calling Jews vampires.

Hitler argued that survival of the Aryan race—that "Prometheus of
mankind, from whose shining brow the divine spark of genius has at all times
flashed forth"—depended upon preventing "all those who are inflicted with
some visible hereditary disease or are the carriers of it" from procreating;
upon halting the "influx of negroid blood on the Rhine" by sterilizing the
Rheinlandbastarde; upon absolutely eliminating sexual contact with Jews, and
thereby putting a stop to miscegenation; and upon reunifying Germany and
Austria, so that "all those through whose veins kindred blood is flowing will
find peace and rest in their common Reich."[41] The German people, Hitler
proclaimed, had to be taught the *völkisch* concept of the world, which "recog-
nizes that the primordial racial elements are of the greatest significance to
mankind." Further, they must know that "the loss of racial purity will wreck
inner happiness for ever. It degrades men for all time to come."[42] To preserve
the race, Hitler declared, any means were justified: "When nations are fighting for their existence on this earth, when the question of 'to be or not to be'
has to be answered, then all humane and aesthetic considerations must be set
aside."[43] He remarked that he had become convinced of the rectitude of a
"ruthless manner" in defending racial purity, and that "in every case where
there are exigencies or tasks that seem impossible to deal with successfully
public opinion must be concentrated on the problem, under the conviction
that the solution of this problem alone is a matter of life or death."[44] To this
end, propaganda was ideal, exciting the emotions of the populace and helping
it "concentrate" on whatever "problem" was at hand.

Mein Kampf makes it clear that even before gaining power, Hitler had
envisioned the programs of mass sterilization, euthanasia, and genocide
which he would execute with complete ruthlessness, aided and abetted by the
German scientific and medical establishment, the courts, the police, the military, and the SA and SS.

On July 14, 1933, the recently convened cabinet of Hitler's Third Reich took the first step on the path to the Final Solution, voting in favor of a Law for the Prevention of Genetically Diseased Offspring, which appears to have been based on Laughlin's model statute and mandated the sterilization of people who were schizophrenic, manic-depressive, feebleminded, epileptic, or alcoholic; who were hereditarily deaf or blind, or suffering from Huntington's chorea.[45] The law provided for the creation of "genetic health courts," to be presided over by a lawyer and two physicians, one of whom was to be an expert in race hygiene. These courts would assess potential candidates and decide whether or not they were to be sterilized. Hitler had calculated that six hundred years of systematic sterilization would restore humanity to robust health—a daunting time frame, but not beyond the realm of feasibility for a Reich that was to last one thousand years.[46] At the behest of some two hundred genetic courts, as many as four hundred thousand people may have been sterilized over the next three to four years.[47]

But it was not sufficient to deal with just those who displayed obvious deficits. Lenz, speaking at the Ministry of the Interior in 1934, announced that his goal was to ensure that the state should inspect all citizens and issue certificates of heredity which would permit or forbid them to reproduce. "Those who do not suffer from hereditary disease within the meaning of the law are not necessarily healthy and fit to breed," Lenz said. "As things are now, it is only a minority of our fellow citizens who are so endowed that their unrestricted procreation is good for the race." He went on, "Safeguarding the hereditary endowment and safeguarding the race are basically the same thing. The biological foundations of the race are the biological foundations of the hereditary endowment and vice versa."[48] By 1937, the Reich had moved to sterilize children of color, including the unfortunate mulatto children from the Rhineland. Next in line came Gypsies, Jews, and "asocial individuals." Historians can only guess at how many thousands of sterilizations were performed on individuals falling into these categories. The sterilizations, usually surgical but sometimes experimentally performed by irradiating the victims, were carried out by physicians and staff at clinics and hospitals, who by this time had been suborned into the service of National Socialism.

Early on, the main difference between the German sterilization program and those undertaken in other countries, including the United States, was one of scale. But certainly by 1938 and possibly before, Hitler had moved to replace sterilization with euthanasia. Once the state had sterilized undesirables, they remained a burden, a drain upon its resources. Distorting the language of compassion, race hygienists contended that people hobbled by birth defects or incurable illnesses should be spared from suffering. Theirs were *lebensunwerten Lebens*—lives unworthy of living. Better that those lives be ended. In a letter dated September 1, 1939, the day on which Germany invaded Poland, Hitler issued a directive on euthanasia, giving "specially designated physicians" the right to review patients and, having judged them "incurable," to "grant" them "mercy killing."[49]

Paid for each evaluation, several dozen professors of psychiatry and medicine had before the year was out exercised their "extended rights." Of 283,000 people evaluated, the experts deemed some 75,000 as *lebensunwerten* and "granted" them death by starvation or lethal injection.[50] Mental patients were also left untreated when sick and allowed to perish. Or, doctors handed them over to storm troopers, as in October, November, and December 1940, when 4,400 mental patients in Poland and 2,000 in Germany were marched out into the cold and shot. A select few hospitals experimented with gassing patients, using carbon monoxide supplied by the industrial giant IG Farben. This proved so efficient an approach that 70,723 mental patients had been gassed to death by September 1941.[51] Although the victims' relatives, a few clergymen and physicians, and assorted others had periodically protested the killings, Hitler did not terminate the euthanasia program until that year—not until he had put in train the Final Solution, borrowing the gassing techniques which had been developed to handle *lebensunwerten* and setting up chambers of mass murder first at Auschwitz, then at other concentration camps.

Numerous scientists offered theoretical justification for such a program. For example, Konrad Lorenz, who would win a Nobel Prize in 1973 and come to be embraced by the public as an avuncular figure for his work with geese, wrote in a 1940 paper that "there is a certain similarity between the measures which need to be taken when we draw a broad biological analogy between bodies and malignant tumors, on the one hand, and a nation and individuals within it who have become asocial because of their defective constitutions, on the other hand."[52] This echoes dozens of other statements by eugenicists, who carried the argument to its logical, but despicable, conclusion: For the sake of the *Volk,* the surgeon had to operate. One propaganda poster showed the Führer bending to gaze into the face of an aproned, blond, plump-cheeked child. The legend read, "Adolf Hitler, Doctor of the German People."[53]

O n another front, Hitler implemented programs aimed directly at middle-class families with the intent of stimulating fertility and enhancing the German germ plasm. To begin with, the Reich proscribed physicians from sterilizing, aborting, or providing birth control to healthy German women, bans which were important ideologically but only minimally effective. Meanwhile, local Nazi organizations mustered youth into race hygiene courses, in which they were exhorted to keep in mind ten essential things regarding sex and marriage:

1. Remember you are a German!
2. Remain pure in mind and spirit!

3. Keep your body pure!
4. If hereditarily fit, do not remain single!
5. Marry only for love!
6. Being a German, choose a spouse of similar or related blood!
7. When choosing your spouse, inquire into his or her forebears.
8. Health is essential to outward beauty as well!
9. Seek a companion in marriage, not a playmate!
10. Hope for as many children as possible![54]

A roomful of youngsters exclaiming these precepts in unison would likely have given Galton great pleasure: Here was eugenics in its most essential form being used to guide the future parents of a nation. Galton and American eugenicists, too, would have appreciated the medals given out by the Nazis for extraordinary feats of motherhood. In the procreative Olympics, four children won you a bronze, six a silver, and eight a gold; eight being, apparently, the best effort the state could reasonably expect, even though Lenz had declared that every pure, fit German woman should ideally produce fifteen children. At the head of the state *Frauenwerk* (Women's Bureau), Gertrud Scholtz-Klink, who had been raised in an anti-Semitic household, supervised a nationwide network of programs promoting motherhood and encouraging women to think of themselves as synonymous with the *Volk*. (Unrepentant when interviewed by historian Claudia Koonz in the 1980s, Scholtz-Klink described her activities as "social welfare" and touted Nazi-style national labor service as a palliative for "the unruly youth of today.")[55] After World War II began, the Nazis in addition launched the *Lebensborn* (Well of life) scheme, whereby SS men were encouraged to impregnate as many single women as possible to increase the country's yield of children of "superior" genetic stock. Many Germans spoke out against this SS privilege, saying that it undermined attempts to inculcate morals in the young.

All the talk of choosing "a spouse of similar or related blood" was of course euphemistically anti-Semitic. Intermarriage between Jews and Germans had been banned at a rally in Nuremberg in 1935, where party members ratified a code "for the protection of German blood and German honour." The so-called Nuremberg Laws also specified who would and would not be considered Jewish. Few race hygienists objected to these regulations; many applauded them. Lenz had offered in Baur-Fischer-Lenz a mild reproach against anti-Semitism, arguing that Jews and Teutons alike had "great powers of understanding" and "remarkable strength of will." In the third edition of the textbook, he had added a footnote reporting that he had been taken to task by anti-Semites for such accommodations to Jews. He responded that "it is a pity that so much enthusiasm and youthful energy should waste itself in the idle clamour of the anti-Semitic movement," but added that "I really must warn my Jewish fellow citizens that they ought not to get the wind up as soon as any one begins to speak about the Jewish race. A tranquil and objective discussion of the Jewish problem would best serve the interests of both sides."[56]

As Lenz well knew, "tranquil and objective discussion" was hardly the approach to offset the hysterical fulminations appearing daily before the German people, like those given vent by Julius Streicher, publisher of the violently anti-Semitic *Der Stürmer (The Storm Trooper)*. Streicher argued that "the seed of a man of another race is a 'foreign protein' " which, during copulation, is "absorbed by the woman's fertile body [literally *Mutterboden,* mother-soil] and thus passes into the blood." Consequently, "a single act of intercourse between a Jew and an Aryan woman is sufficient to pollute her for ever. She can never again give birth to pure-blooded Aryan children. . . . " Streicher contended that this fact was well known to scientists, and to Jews, who used their knowledge to intermarry with, and destroy, the people of "superior" nations.[57]

Although Streicher blamed "science" for "suppressing the truth" and keeping Germans from learning about the perils of "race mixing," many scientists busily sought to ingratiate themselves with the Nazi leadership and to win government funding for research on race hygiene. Eugen Fischer, speaking at the University of Berlin, took pains to thank Hitler for making it possible for geneticists to put their understanding into action through the Nuremberg Laws.[58] (His sentiment was echoed in the American periodical *Eugenical News,* which lauded Hitler for having installed eugenics as state policy.)[59] Fischer would proclaim in 1943 that "it will always remain the undying, historic achievement of Adolf Hitler and his followers that they dared to take the first trail-blazing and decisive steps towards such brilliant race-hygiene achievement in and for the German people."[60] Rudolf Hess would declare National Socialism nothing more nor less than *"angewandte Rassenkunde,"* or applied racial science.

The physician, so said the propaganda, was to be the Führer of the *Volk.* In the name of the *Volk,* the top physician, Adolf Hitler, brought about the murders of 5 to 6 million Jews and unknown thousands of Gypsies, homosexuals, criminals, and mental patients; of anyone perceived as "undesirable," anyone whose existence could be considered "dysgenic." Before, during, and after the war, German researchers could claim that they had pursued worthwhile knowledge, had sought only "objective" fact, while others had prosecuted a distorted version of race hygiene. But the evidence shows that German scientists and physicians were, for the most part, fully cognizant of Hitler's genocidal actions and fully culpable. Scientists, with their specialized argot, their battery of mathematical truth-revealers, and their supreme powers of self-justification, have too often served evil ends. In Germany, their guiding conceits of the corrupted germ plasm, racial purity, and eugenic supermen may not have caused the Final Solution but certainly provided an authoritative-sounding rationale for it. Saddled with their own ignorance, prejudices, ambitions, and fears, race hygienists extolled actions which were monstrous and unforgivable. In mid-1944, when German troops shipped some fifty thousand Russian children to camps and killed them, a memo regarding the action stated, "The operation is planned not only to reduce the direct growth of enemy strength, but also to impair its biological strength in the distant future."[61]

From a purely scientific point of view, eugenics did not go entirely unchallenged from the 1880s through the end of World War II. In England and the United States, even as eugenicists succeeded in bending public policy to their ends, critics occasionally attacked the philosophy, exposing its flawed premises. Advances in genetics had revealed that genes and environment interact in complex ways, so that no strict equation can be drawn between genotype and phenotype. Population biologists had mathematically modeled what would happen if one attempted through selective breeding to eliminate a single gene from a given pool of individuals, assuming that that gene was recessive and responsible for one trait. If the gene was found in 1 in 100 people to start, it would take 22 generations to reduce its frequency to 1 in 1,000. To further reduce it to 1 in 10,000 would require 90 generations more, and another 700 generations on top of that to bring its occurrence to 1 in 1 million—for a total of roughly 16 millennia.[62] Since many geneticists had by the 1930s concluded that virtually all the higher cognitive functions of humans, in addition to many physical traits, were produced not by one gene but by the interactions of arrays of genes, eugenicists' claims that any trait, positive or negative, could be selectively bred into or out of a population seemed ludicrous at best, downright deceptive at worst.

In light of a panoply of mounting evidence, some prominent scientists, including anthropologist Franz Boas and biologists T. H. Morgan and Herbert Spencer Jennings in the United States, and biologists Lancelot Hogben and J. B. S. Haldane in Britain, drew a bead on the doctrine in its "hard-line" form. Haldane, an Oxford-educated professor of biometry at the University of London—who was then active in Marxist politics and had condemned the Nazi's expulsion of Jews and called Streicher's justifications for state anti-Semitism in *Der Stürmer* "obscene"—spoke out against the excesses of the international eugenics movement in a short 1938 book for the general audience, *Heredity and Politics*. While long in sympathy with eugenics in principle, Haldane, who carried out important studies in population genetics, argued that there was little support on a practical basis for legislation which attempted to improve the species. He deemed forced sterilization programs in the United States worthless. They exposed women especially to unnecessary risk of death from surgery without saving them from some greater personal danger. He wrote, "I personally regard compulsory sterilization as a piece of crude Americanism like the complete prohibition of alcoholic beverages."[63]

As a corrective, Haldane offered his readers a short course in genetics and demonstrated the relative roles of heredity and environment using the example of four lines of guinea pigs bred to produce extra toes. Haldane explained that within itself, each line of guinea pigs might exhibit a hereditary tendency to produce a certain percentage of extra-toed young; however, when comparing all four lines, one could see that the prime factor determining the percentage of congenitally "defective" offspring born was the age of the mother. The genes of the guinea pigs in each line code for a range of variability, but the actual variability that is seen in offspring results from an interaction between genes and the environment. The argument holds for humans and other animals, too. In the end, both nature and nurture shape the outcome.

Haldane accused eugenicists of routinely inflating the percentages of "defective" children born to "defective" parents (one researcher had put the figure as high as 75 percent), and deployed solid evidence against such claims by citing actual rates from the city of Birmingham and from East Suffolk, where 7.5 percent and 6 percent, respectively, of children born to those incapable of holding a job themselves proved incapable of holding a job. Haldane, seriously but somewhat wryly, remarked that if defects were owed to inbreeding, then "it is likely that the introduction of motor omnibuses to our rural areas will prove to be a eugenic measure quite as valuable as sterilization."[64]

Because in popular parlance evolutionary "fitness" had become misconstrued as referring to a general physiological robustness, Haldane was at pains to point out that Darwin had used the term only to signify "individuals of such a constitution that they are likely to propagate themselves in larger numbers than their fellows, either as a result of being better adapted to their environment or more fertile, or both."[65] By no means could Darwinian fitness be used as a justification for sterilizing those with mental ailments like schizophrenia, or those with a "grave physical disability" that might or might not be genetically transmissible. Haldane suggested that the zeal for sterilization, while supposedly motivated by scientific concerns, had far more to do with psychological aversion to those who were not perceived as "normal," and with race and class prejudices. Instead of depriving people of their reproductive powers, Haldane advised societies to ensure good prenatal care for women, to clean up workplaces that subjected workers to chemicals which might be damaging to their reproductive systems, and to reshape society to accommodate everyone, regardless of handicap, and give them a chance to serve in some productive capacity.

However, for every Haldane, for every scientist who challenged the paradigms of eugenics, there was one who, while deploring the Germans' more extreme methods, insofar as they were known, retained a belief in the worthiness of attempts to improve the human line. Researchers attending the Seventh International Genetics Congress in Edinburgh at the end of August 1939, the week before the United States entered the war, formally denounced

Hitler. But they also promulgated a "Geneticists' Manifesto" which proposed a recast eugenics. It argued for a leveled social, economic, and political playing field in which all individuals, regardless of ethnicity, gender, class, intelligence, or physical makeup, might have a chance to lead fulfilling lives. The manifesto specified that greater economic security for workers, with improvements in housing and adequate medical and educational benefits; financial assistance for families; the dissemination of birth control and of information regarding the biology of reproduction—all must play a role in any attempt at genetic improvement. This was indeed a radical departure from hard-line hereditarian eugenics and race hygiene. Yet while embracing values expressed by the labor, women's, and socialist movements, the manifesto still maintained that there could be biological solutions to social problems.

Despite claims that eugenics was overcoming its classist and nationalist shortcomings due to recognition of the "truth that both environment and heredity constitute dominating and inescapable complementary factors in human wellbeing," the established organs of eugenics continued to spread what was, for all intents and purposes, the same old message.[66] Surveying the editorial slant and content of the *Journal of Heredity* from 1939 to 1946, when it was a joint publication of the American Genetics Association and the American Eugenics Society, one discerns no marked retreat from the fundamental tenets of eugenics and encounters only mild repudiations of Hitlerian policies. In 1939, the journal heartily recommended a sampling of German genetic literature; in 1940, it ran praiseworthy pieces about Alfred Grotjahn, a physician and race hygienist who believed in compulsory sterilization of "defective antisocials."

During World War II, only one piece in the *Journal of Heredity* took a strong stand against racism and against policies based upon it. The piece was a reprint of a speech given by Ashley Montagu, a Philadelphia anatomist, at a 1941 meeting in Chicago of physical anthropologists. Montagu indicted racism as a "whitened sepulchre" and called upon anthropologists to cease and desist in their misbegotten attempts to differentiate between races through anatomical measurements. The anthropologist, largely ignorant of the basic principles of genetic variability, mutation, and inheritance, had taken a crude eighteenth-century notion and having "erected a tremendous terminology and methodology about it, has deceived himself in the belief that he was dealing with an objective reality," Montagu wrote.[67]

The overall editorial slant of the journal, however, tended to reinforce that status quo. The same year it ran Montagu's piece, the journal also offered readers a minute analysis of race crossing in Hawaii, complete with photographs showing the assorted features produced by the mixing of Chinese, Hawaiian, and European peoples on the islands. And in 1942, it printed an account by one Tage Ellinger of a visit to Germany in the winter of 1939–40, during which he met with Fischer at the Kaiser Wilhelm Institute of

Anthropology. Ellinger wrote that the Nazi treatment of Jews belonged "exclusively in the shameful realm of human brutality," but added that the breeding program implemented by Hitler's Reich wasn't bothersome per se and felt confident that "biological science can assist even the Nazis."

Eugenics as a philosophy survived the war relatively intact. A few zealots like Laughlin were hurried into retirement as genetics attempted to clean house and give itself new legitimacy. But by the 1950s, geneticists could be heard making obligatory references to the "lurid and disquieting history" or "spurious use" of eugenics and bemoaning the atrocities committed in the name of the philosophy, while going on to maintain that the basic goals of eugenics were laudable.[68] The *Eugenics Quarterly* would crow in 1954 that

> since World War II the reconstructed eugenics movement has succeeded in gaining the scientific respect of most geneticists, regardless of their individual moral views about eugenic programs. One sign of its newly found scientific acceptance is that leading biologists, such as Joshua Lederberg and, later in life, H.J. Muller, have once again become leading proponents of eugenic schemes.[69]

In Germany, where geneticists like Lenz had been rehabilitated and given university positions, a reconstituted eugenics movement preached pronatalism and even nervily campaigned for the reinstitution of sterilization programs. There, as elsewhere, a whole new generation was shaping eugenics to suit its own values and expectations, reframing its tenets to exploit a growing knowledge of the molecular foundations of heredity. Galton's legacy, despite the atrocities committed in its name, refused to go away.

9

A Crisis in
Paternity

I n the history of assisted reproduction, the Philadelphia physician William
Pancoast, who distinguished himself during the Civil War as surgeon in chief
to the Union army, stands as a key figure. Pancoast is notable not only for having performed one of the first artificial inseminations using donated sperm but
also for having done so without gaining the permission of either the woman being
inseminated or her husband. After a wealthy couple came to Pancoast in 1884
complaining of childlessness, he and a group of medical students at Jefferson
Medical College determined that the problem lay with the husband, who
apparently had been rendered sterile by gonorrhea. They attempted to reverse
the man's condition for some months, to no avail, then hit upon a novel idea:
They would inseminate the woman with the sperm of a substitute, of a "hired
man." Accordingly, one day they anesthetized the woman, collected semen from
"the best looking member of the class," and with a rubber syringe deposited his
untreated ejaculate into her vagina.

Pancoast did not tell the couple what he had done. Not until after the
woman gave birth did he confess his medical ruse to the husband, who purportedly took no offense but asked that his wife be kept in the dark about her
son's true origins. In 1909, several years after Pancoast's death, one of the students who had participated in the insemination, Addison Davis Hard, confessed all in the publication *Medical World.* The veracity of Hard's tale was
challenged, and a physician who had known Pancoast insisted that his colleague would never have stooped to such a "ridiculously criminal" act, which
essentially amounted to rape.[1] Outraged letters to the editor lambasted Hard
for his immorality. In return, other letter writers rose to the young man's

defense, proclaiming morality the business of theologians, not scientists. One correspondent in Oregon contended that "doctors have enough of the laws of God when they are young," and declared it entirely right and proper that physicians undertake to "modify creation and improve it with intelligence."[2]

Hard, who may in fact have been the provider of that essential ingredient for Pancoast's experiment, persisted in defending artificial insemination by donor (AID), in particular as well as in general. He suggested that AID represented a potential boon to women, since so many American men had contracted venereal diseases and would therefore spawn "defective" children. Women having access to supplies of sperm acquired from men who were free of infection and accomplished to boot would be spared a great deal of worry and could rest assured of producing wellborn offspring.

Here, at the inception of the age of assisted reproduction, the dangers of technological incursions into this intimate realm were already drawn large. Pancoast's hubris and dishonesty would seem to have amply justified fears that in addressing infertility, physicians might violate women of their bodily integrity and rob them of control over their own reproductive processes. Further, the news that a woman could conceive entirely without the assistance of her spouse aroused fears among men that medicine would somehow render them obsolete. Artificial insemination separated sexual congress from reproduction in a more profound way than either contraception or intentional abortion did. Indeed, it removed all need for bodily contact. How could one even begin to comprehend this cleavage; how assess its potential effects upon people's attitudes toward, and treatment of, their spouses and offspring? Without forethought or intention, Pancoast helped launch the age of assisted reproduction, ushering in a host of ethical, legal, and social concerns which even today remain largely unresolved.

A t its simplest, artificial insemination barely qualifies as a technology. No complicated instruments, drugs, or physiological calculations are involved, merely the collection of sperm and insertion of them into a female's vagina or uterus at a moment conducive to conception. There are indications that fifteenth-century Arabian horse breeders may have gathered the semen of prized stallions on wads of cotton which they then placed inside brood mares, performing a rudimentary but effective type of insemination.[3] Certainly, the early attempts by London physician John Hunter and others required little more than a syringe and a sense of timing.

This same, plain pipe-rack approach sufficed when Ilya Ivanovich Ivanov set out to improve domesticated stocks.[4] A microbiologist who had studied and worked at several institutions in Russia, as well as in Geneva and at the Pasteur Institute in Paris, Ivanov conducted a thorough survey of artificial impregnation in 1899, scouring the veterinary and medical literature and consulting the annals of fish, dog, and horse breeding for any mention of the subject. The evidence he amassed convinced him that artificial impregnation of mammals "is not only possible but also must become one of the powerful forces of progress in the practice of livestock breeding."[5] Just as social scientists worried about degeneration of the human line, Ivanov and others of his day were concerned that the overall quality of animal stocks was worsening over time. Artificial insemination seemed to offer a way of reversing the negative trend by allowing a single "superior" male to bestow his sperm within a very short time upon not dozens but hundreds, and potentially—if a method could be developed for preserving sperm for longer than a few days—thousands, of females.

Ivanov grappled with concerns over whether eliminating intercourse might somehow drain future offspring of vigor, and upon returning home from France enlisted the manpower of several Russian laboratories to delve into a host of issues involving reproduction. He soon concluded that the whole range of male physiological alterations that went along with arousal and copulation were just so much *Sturm und Drang*. Their main function was to ensure that egg and sperm collided, yet at the cellular level, they made no difference. Sperm, Ivanov believed, was perfectly well equipped to fertilize an egg prior to ejaculation. Accordingly, he launched a pilot program for artificial insemination of horses, setting up shop in a village about two hundred miles south of Moscow.

It is important to note here that breeders of Ivanov's day assumed that the qualities, whether physical or behavioral, of an adult male animal could be transmitted virtually in toto to his offspring. This suggests the Aristotelian belief that males contribute the motive force and are the dominant influence upon the quality of offspring. At the same time, it reduces the complexities of heredity to a single factor, the "superior" germ plasm, which supposedly is conveyed in unitary fashion down the generations, by strict Mendelian inheritance. Although entirely bankrupt from a strict genetical perspective, this assumption has proved remarkably resistant and surfaces repeatedly in the history of modern assisted reproduction, with regard not just to sperm but also eggs.

Ivanov scored such impressive successes with artificial insemination that soon no forward-thinking animal breeder in Russia would consider mating his animals conventionally. By 1932, veterinarians using the method at several experimental stations operating under Ivanov's direction had inseminated 650,000 mares, 2 million cows, 3 million ewes, and 200,000 sows.[6] Breeders were sold on the technique not only because it allowed them to "improve" the overall genetic stock within the span of a few generations, but also because they believed it eliminated the problem of telegony. According to this ancient supposition, since debunked, the first male to impregnate a female leaves an indelible mark upon her, and subsequent offspring borne by her, regardless of their sire, will resemble the original male. If a less than exalted first mate might ruin the entire reproductive life of a cow or mare, a first insemination with sperm from a high-caliber male would guarantee the quality of her future output.

During the same period, British animal scientists carried out independent investigations into artificial insemination. Most notable of these was F. H. A. (Francis Hugh Adam) Marshall, a lecturer on animal and human reproductive physiology at the Cambridge University agriculture school, who for a time resided in the very Christ College rooms once occupied by Charles Darwin. Marshall had, while a postgraduate student at the University of Edinburgh, carried out a frustratingly uninformative study of hair length in horses at the behest of a professor fixated on telegony. Casting aside that line of inquiry, he looked into a subject of far greater interest to him, the estrous cycle, which he studied in sheep, ferrets, and dogs. His careful tracking of estrus revealed that ovaries do not merely produce eggs; rather, they also secrete substances which alter the state of the uterine lining. Marshall soon gained a reputation as the era's preeminent expert on animal reproduction, and his groundbreaking 1910 overview of the field, *Physiology of Reproduction,* went through multiple editions and continued to be revised by others after his death in 1949.

Around the time of his textbook's publication, Marshall and a student named John Hammond performed a series of experiments with horses in hopes of boosting their fertility. Marshall looked for inspiration to the work of Walter Heape, not only a versatile biologist but also inventor of a high-speed camera known as the Heape and Gryalls Rapid Cinema Machine. At Cambridge, Heape had, in addition to exploring the mechanics of ovulation, fertilization, and gestation, carried out numerous artificial inseminations of rabbits. Marshall had duly noted this work, and it came to mind when, around 1912 or 1913, he overheard horse breeders complaining that many of the stallions that were being carted around the English countryside to serve mares were achieving less than stellar success rates. Marshall suggested to Hammond, "Now look here, these travelling stallions are very infertile. Why not let's try some artificial insemination, as Heape did?"[7]

Hammond eagerly undertook the project, although he was not quite sure where he would find the needed semen. He soon realized that he had a handy nearby source. Theretofore, when researchers at the agricultural school wanted to have mares impregnated, they had taken them to a farm not far from the university and had them serviced by studs. On occasion Hammond had lent a hand in this chore. Frequently, if the mare's cervical opening was small, a good deal of semen would run out after the male had finished mounting her, and so Hammond would gather the overflow in a pipette and place it back into the cervix. "Then I said to myself, 'Hello, why shouldn't we keep this stuff?'" Hammond recalled in a 1961 interview. He went on,

I can well remember that in the early days I thought it was necessary to keep the semen warm, so I fastened a tube of it under my armpit with elastic. And I used to cycle out to the University Farm to inseminate mares from this tube of semen under my armpit. It wasn't very successful![8]

World War I took Hammond away from his research, but in 1920, after serving as a captain with the Norfolk Regiment, he returned to Cambridge and, perceiving an acute need to rapidly rebuild livestock herds, set about expanding his investigations into artificial insemination. His goal was to augment the ranks of high-yield dairy cows, but because money for scientific undertakings was tight, he had to settle for experiments with rabbits, which were considerably cheaper to keep than cattle. Undaunted, he joined forces with Arthur Walton, who had been probing the biomechanics of sperm motility at Edinburgh University. Walton moved to Cambridge, and for a first project the pair endeavored to improve upon Hammond's armpit method of storage.

Rather than heat, cold turned out to be more conducive to sperm survival. "We found we could best keep semen cool, not ice-cold, but somewhere about 4°C. [39°F], and we sent a thermos of rabbit semen off to [Walton's mentor] Crew in Edinburgh. I think that was about 1922, and he inseminated some animals and got young born," Hammond later recalled. Spurred on by this achievement, Hammond and Walton expanded to other species. "By then we had a few sheep about the place and we sent some ram semen out to Poland, where I had a friend, and that was successful," Hammond said.[9] The Cambridge duo communicated their results through formal scientific channels as well as corresponded directly with researchers in Russia. They also demonstrated their techniques to cattle breeders who made pilgrimages from Denmark and the United States to the grassy banks of the River Cam and then returned home to proselytize for artificial insemination.

Meanwhile, the British Ministry of Agriculture could not be convinced of the technique's commercial potential and refused to fund large-scale tests. The outbreak of World War II, though, changed everything. Hammond recalled that in the late spring or summer of 1940, around the time that British troops were forced to decamp the Continent in a massive evacuation from the North Sea port of Dunkirk, Robert Spear Hudson, the minister of agriculture and fisheries, arrived at Cambridge. Heir to a soap fortune, Hudson was a dynamic man who had been charged with boosting Britain's sagging wartime food production. He was looking for additional ways to expand output. Hammond later told an interviewer that Hudson "was a very blunt man. He said, 'Now look here, what have you chaps got for today? No good thinking about tomorrow—may not be a tomorrow!'" Hammond immediately suggested a program of artificial insemination, which had been rejected several times by the ministry "on the grounds that though very spectacular, it had no commercial use." But Hudson thought otherwise, and allocated money for a pilot program.[10]

Soon, the ministry moved to establish the practice countrywide. Encountering opposition first from the Cattle Breeder's Association, whose members feared that they would be driven out of business by a severe drop in demand for bulls, those in favor of artificial insemination succeeded in quelling any worries on this score by pointing out that while the number of bulls required might be less, the price commanded by each bull would rise compensatorily. Rather than selling ten bulls at fifty pounds, breeders might sell one bull at five hundred pounds. Breeders bought the argument, and were gratified when prices did indeed spiral upward.

Other objections came from the Church of England, and under the minister of agriculture's urging, Hammond faced down a steering committee of the church, whose members firmly opposed the technology. "They attacked me first by saying that it was artificial, and nothing artificial could be of any use. It was going against nature," Hammond said of the meeting. "I had my answer ready for that one. I said, 'Do you drink milk?' They said they did and that it was very good for them. I said they were being artificial—milk was meant for the calf, not for them. That settled artificiality!"[11] This left the committee's moral reservations, which Hammond reported that he countered by arguing that normally, when a cow was mated in a village, "all the small boys come and watch." If artificial insemination were adopted, the cow's "owner would phone up, a car would appear, and a man with a little black bag would do an insemination before any boy in the village knew anything about it."[12] Hammond thus convinced the committee that the ancient customs of the land had perverted the morals of rural English lads all along. Technology, by removing the chance for prurient interests to be aroused, would serve the interests of Christian civilization. Hammond had won the battle and the war.

Over the next decade, scientists continued intensive investigations of animal reproduction, studying estrous cycles, discovering the role hormones play in sexual reproduction, determining the best methods for collecting semen and winnowing sperm from it, and learning to refine the timing of inseminations and thereby improve success rates.

The main impediment to more widespread adoption of artificial insemination was the short shelf life of fresh sperm, which degraded after a few days. Spallanzani had observed that exposure to cold caused sperm to turn sluggish, but that upon being warmed, they regained motion. In the 1860s, another Italian, the physician, anthropologist, and romantic novelist Paolo Mantegazza, had cooled frog sperm down to -15°C (5°F), with no ill effects. Mantegazza, a prolific popularizer of science, not only predicted that frozen sperm might one day be used in animal husbandry, but also spun a heroic fantasy about a soldier dying upon the battlefield and leaving a legacy of frozen sperm, which would enable him, through the agency of a physician, to engender children after his death.

Charles Davenport, in 1897, had observed that sperm could survive freezing to -17°C (-1.4°F), although few scientists took note of his finding. During the 1930s, several researchers, including Landrum Shettles of New York City's Columbia-Presbyterian Sloane Hospital for Women, tried different freezing regimes, sometimes plunging sperm into baths as cold as -269.5°C (-453°F), sometimes taking them down to that temperature step by step. They suspended sperm in various solutions, and thawed them both rapidly and gradually. But the die-off always hovered in the 60 to 90 percent range, and it was unclear whether those sperm that survived had been damaged somehow, in a manner that would only become obvious later, after they had been used to create an embryo.

In 1945, British scientist Alan Parkes, at the National Institute for Medical Research at Mill Hill, north of London, undertook a comprehensive review of the work done with sperm up until that point, and concluded that freezing should pose no genetic hazard. He also decided that the rate of freezing and thawing was far less crucial than the vessel in which sperm were stored. Earlier attempts at cryopreservation had failed because the sperm were frozen in thin films or in very fine capillary tubes. When Parkes tried larger capillary tubes and ampoules, an abundance of sperm revived after being rewarmed. Over the next several years, Parkes and his lab concentrated on concocting a solution that would convey additional insulation for sperm. After repeated frustration, they accidentally hit upon the right recipe in 1949 when someone in the lab unwittingly contaminated the solution in which sperm were being frozen with glycerol. This ingredient turned out to provide ideal protection. When Parkes suspended sperm in a solution containing about 5 percent glycerol, the proportion of sperm surviving storage on dry ice increased.

As of the early 1960s, researchers made liquid nitrogen, which kept sperm at -196°C (-321°F), their refrigerant of choice. Animal sperm banks proliferated, and around the world thousands of squat, hollow metal drums bristled with thin glass rods, called straws, filled with semen. From these icy arks would issue whole herds. In the late 1950s, the British Milk Marketing Board feted dairymen at a banquet in honor of the country's 10 millionth official insemination.[13] A 1964 survey found a one-year worldwide total of 59 million cows, 47 million ewes, 1 million sows, 125,000 mares, 56,000 goats, and 4 million turkey hens inseminated artificially.[14] By 1969, some 60 percent of all cattle matings in the United Kingdom were by artificial insemination. American use of the practice reached a commensurate level within the next decade, when about 6 million calves were being produced annually from thawed sperm, some of which had sat in the deep freeze for almost twenty years with no apparent ill effects.[15]

Although the rigors of freezing and thawing sperm robbed them of a degree of potency and motion, or motility, the benefits of cryopreservation were manifold. Without the difficulty and expense of trucking a stallion, bull, boar, or ram around, breeders could mate him with females in another part of the country or elsewhere in the world. Long-distance mating became the vogue, and was especially appreciated in less-developed countries like India, where breeders considered their native stocks of animals inferior and incapable of raising the overall quality of herds. Shipping proven adult males halfway around the world was not only expensive but a gamble, since the animals ran the chance of becoming sick or dying en route. Through shipments of frozen sperm, breeders in the Third World acquired the "superior" genetic material they desired at a far reduced cost, and without having to assume a major financial risk.

Artificial insemination spelled increased profits and productivity. It is generally agreed that the technology has during the past half century enabled the dairy industry to boost milk production per cow several times over. Furthermore, in conjunction with the rise of large-scale feedlots, it has facilitated the rise of the factory farm.

Using banked sperm, breeders can greatly increase a prize male's total lifetime progeny. For example, with each ejaculation, bulls issue billions of sperm, far more than necessary for the job in question; diluted, a single sample suffices for as many as six hundred inseminations. Among swine, a boar annually yields enough sperm to artificially inseminate some two thousand sows, about ten times as many as under an old-fashioned pen system in which a boar mates in turn with females in estrus. When a sire's sperm is collected and frozen over a period of years, his line can be perpetuated for several decades, long after his death.

Breeders like the consistency afforded by artificial insemination: It allows them to minimize the introduction of new genetic material which might render herds more susceptible to disease or to the vagaries of a given climate. Too, it provides cattlemen with a more marketable product, since most feedlots today will pay top price only for animals possessing a highly homogeneous genetic makeup. Feedlots, out to maximize profits, prefer to buy lots of genetically similar animals of roughly the same age for reasons of predictability. Such animals tend to enter the lot at a fairly uniform size and weight, and to fatten at the same rate.

However, while this approach has yielded short-term gains, some biologists suggest that decades of close inbreeding which have narrowed the genetic pool of cattle have produced a situation in which a single, virulent strain of disease might have the power to decimate the industry. Disaster of this sort has repeatedly struck modern agriculture, most recently in 1970 and 1971, when a blight that rotted corn ears in their husks destroyed 15 percent of America's national crop—up to half the harvest in some southern states. The enormity of the loss was owed entirely to genetic homogeneity. Hybrids deriving from a line known as Texas male-sterile cytoplasm, or T-cytoplasm, corn had been planted from the Gulf of Mexico up into Canada. The only problem was that T-cytoplasm was susceptible to southern corn-leaf blight, caused by the fungus *Helminthosporium maydis*. Proliferating during the hot, wet summer of 1970, the fungus had a fifteen-hundred-mile swath of corn to infect. It spread north at an incredible clip, felling whole fields as it went.

It is but a short leap from barnyard to babies. The innovations in animal husbandry led inexorably to new approaches to human reproductive problems. As artificial insemination was becoming the norm on farms around the world, a growing number of physicians were contemplating its possible role in circumventing male infertility.

Normally, with each ejaculation, a man releases huge numbers of spermatozoa, in the range of 20 to 100 million per milliliter of semen. These sperm, containing twenty-three chromosomes, or one-half the genetic component of other bodily cells, are produced in the testes through a series of hormonally triggered steps, and then undergo further modifications within a crimped duct called the epididymis, which empties into the vas deferens. The transit through the epididymis prepares the sperm's outer membrane and nucleus and gives its tail the biochemical wherewithal to beat vigorously over the course of the long swim up through the female reproductive tract to the fallopian tubes.

Glitches can occur at any point in the manufacture or maturation of sperm, leaving them chromosomally faulty or poky or slow to shed their protein cap. Furthermore, blockages within the epididymis may allow only a few sperm to trickle out, or entirely prevent their exit, making for near or total sterility. Scientists now know that when the number of sperm crowding a given volume falls below a certain level—generally pegged at about 20 million per milliliter of ejaculate—fertilization is less likely to take place after regular intercourse. A consistently low sperm count is technically referred to as oligospermia. In addition, men may suffer from asthenospermia (a word derived from the Greek for "weakness of the seed"); their sperm lack the motility to perform effectively. Some men also display a tendency to turn out malformed sperm, with one too many heads, or heads that are too small or overly inflated, or that are pinched, pointy, or amorphously shaped rather than oval. The manufacture of sperm with poor morphology, or outward appearance, is classed as teratospermia and often has genetic causes. Infertility may also result because a man's immune system runs amok and mounts an attack against his own sperm. Or, as in the case of the London draper who visited Hunter in 1793, congenital anatomical flaws may interrupt or deflect the flow of sperm from the testes.

Around the turn of the century, physicians in Europe and the Americas concluded that poor sperm quality or low counts were responsible for the problems of a sizable fraction of couples who came into their offices complaining of having trouble conceiving. Given that infertility proved a torment to so many couples, why not employ already ejaculated sperm to impregnate women, as was being done with brood animals? Although this made perfect sense from a medical standpoint and in terms of alleviating the psychological pain of infertility, specialists intuited that the societal response might be less than enthusiastic. Thus, gynecologists quietly administered donor sperm within their practices, preferring for the most part not to publicize their results.

Only a handful of case studies of artificial insemination surfaced in the medical literature from the turn of the century through 1940. Accordingly, it is almost impossible to document how many women underwent the procedure with their spouses' sperm, much less with that of strangers, during this period. In 1924, one researcher, having made a concerted effort to track down every citation in the literature, tallied 123 artificial inseminations, of which 47 led to pregnancy. A separate study carried out four years later located evidence of 185 attempts worldwide, with 65 pregnancies ensuing. These assessments apparently jibed with the medical community's anecdotal sense of the procedure's prevalence.

In 1934, a New York gynecologist, Hermann Rohleder, wrote what amounted to a sales pitch, albeit a dignified one, for artificial insemination. The book, *Test Tube Babies: A History of the Artificial Impregnation of Human Beings,* came out in a limited edition of two thousand copies. (The term "test-tube baby" referred originally to infants produced by artificial insemination; in the 1970s, it was adopted to describe children born as a result of in vitro fertilization.) The title page promised readers "a detailed account of [artificial insemination's] technique, together with personal experiences, clinical cases, a review of its literature, and the medical and legal aspects involved." Rohleder had tried artificial fecundation in nineteen cases, and had had five successes. He advised physicians to carry out detailed physical examinations of patients before undertaking the procedure, then to attempt the insemination during the final days of a woman's menstrual period. (Since this is a woman's *least* fertile time, it's astonishing Rohleder achieved any pregnancies.)

Rohleder believed that female orgasm was necessary if conception was to occur; thus, he performed the procedure directly after couples had copulated, or, in cases where the man was impotent, after the husband had performed "external irritation of the clitoris."[16] Realizing that even physicians might balk at this, Rohleder chided them:

> We should not be too squeamish about this. I can assure some of
> my colleagues that many a sterile woman will, in order to get a
> child, gladly excite herself sexually through masturbation before the

operation is carried out, or will suffer herself to be so stimulated by her husband.[17]

Rohleder admitted, though, that at least one couple he treated refused to undergo a second insemination because the woman had found the first experience too embarrassing.

The technique Rohleder advocated was straightforward. Shortly after the woman has been aroused, and while she remains in bed, the physician "appears" and with the husband's assistance—"indeed," Rohleder wrote, "I would not advise carrying out this operation save in his presence"—introduces a vaginal speculum.[18] Working quickly, he places a syringe filled with the husband's semen a few centimeters into the cervical canal and injects a few drops slowly. He lets the syringe sit for about fifteen seconds, then withdraws it. He ties the woman's knees together with a towel, and she remains in bed for several hours. For the next month, she must avoid riding, dancing, and other strenuous activities.

As for the morality of artificial insemination, Rohleder found it acceptable, despite the fact that an 1877 encyclical of Pius IX had condemned it as "abominable." He believed that the "morality of reason" argued that the physician had an obligation to provide therapeutic treatments for patients. Although artificial insemination was not a cure, it did answer the desires of infertile couples, and thus was an estimable act. Interestingly, in discussing the effects of infertility, Rohleder focused upon the "childless wife," who would supposedly go to unusual lengths to overcome her condition, including stealing other women's children.

Notwithstanding his general approval, Rohleder did feel chary about performing an insemination upon a woman using the semen of a "stranger." Apparently, a woman being treated by "Professor Semola of Rome" had requested, after several failures with her husband's sperm, that the doctor try another man's sperm. Semola reportedly "told his patient the evil contained in her suggestion and pointed out to her that the artificial introduction of the semen of a strange man would be just as much of a sin as if she had herself consorted with a strange man."[19] Rohleder felt that there might be occasions when using another man's sperm would be all right; however, he expected that he would agree to do so only if the couple signed a full release and allowed him to perform a physical exam on the donor.

Despite the enthusiasm of Rohleder and a few others, most gynecologists retained their reservations about artificial insemination throughout the depression and early phases of World War II, and seem to have adopted it only haphazardly.

In 1941, a report in the *Journal of the American Medical Association (JAMA)* asserting that "9,489 women had achieved at least one pregnancy by this method," drew considerable fire.[20] Several prominent researchers took issue

with the report's data as inflated. Their suspicions were raised by the disproportionately high number of males among the total births, and by the
extremely high rate of viable pregnancies—97 percent—for the given number
of conceptions. Customarily there is tremendous attrition among fetuses. Of
all conceptions, perhaps half or more end in the resorption, or dissolution, of
the embryo into the uterine lining at a very early stage, or in spontaneous
abortion at a later date. It seemed unlikely, then, that the use of a mechanical
means of fecundation would so substantially boost the fraction of fetuses carried to term. Dismissing the *JAMA* study a few years later in the *American
Journal of Obstetrics and Gynecology*, physician Clair Folsome commented with
some sarcasm that if one were to accept the figures it gave, one would con-
comitantly need to recognize "that the old-fashioned connubial techniques,
as practiced to effect conception, are distinctly deleterious as a process involving perpetuity to our race."[21] Folsome concluded that the study was unscien-
tific and misleading, and that artificial insemination should be considered
only as an option of last resort to overcome male sterility.

However, other physicians had come around to the view that the technique deserved broader application. For instance, Mary Barton, who worked
at the fertility clinic of the Royal Free Hospital in London, along with surgeon Kenneth Walker and biologist B. P. Wiesner, both at the Royal Northern Hospital, also in London, maintained that a range of sexual dysfunctions
warranted its use. In a pithy 1945 *British Medical Journal* report, the three
declared that impotence, intercourse which caused pain for either party, failure to ejaculate, and scarce or poor-quality sperm all called for artificial
insemination. They themselves had prescribed it in such instances, and had
employed donated sperm in cases of both temporary and permanent sterility,
as well as in two cases where heredity ailments ran in the husband's family.
Barton and her colleagues had also discovered that some men who were
unwilling to undergo treatment for sterility actually took the initiative and
requested AID, their wives readily assenting. Also, some women asked that
they be given donor sperm without their husbands' knowledge, "claiming that
paternity would save his self-esteem." The clinicians considered this a legitimate reason for AID, stating that "such women are usually good and devoted
wives who, having longed for children for many years of marriage, have
rejected other ways of becoming pregnant."[22] However,

> [other women] are impelled by different motives: e.g., they plan to
> force marriage by what they consider to be a guiltless conception.
> Neither these nor other more dubious reasons have been accepted
> by us; for while we do not feel in a position to suggest the genetic
> principles which should govern the application of AID, we recog-
> nize the need for the greatest caution. In our work we have taken
> into account not only biological and medical factors, but also the

suitability of the couple for parenthood, so far as this can be reasonably estimated.[23]

In other words, while Barton et al. shied from outright eugenic judgments, they felt more than justified in withholding their services from women whose motives they considered "dubious," or from couples who did not appear "suitable" candidates for the all-important chore of raising children.

That said, the specialists at the Royal Free Hospital fertility clinic appear to have been fairly liberal in disbursing their expertise. Except where donated semen was involved, the clinic relied on wives to carry out the insemination procedure at home. Women were taught how to monitor ovulation according to fluctuations in their morning temperature and other signs. Then they were instructed in the art of impregnation. Barton detailed the procedure that a woman would be taught to follow:

> She first douches with warm water (1 pint); half an hour later the husband passes semen into a cold dry glass container and allows it to liquefy at room temperature (about 10 minutes). The wife then draws up the semen into a clean dry urogenital glass syringe, and, lying on her back with knees drawn up, she passes the syringe into the vagina and very slowly expels the semen. The prone position should be retained for about half an hour.[24]

As opposed to their British counterparts, physicians in the United States insisted that artificial insemination be kept within a strictly medical purview. So strongly did the profession wish to limit access to the technology that the American Fertility Society, the largest organization of infertility specialists in the world, would eventually recommend legislation making artificial insemination by nonmedical personnel a criminal offense punishable by fines and imprisonment.[25] The society's ostensible reason was the higher success rates achieved with cervicovaginal inseminations, whereby semen was instilled not only into the vagina but also into the layer of mucus surrounding the cervix; as well as with intracervical and intrauterine inseminations, in which a cannula, or small tube, was threaded through the cervical opening so that semen could be deposited either halfway through the passage to the uterus or into the uterus itself. These approaches, which require a fair degree of skill and care, obviously could be performed with greater safety in a clinical setting. However, one must also suspect a pecuniary motive behind the society's recommendation.

Immediately following World War II, as birthrates rose steeply worldwide, AID underwent a boom of its own, especially in the United States. For example, it is thought that only 10 conceptions were owed to AID in Holland between 1948 and 1960, and only about 1,150 in Britain from 1940 to 1960.

However, one expert claimed that in the United States by 1957, a total of 100,000 babies had been born as a result of artificial insemination, the bulk of them from donor sperm. A second researcher claimed that by 1960, some 5,000 to 7,000 babies delivered annually were being conceived with fresh sperm, which was provided anonymously, usually by medical students.[26]

By then, a handful of babies born each year also derived from sperm that had weathered a period of suspended animation at subarctic temperatures. Although animals created from frozen sperm seemed to show no ill effects attributable to the process, many physicians had continued to worry throughout the 1940s that freezing human sperm might somehow damage the genetic material contained within its nucleus, and that fetuses would as a result be deformed or physiologically incompetent. Pushing ahead with the conviction that the animal results would hold for humans, Jerome Sherman, while a doctoral candidate at the State University of Iowa, compared a range of freezing methods and devised a simple approach which he believed was both safe and efficient. When a sample was cooled slowly and stored on dry ice, about two-thirds of the sperm survived. Moreover, they retained the capacity to fertilize eggs and spark normal embryonic growth. In 1954, Sherman and two physician colleagues announced that a woman in Iowa City had given birth to the world's first baby from frozen sperm, and fifteen other such births took place over the next five years. Japanese physicians boldly adopted the use of frozen sperm, and reported two dozen pregnancies by 1965. However, not all of these went to term, and at least one baby was stillborn. Just eight years later, five hundred babies had been delivered worldwide thanks to frozen sperm, and medical investigators had satisfied themselves that the technique was safe and produced rates of anomaly no worse than natural conception.

In the 1930s, when American experiments with artificial insemination were taking place with regularity, Walter Gross, head of the German National Socialist Office of Racial Policy, opined that the technique was not only "tasteless" and "unnatural" but also fundamentally misanthropic and sure to be employed by women "who have an ineradicable hostility toward men."[27] As head of a state-directed euthanasia program, Gross was hardly one whose squeamishness should be taken seriously, yet his comments fell squarely in the mainstream of criticism of artificial insemination in the first half of this century.

Many commentators realized that, unlike most other medical procedures, this one challenged a whole raft of personal and societal assumptions. It not only removed physical contact from the act of conception but also replaced the intimate emotions of intercourse with the dispassionate ministrations of the technician. Caught up in the excitement of research, and rationalizing artificial insemination as a godsend for infertile couples who could not previously be helped, physicians by and large ignored the effects it might have upon men's and women's psyches—not to mention those of the children born as a result of the procedure. But public commentators were quick to seize upon such issues. Manhood had long been equated with the ability to father children; now that the male sexual function could be replaced by a syringe, what did that imply about men's virility? As for women, shouldn't they feel ashamed to undergo such treatment?

Far more worrisome to some critics were the larger social implications. Artificial insemination in a sense represented a violation of strictures concerning who might acceptably impregnate a woman. If a woman was inseminated heterologously, or with another man's sperm, as opposed to homologously, with sperm from her husband, should she be considered to have committed adultery? Morally offended critics proclaimed that indeed she should. And even if AID in itself did not constitute adultery, some said it was pernicious insofar as it served to cover up infidelity: Abetted by physicians, wives of sterile husbands could mask affairs in which they had carelessly become impregnated.[28] This farfetched scenario did not seem to excite widespread alarm; however, many commentators agreed that the status of a child produced through AID was equivocal. Who, in fact, should be considered the "real" father of a child who had been engendered with donated sperm: the donor or the man who helped feed, clothe, shelter, and educate the child?

The issue of legitimacy disturbed medical practitioners to the extent that at least one went to some lengths to develop a method of disguising the use of nonspousal sperm and giving artificial insemination a component of "naturalness." The surgeon C. T. Stepita in 1933 recounted his strategy: He had first operatively inserted catheters into a sterile man's seminal vesicles, tiny glands then thought to store sperm. Through these implanted conduits Stepita then periodically injected sperm collected from a blood relative of the man. The idea was that if the man hurried home after receiving a consignment of freshly donated sperm and engaged in coitus with his wife, he might succeed in impregnating her. The medical justification for such a procedure was nil; it was motivated entirely by the desire to circumvent unsettling psychological (and possibly legal) issues.

Later, in the 1950s and 1960s, driven by the same impulse to camouflage the procedure, physicians habitually mixed a husband's sperm, however incompetent, with the donated sample. They also advised couples to continue having sexual relations before and after inseminations. That way—one never could tell—any baby who came along might actually be the husband's. This incitement to self-deception cannot be considered terribly constructive or healthy, but as late as 1979 Britain's Royal College of Obstetricians and Gynaecologists officially endorsed the tactic as a way of fudging the legal status of AID children, who technically should have been registered as "father unknown," and therefore illegitimate. (In Europe during the Middle Ages, to

be of "unknown" birth was to be either a bastard or a foundling, and the epi-
thet was strongly associated with a conviction that such a person was ipso
facto "low-born." The notion that a label of illegitimacy revealed a person's
legal, social, and moral character persisted into this century, a singular case of
visiting the sins of fathers and mothers upon the children.)

Should it be kept secret from everyone that a woman had undergone artificial insemination? Folsome in 1943 anticipated that opprobrium would
accompany any revelation that a woman had borne a child not biologically
her husband's:

> The measuring stick of the family continues to be social propriety.
> The happy wife, contented through attaining a baby by means of
> homologous artificial insemination, may give voice to her joy and
> win approbation. But the woman, made pregnant by use of donor
> semen, who even whispers out of turn, on a single occasion,
> becomes a medical curiosity. She is envied by the primitive and
> wanton-minded, pitied by those gifted with easy fertility, shunned
> by her relatives and perhaps unfortunately by her own child. The
> so-called veneer of civilized culture is thin, none the less it is of an
> oppressive nature to the woman willing to overstep the bounds of
> her environmental social mores and wedlock in her real desire to
> have children of her own.[29]

Folsome's analysis appears to be based on nothing more than his own perception of British mores. Nevertheless, one can infer from first-person accounts
later provided by couples who opted for AID during the decades after World
War II that Folsome fairly accurately took the pulse of his countrymen. In
Britain, as elsewhere, the shame attached to infertility was such that couples
often displayed reluctance to discuss their inability to conceive with trusted
family doctors, so it is not surprising that they would feel uncomfortable disclosing that they had substituted an emotionless mechanical technique for
the customary intimacies. In many cases, people opted to lie to their parents,
siblings, and other relatives, and to their friends and neighbors, and to hide
from their children the truth of their origins.

This conspiracy of secrecy struck many observers as problematic. How
would a child be affected to learn that he or she was a test-tube baby, conceived with sperm belonging to a stranger? Wouldn't a person who found out
that he or she had been lied to for years and that his or her father was a
"sham" feel deeply betrayed? Might not the shame eat away at marriages from
the inside, leading to estrangement or divorce?

Jealousies or resentments might build up in men who were raising children
not biologically their own. Women might fantasize about the sperm donors,
thereby creating tensions in the household. Worse, from the point of view of

most early critics, if the government did not control artificial insemination, women might decide to have children on their own, without marrying. Didn't society have the right to intervene in the reproductive decisions of its members if it meant ensuring the viability of the family as an institution?

From the 1920s on, courts in many countries were forced to deal with the consequences of artificial insemination. For the most part, the legal debate centered on the status of the children conceived through AID: Were they or were they not legitimate? Secondarily, there was the matter of consent: Was AID performed without a husband's specific go-ahead adultery? In one of the earliest rulings, in 1921, a judge for the Supreme Court of Ontario, Canada, asserted in *Oxford* v. *Oxford* that it was, and rendered an opinion that AID might even be considered a form of sexual intercourse. However, few courts in the United States bought this argument, with the notable exception of an Illinois judge who in a 1956 case, *Doornbos* v. *Doornbos,* declared that AID was adultery even in the presence of spousal approval. Generally, courts, as in a 1943 Pennsylvania case and an unreported 1945 Illinois case (*Hoch* v. *Hoch*), recognized that AID, involving no bodily union between the woman and sperm donor, fell outside the customary definition of adultery. Furthermore, it could not be held as grounds for divorce.

As late as the 1970s, no laws applying to artificial insemination existed in France, but legal experts looked to the criminal and civil codes governing medical contracts for general guidance. Artificial insemination by donor could not be considered adultery, although a woman who underwent the procedure without her husband's knowledge could be held to have wronged him sufficiently that he might successfully petition for divorce. Similarly, a man who conspired with a doctor to inseminate his wife with donor sperm without her knowledge could be criminally liable. As for AID children, those born to unmarried woman would technically be considered illegitimate; however, few French physicians would deign to inseminate unmarried women. As long as a woman's husband accepted paternity, an AID child would be deemed legitimate. Hypothetically, a man might sue within six months of such a child's birth to deny paternity; however, he would have to prove that another man had fathered the child, which was at the time (before the development of DNA testing) extremely difficult. Elsewhere in Europe, courts gave men enormous leeway to change their minds. A German father might petition to deny paternity even after having approved AID.

In Britain, the archbishop of Canterbury, heading an inquiry on artificial insemination in 1948, recommended that Parliament make AID a criminal offense. The following year, speaking in the House of Lords during a debate over the legitimacy of AID babies—a touchy issue in a nation where inherited titles mean so much—the archbishop reframed his reservations, arguing that caution was justified because "there are too few observed cases, and too few cases observed for a sufficient length of time" to assess either the psychological impacts which AID might have upon children or its broader social ramifications.[30] A decade later, in the wake of *Maclennan* v. *Maclennan,* a Scottish case involving AID, the British home secretary and Scottish secretary of state jointly drafted a committee to examine whether changes in the laws regarding marriage or inheritance were warranted.

Chaired by the earl of Feversham and composed of nine physicians, legal experts, and citizens, the committee released its recommendations in the summer of 1960. To begin with, its members held that a successful artificial insemination, either with homologous or heterologous sperm, eliminated the right of a couple to be granted a nullification of marriage on the grounds of sterility. In effect, AID operated as a surrogate for the husband's missing sexual potency. (However, a man's donation of sperm could not be construed as the equivalent of a sexual act, and wives should not be allowed to file for divorce if fertile husbands donated sperm without their knowledge or consent.) On the financial obligation of men to children born as a result of consensual AID, the committee took a firm stand. It advised that in Scotland, AID offspring should be treated by the same lights as adopted children, while in England, they should be entitled to receive support from the estate of their social father in the event of his death. Notwithstanding, the committee held that AID babies should not be granted outright legitimacy at birth but should be registered as "father unknown," and therefore illegitimate. Physicians who falsely attested otherwise perjured themselves. The committee hesitated to press the issue further, however. Although "[i]t had been suggested that persons assisting in AID are guilty of a criminal conspiracy to produce an illegitimate child," the members were reluctant to accept such a construction except in the remarkable eventuality that "the intention of the parties was to defraud."[31]

The possible detrimental effects of AID upon the family concerned the committee greatly, and its members expressed "unqualified disapproval" of granting single women access to the technology. "[Artificial insemination by donor] is undesirable because it is a danger to the institution of marriage and to the resultant children," the committee decided, because it dissociates "responsibility for rearing children from responsibility for their procreation."[32] So little solid research had been done into the psychological riptides which AID might set in motion, that the committee urged physicians to exercise extreme caution in picking candidates for the procedure. It also saw a "pressing need" for comprehensive follow-up studies of families and AID children in order to better assess the detriments and benefits of the practice.

One possibility—that couples would employ artificial insemination for the purpose of selecting the sex of offspring—particularly bothered the Feversham Committee. Although researchers had not yet succeeded in separating human sperm according to whether they carried a male-determining Y chromosome or female-determining X chromosome, laboratories around the world were busily seeking efficient methods of doing so. Should such techniques become available, the committee feared that people's preference for males might "upset the balance of the sexes" within the population. For this reason, the government should keep a keen eye on developments in this area, the committee advised.

Notwithstanding its array of reservations, the committee preferred to leave the ultimate choice on whether to use AID to individuals and their doctors, and did not recommend "its regulation, either by statute or by the profession." By taking this laissez-faire stance, the committee explained, it actually hoped to send a signal of disapproval not encouragement, reasoning that "regulation might be understood to imply a degree of official recognition which would be undesirable."[33]

The Feversham report heavily influenced British policy on artificial insemination, and its opinions were essentially echoed thirteen years later by a British Medical Association committee engaged to revisit the issues. Once again, a group of panelists undertook to examine the extent and value of the practice, and to weigh its legal implications, this time led by Sir John Peel, an eminent obstetrician at King's College Hospital in London, former president of the Royal College of Obstetricians and Gynaecologists, and surgeon gynecologist to Queen Elizabeth II. Peel, long a staunch supporter of maternal and child welfare, held that a primary charge of governments must be to increase expenditures on prenatal care and childbirth, and to ensure that every child received adequate health services into adolescence and beyond.

Peel's panel, reflecting upon the "considerable changes" in social attitudes ushered in by the 1960s, came to a more sanguine view of AID than had the Feversham Committee. Although sketchy, the statistics pointed to a rise in the prevalence of AID births, and a corresponding diminution in public disapproval for the procedure. With fewer children available for adoption due to the Pill and other contraceptive measures, AID seemed a reasonable option for childless couples. In fact, providing that careful screening programs were implemented, the panel advised that the National Health Service should offer AID to suitable candidates, that is, couples who would be unable to have a family in any other way. The panel suggested that five centers, two in England, one each in Wales, Scotland, and Northern Ireland, could adequately meet current demand for the service. Directors of these centers would recruit donors or rely on sperm banks at their discretion, and might employ frozen sperm (then generally eschewed in Britain) "for reasons of accessibility and the maintenance of confidentiality."[34] Notwithstanding the need to protect anonymity, the panel did feel strongly that centers should maintain extensive records on donors for the sake of genetic research. In addition, it countered prevailing opinion by insisting that long-term follow-ups on AID children were essential: "Information must be obtained on the genetic effects, especially where frozen semen has been used, and it is important to learn the effects, in human terms, on the development of personal relationships in families resulting from the use of AID."[35] To ease this task, the law should extend legitimacy to AID children, thereby removing one major incentive couples had for hiding the truth about a child's origins.

Meanwhile, in the United States, a nationwide consensus regarding legitimacy failed to emerge. In the mid-1960s, Georgia and then Oklahoma passed laws declaring any offspring of AID to be legitimate if the procedure was undertaken with the husband's agreement; however, state legislatures in several other states, including Indiana, Virginia, and Wisconsin, voted down similar provisions. Many states, like New York, consigned AID babies to the same marginal legal status as those born out of wedlock or put up for adoption, and would assign parental rights to the social father only if he went through formal adoption proceedings. Accordingly, doctors and patients continued to collude, covering up the truth of AID babies' paternity in order to avoid legal hassles and expense.

On the other hand, court rulings in some places compelled men to take responsibility for AID children conceived with their cooperation during the course of a marriage. In California in 1968, a case was successfully brought against a man who failed to support a child borne by his wife after AID. The decision required the man to provide for the child.[36] That such financial obligations might be counterbalanced by paternal rights was determined in the 1977 case *CM* v. *CC*, which was heard in Cumberland County, New Jersey. Here, the judge ruled that CM, a donor not married to CC, would have visitation rights to the child who had been conceived with his sperm.

Lawyers, social scientists, physicians, reproductive physiologists, clerics, and ethicists from around the world continued to discuss the legal and ethical aspects of AID well into the 1980s, attending symposia and giving lectures, sitting on government panels, testifying before politicians and bureaucrats, penning overviews for the medical and popular press. On the whole, those within the medical and scientific communities downplayed the possible ill effects of the practice, while those within the philosophical, religious, and psychological communities continued to voice misgivings about it.

10

Shopping for the Future

Pancoast's student Hard had, in revealing his professor's malfeasance, defended the eugenic benefits of artificial insemination, and so, too, did certain leading scientists from the 1930s on. While a visiting researcher at Moscow's Institute of Genetics in 1935, Hermann Muller, geneticist, socialist, and sometime board member of the American Eugenics Society, had realized the benefits that a program of artificial insemination with sperm from "superior" men might yield. The concept of "germinal choice" became for Muller a thirty-year passion, upon which he issued occasional broadsides. In September 1961 he once again took his case to the public in a *Science* magazine article titled "Human Evolution by Voluntary Choice of Germ Plasm."

Muller proposed that special sperm banks be set up and stocked with the frozen sperm of "superior" men. He suggested that each couple in the United States, after having a child of their own, should proceed to have at least one additional child, this one to be conceived with the banked sperm. If followed religiously, this program for voluntary choice of germ plasm—VCOG, for short—would, he purported, move humans out of the unfortunate "genetic cul-de-sac" they currently found themselves in and fend off an "ultimate disaster" which might lead to the extinction of the species. Muller was savvy enough to suspect that not everyone would appreciate his plan; however, he believed it would be eagerly adopted by "tiny groups of the most idealistic, humanistic, and at the same time realistic persons," who shared in "our great common endeavor: that of consciously controlling human evolution in the deeper interests of man himself."[1]

The well-known British biologist Julian Huxley seconded Muller's one-

child-of-your-own/one-with-donor-sperm idea in a speech given in 1962.
(That same year, in a utopian novel titled *Island,* Julian's brother Aldous
Huxley imagined a society in which people opt for AID, at least when having
their third child, in order to improve the overall intelligence of the population.) Muller's idea was also touted by Jerome Sherman, who contended that
a true program of population control called for "not only the reduction of the
number of births in our world's population explosion, but also the genetic
improvement of the population."[2] This, of course, was the familiar logic of
mainline eugenics.

Standing before his peers at the eleventh annual International Congress of
Genetics at The Hague in 1963, Sherman outlined the future of sperm banking.
He foresaw the emergence of national or international networks of infertility
specialists drawing upon centralized banks that would sort sperm according to
blood type and other characteristics. Healthy men might store their own
sperm in the event that they one day lost their potency or wished to father children in old age or posthumously. "Thus, the reproductive effectiveness of
husband or desirable donor can be extended indefinitely. Children in one
family can be fathered from the stored semen of the same desirable donor independent of geographic location or his vitality," Sherman said.[3] Sperm banking
was a hedge against the threat of genetically harmful radiation, from nuclear
fallout, say, or space travel. It would also, counterintuitively, contribute to
efforts to control population, Sherman claimed, because men would more
readily submit to vasectomization, since they could always reverse their decision
at a later date.

Moreover, sperm banks offered an unparalleled opportunity for eugenic
endeavors:

> Thousands of donor inseminations are requested by couples and
> performed each year in the United States as an endorsed practice of
> the American Medical Association. The question may be raised,
> "Why not make available for the recipients' choice, not the physi-
> cian's choice as it is now, a selection of donor semen bearing the
> characteristics of mind and body which are deemed ideal by the
> recipients?" Professor H.J. Muller believes that this concept of ger-
> minal choice or parental selection may extend beyond that couple
> with its infertile male member to fertile couples wishing to improve
> upon the genetic constitution of their offspring. Frozen human
> semen banks can offer a wide range of genetic material for selec-
> tion. . . . Also, a program of progeny testing with donor semen will
> become feasible to evaluate donors. We have only to look at
> improvements in our dairy cattle and their products to appreciate
> the merits of the utilization of select donors. Frozen semen banks
> perhaps, for the first time, will permit scientists to evaluate genetics

in man on an experimentally controlled basis from generation to generation.[4]

In other words, through the fruits of the donor, the donor's fitness would be judged—although Sherman failed to specify how, precisely, progeny would be "tested"—and then his sperm either retained or, presumably, discarded. Once again, the misguided notion that "characteristics of mind and body" reside in sperm (like little homunculi?) had surfaced.

By alluding to the prospect of human experimental genetics, Sherman unabashedly upped the scientific ante. The overreaching ambition of legions of researchers for generations has been to resolve, once and for all, the debate over the relative contributions of nature and nurture to human character, behavior, and achievement. Here Sherman was claiming that artificial insemination offered investigators a chance to run a massive experiment in heredity which—best of all—was being willingly entered into by individuals. On their own, people who desired children were employing donor sperm; if the "qualities" of the sperm were known, and the outcomes—that is, the children engendered by that sperm—could be monitored, something of moment might be learned about the relative contributions of inheritance and environment.

To this end, Sherman encouraged existing sperm banks to expand their screening of sperm samples, carefully documenting "volume, liquefaction, density, motility, abnormalities, and the like," as well as taking detailed information on the physical and psychological traits of donors, and their genetic backgrounds. Sherman envisioned a time when banks would band together in a self-monitoring organization with a single coordinating physician who would undertake systematic reviews of freezing and thawing procedures, in an attempt to appreciably improve sperm survival rates and clinical outcomes for insemination. In addition, "A 'clearing house', with all pertinent information on stored semen should be established. It should function in close association with consultants in medicine, genetics, psychology, and social work for the necessary advice and guidance relative to such programs as germinal choice."[5] In the best of all possible worlds, sperm banks would merely be an "interim vehicle" for genetic improvement of the species, for Sherman's dream was that one day scientists would usher in "man's biological destiny," and through "genetic surgery" succeed in manipulating "the coding elements [of DNA] to achieve human betterment of mind and body."[6]

Notwithstanding their elevated standing within the scientific community, Muller and Sherman advanced schemes that had only a tenuous theoretical footing. In fact, as population geneticists knew by then, even minor "improvements" across the entire gene pool would either require thousands of years or prove altogether impossible. On this basis, John Maynard Smith, writing in the quarterly *Daedalus* in 1965, took issue with both Muller and Huxley. Smith assessed what might happen were Muller's VCOG program implemented using sperm from donors possessing a mean IQ one standard deviation—or about fifteen points—above average. Supposing that intelligence might be about 50 percent heritable, Smith estimated that after a century, the mean IQ of the population would have edged up to match that of the donors. Such a shift, Smith wrote, "seems hardly sufficient to justify the establishment of a new religion," especially since a concerted program of breeding with donor sperm might yield a society in which inordinate burdens were placed upon children. Probably to a greater degree than with normal births, people would expect offspring who had been conceived with sperm from intelligent, musically gifted, or highly athletic men to excel in the same ways that their proxy fathers had.

Smith concluded that the main error made by those who wished to exploit the eugenic side of AID lay in their presumption that an increase in the proportion of high-IQ people in a given population "would necessarily, or even probably, be associated with an increase in the number of people of outstanding ability as judged by their achievements."[7] However, in the end, he supported what he deemed "transformationist eugenics" and "biological engineering," that is, the development of techniques for correcting faulty genes in a person's hereditary makeup and for surgically or chemically enhancing embryonic development so as to eliminate disease and abet the evolutionary process. (Clearly, he here makes the faulty assumption that the "proper" direction of evolution is "upward" toward some state of "perfection.") These approaches, Smith believed, were far more worthy of scientists' time than clumsy attempts at selective breeding through artificial insemination.

Critiques like those made by Smith barely fazed AID's true believers: Many persisted in touting the technique's power to deliver qualitatively better babies. Japanese physicians Rihachi Iizuka and Yoshiaki Sawada, among other Tokyo trailblazers, went so far as to contend in the *International Journal of Fertility* in 1968 that babies produced from frozen sperm outstripped normally conceived babies in intellectual development, routinely exhibiting IQs above the mean. Although admitting that this might be owed to environmental factors, since couples requesting AID in Japan tended to hail from higher socioeconomic groups and to be employed in professional fields, Iizuka and Sawada nonetheless considered AID a telling factor in the development of the fifty-four infants they examined.

As for Muller's VCOG scheme, it inspired a scientist at the California Institute of Technology, Robert Klark Graham, to found a special sperm bank in 1976. Graham was a devotee of Muller, of the discredited British psychologist Sir Cyril Burt (another believer in the "great common endeavor" of eugenics), and of Nobel laureate William Shockley, who contended that blacks were genetically inferior to whites in terms of intellect and insisted that anyone with an IQ under 100 should be paid to undergo sterilization. Graham's seminal idea, as it were, was to collect sperm from Nobel laureates and bank it so that a wide array of women might have the opportunity to bear the sons and daughters of "geniuses." Graham initially named his Pasadena-based enterprise the Hermann J. Muller Repository for Germinal Choice, but in the face of objections from Muller's widow changed the name. Today, Graham's Repository for Germinal Choice, based near San Diego, is often referred to in the press as the "Nobel sperm bank," but in fact policy has changed, and sperm is now accepted from "outstanding" men, whether or not they have mounted the podium in Stockholm.

For many decades in the United States, the legal, medical, and ethical debates about artificial insemination were conducted virtually in private, within the confines of courts, clinics, and universities. A Harris poll conducted in 1969 revealed that just 3 percent of Americans had heard of artificial insemination.[8] Around the time of the poll, the British researcher Alan Parkes—an active member of the Galton Society who also served as honorary secretary of the Eugenics Society and had organized several conferences on the genetic aspects of social problems, human ability, and ethnicity—could be heard bemoaning the fact that the general public had not more readily accepted the new technology, a reticence he attributed to their "lack of enthusiasm for the idea of abolishing in human reproduction the need for contemporaneous and contiguous action on the part of the two sexes."[9]

Parkes acknowledged that "fearful crimes [had been] committed allegedly for eugenic purposes" but also felt strongly that the genetic aspects of modern social problems must be investigated. Also disturbed by the burgeoning global population, he proclaimed the urgent need for family planning, especially in the developing world, and for eugenics. "Certainly," he wrote, "by some means, the parental urge must be further diverted from quantity to quality, with all that quality implies."[10]

Public awareness of artificial insemination would soon increase, as Jerome Sherman's hoped-for sperm-banking network gradually emerged. In the United States, as more and more couples who experienced difficulty in having children eschewed "contemporaneous and contiguous action" in favor of repeated inseminations at a fertility specialist's office, banks run frankly for profit arose, catering to the expanding demand for sperm. Since roughly one-third of infertility among couples can be traced to the male partner, sperm banks were looking at a potential market of some 3 million couples annually in the United States alone.[11] In addition, a growing number of men worldwide were opting to undergo vasectomies as a birth control measure. Enterprising physicians saw that sperm banking might satisfy a need that these men might not even realize they had: It could provide "fertility insurance." If a vasectomized man divorced and remarried and wanted to father children with his new wife, he could do so if he had taken the precaution of freezing sperm before his operation. If, having reached a certain position in life, he decided that he regretted his earlier surgery and wanted to have a "second" family with his original wife, he could do so. Furthermore, men undergoing treatment for prostate or testicular cancer could stockpile healthy sperm in case the doses of radiation or chemotherapy rendered them sterile or incapable of manufacturing competent sperm.

Using selling points such as these, several commercial banks opened during the 1970s. The first, Genetics Laboratories in Minneapolis, was founded by Arthur Beisang, who had gained his expertise in cryopreservation working with bull semen. A second privately held company, IDANT Corporation, began taking deposits in New York City. IDANT, one of whose scientific consultants was Sherman, would survive a poor opening decade, which brought down all the other banks, and grow into the largest sperm distributor in the world. By the end of the decade, the pecuniary nature of the business would lead Boston University professor of law and medicine George Annas to point out that the term "donor" wasn't exactly applicable any longer: "Sperm vendor" was more like it, since at the larger banks men earned anywhere from $20 to $35 for each ejaculation.

Americans were far and away the pacesetters in commercial sperm banking. Because in Britain physicians preferred to use fresh sperm, banking did not immediately catch on there. In France, banking was done under the aegis of a group of clinics scattered across the nation and federated under the name Centres d'Etude et de Conservation du Sperme Humain (Centers for the Study and Preservation of Human Sperm), or CECOS. Partly funded by the central government and operating under consistent protocols, these centers both stored sperm and inseminated women. Between 1973 and 1978, CECOS performed AID on 4,253 women. Its physicians were thus uniquely positioned to carry out long-term studies of the efficacy of the procedure, which revealed that the older women got, the longer it took them to conceive and the more likely they were to fail altogether to become pregnant.

New York City, through its Sanitary Code, had regulated the collection of sperm as early as 1947, requiring donors to undergo a complete physical examination, with tests for syphilis, gonorrhea, tuberculosis, and Rh factor, and to possess family histories free of any known hereditarily transmissible diseases or defects. But standards at many smaller clinics and doctors' offices elsewhere around the country, subject to no such regulations, were often less rigorous. Here and there in the medical journals reports surfaced of women contracting gonorrhea, trichomoniasis, or chlamydia after AID with fresh sperm. Physicians frequently maintained stables of just a handful of donors who appeared healthy and intelligent, and might keep dilatory records on them or none at all. Often, if a donor wasn't handy, physicians themselves provided the needed sample, without informing anyone. (Cecil Jacobson, the

Virginia physician who clandestinely inseminated seventy-five women with his own sperm from 1976 to 1988, differed from his colleagues only insofar as he misled women into thinking that they were receiving sperm selected from a commercial source [see page 258].)

Although unregulated, most commercial banks instituted rigorous screening of sperm. An influential 1979 study on the practice of AID in the United States by Martin Curie-Cohen, Lesleigh Luttrell, and Sander Shapiro of the University of Wisconsin in Madison had taken physicians to task for their lax approach to this matter. The researchers had mailed out 711 questionnaires to members of the American Fertility Society who were known to perform AID, to medical school obstetrics and gynecology departments, and to medical researchers. Of these 471 were returned. Answers revealed that few practitioners took more than cursory family histories from donors, and almost none displayed a real comprehension of genetics. For instance, "71.4 per cent [of respondents said they] would reject a donor who had hemophilia in his family, even though it would be impossible to transmit this X-linked gene unless the donor were affected."[12] Even when they rejected donors for sound reasons, for example, the suspicion that they were carriers of a defective recessive gene known to produce disease, physicians rarely bothered to confirm this through biochemical tests.

Curie-Cohen et al. advocated that well-trained genetic counselors carry out screening of sperm donors "to ensure that children born through artificial insemination will have a minimum number of defects." The risk had already been diminished somewhat, these researchers asserted:

> Donor selection tends to promote positive eugenics (genetic improvement) as well as negative eugenics (prevention of genetic diseases), since donors are usually healthy university or medical students. This restricted donor pool may be responsible for the low frequency of congenital abnormalities and spontaneous abortions reported among pregnancies resulting from artificial insemination.[13]

The notion that medical students are somehow healthier and more exalted than the man in the street is gospel in the literature of artificial insemination.

Lest anyone worry that physicians might perpetrate the sort of eugenics which had so marred the past, Curie-Cohen and his colleagues quickly added that "the practice of matching the donor to the husband's phenotype reduces the dangers inherent in positive eugenics by limiting the number of characteristics that physicians are free to select." In other words, eugenics imposed from above would be unacceptable, but eugenics enacted by couples of their own free will was to be applauded.[14]

Many of the four hundred or so banks in the United States now follow screening guidelines set by the American Association of Tissue Banks, the American Society for Reproductive Medicine (formerly the American Fertility Society), and the Centers for Disease Control and Prevention. Reputable banks scan donors for chlamydia, cytomegalovirus, gonorrhea, hepatitis B and C, human immunodeficiency virus (HIV, the cause of AIDS), human T-cell lymphotropic virus type 1 (HTLV-1), and syphilis. Because HIV infection may not show up on initial tests, health regulations today require all donated sperm to be quarantined for six months and retested before being sold. Banks may also look for carriers of the genes responsible for the wasting diseases sickle-cell anemia, Tay-Sachs disease, and beta-thalassemia, which tend to strike blacks, Eastern European Jews, and those of eastern Mediterranean ancestry, respectively.

In addition, commercial sperm banks have on the whole improved record keeping. The larger ones assign donors coded identification numbers, maintain files on their medical and personal history, and track distribution of their sperm. Some put a cap on the number of babies to be conceived with a given donor's sperm, something that doctors with modest practices often neglected to do in the past. The Curie-Cohen study had singled out excessive use of a single donor as especially troubling. If the physician practiced in a small community, a whole crop of AID children who were siblings but did not know it might wind up going through school together, and possibly even dating or marrying. Curie-Cohen and his colleagues opined that "present records on artificial insemination are woefully deficient. At a minimum, records should be maintained on the outcome of these pregnancies and on paternity."[15]

This potential problem could, of course, have been eliminated by opening up the system and making sperm donors known to recipients. Sherman had seen a scientific pretext for openness, but mental health professionals who began looking at AID in the 1970s suggested a more important reason for it: the psychological well-being of AID children, who would seem to have a compelling right to know their own genetic heritage, if only for medical reasons. As Houston psychologist Patricia Mahlstedt and social worker Kris Probasco have written, "a young woman whose mother and aunt had breast cancer will be encouraged by her physician to have more frequent examinations and mammograms than the woman with no history of such a disease."[16] So, too, those coming from families riddled with heart disease, alcoholism, and other ailments have a claim to such information. Without it, they lack "a powerful tool for ensuring a healthy life."[17] The privacy rules that serve the best interests of donors do not in this regard benefit the children born as a result of AID.

As judicious as revealing a child's family history to him or her might sound, most within the infertility industry offered staunch opposition to suggestions that they move toward greater openness. Sperm banks and physicians argued that their recruitment efforts would be severely hampered if men knew that they might one day be approached by children claiming them as their fathers. It was also widely suggested that nonanonymous donors might encounter legal difficulties: Although case law was lacking, no one wanted to find out whether courts might require donors to provide financial support for any children conceived with their sperm. British law professor Glanville Williams offered two other justifications for keeping a donor's identity shrouded: "the desire to protect the donor's reputation (think of the repercussions for his family if his adventures in paternity become common gossip!), and to eliminate the risk of the [recipient] wife transferring her affections to the donor."[18]

Even as Williams and others were insisting upon the need for donor anonymity, a grassroots movement to give a greater variety of women access to sperm banks was coalescing. Single women and lesbians objected to the fact that the few existing sperm banks had adopted the restrictive policy of dealing only with married women. Some women, realizing the ease with which artificial insemination might be accomplished, took matters into their own hands. For example, Francie Hornstein, a gay woman active in women's health collectives in Los Angeles, gave birth to a son in 1978 who was the product of self-insemination, and she happily shared the details of the process with other women. The multiple purposes of the turkey baster became an inside joke.

Within the next few years, women's health centers in California, Vermont, and England began offering clients AID. In London, a group of lesbian feminists, drawn together by an ad in a women's liberation newsletter, formed the Feminist Self-Insemination Group, which in 1980 published a fifty-page how-to guide on AID. While feminists began spreading the word on AID within the network of women's health centers, other women used the courts to obtain access to conventional clinics. One suit, pressed in 1980 against Wayne State University School of Medicine, forced a reversal of the school's policy to deny single women AID.

Perhaps the most enlightened sperm bank was founded in Oakland by the Feminist Women's Health Center, which in 1982 opened the Sperm Bank of Northern California (now simply the Sperm Bank of California) in a building that had once housed a funeral parlor. The sperm bank provided its product to any woman, single or married, straight or gay. Like London's Royal Free Hospital, it encouraged women to perform self-insemination, providing homey consultation rooms for that purpose. Without giving the matter a great deal of thought, its administrators—all women—adopted an open donor policy, whereby men were asked before giving sperm whether they would allow their names to be revealed to any potential offspring seeking that information after reaching age eighteen. Over the years, some 40 percent of donors have agreed to allow release of their name, place and date of birth, and Social Security and license numbers to those of their biological children who ask.

This sperm bank was also the first to provide extensive, nonidentifying information on the health, habits, appearance, and social standing of donors before insemination. Gradually, the approach put together by the Sperm Bank of California, today located in Berkeley, has come to set the standard for client relations in the $160-million industry, forcing other concerns to expand the services they provide.

The top sperm banks compete vigorously for business, showing up at all the major convocations of gynecologists and fertility specialists, their booths stocked with glossy literature and gimmes. Like drug and biochemical supply companies, they are selling a product, one that just happens to be manufactured in the testicles of men rather than in factories crowded with fermentation vats and automated equipment.

A concern like California Cryobank, a privately held company founded in 1977 by physician Cappy Rothman and based in Westwood just a block from the gates of the University of California Los Angeles campus (its other branches, in Palo Alto and Boston, are also located hard by universities—and thus close to the source of their raw material), rejects as many as 90 percent of those volunteering to provide semen. This is typical of other large sperm banks which sell their products worldwide. Some would-be donors are eliminated because of previous or current infection, or a known incidence of hereditary disease in the family. The preponderance, though, are knocked out of consideration because they do not consistently produce a large enough number of highly motile sperm with each ejaculation.

There are other unspoken standards employed by sperm banks which have little medical rationale and are based in large measure on eugenic beliefs. Genotype is judged not only by phenotype but also by personal performance: Looks, good grades, and outstanding achievement are all definite pluses for the candidate hoping to earn a little extra cash making weekly visits to a tiny cubicle, sometimes called a "masturbatorium" or "blue room," stocked with erotic movies and magazines. Such cubicles are standard features of sperm banks and can also be found at leading fertility clinics around the country.

Flipping through the September 1994 Donor Catalog of California Cryobank, a photocopied 8 1/2-by-11-inch, eight-page document which lists available specimens and is sent to physicians who provide it to those seeking AID, one finds donors broken out by race (Asian, Black/African-American, Caucasian, Mixed) as well as by ethnicity (eighty "ethnic origins" from African to Flemish to Polynesian to—perhaps inappropriately given the ethnic cleansing which has been prosecuted by former citizens of this nation against one another—Yugoslavian). Four and two-thirds pages of the catalog enumerate and describe Caucasian donors; two-thirds of a page, Asian donors; two-thirds of a page, donors "who belong to TWO OR MORE RACIAL GROUPS OR have UNIQUE ANCESTRIES such as American Indian, East Indian, or Mexican"; and less than one-third of a page, Black/African-American donors.

Of Caucasian donor 728, for example, it is recorded that he is of Lebanese, Irish, and American Indian descent, has dark brown wavy hair and hazel eyes, stands 6'2", weighs 200 pounds, has O-positive blood, "medium" skin, and, at the time of his sperm contributions, had completed five years of higher education, majoring in psychology and law. Asian donor 514 is Chinese, with straight black hair and brown eyes, stands 6'2" (Steve Broder, the lab director, said that taller Asian donors are very much in demand by clients), weighs 195 pounds, has O-positive blood, "medium" skin, and had completed nine years of college and graduate school to become a mechanical engineer. Black/African-American donor 714 is of African-American and French Creole extraction, has dark brown curly hair and brown eyes, stands 6'1", weighs 180 pounds, has A-positive blood, "medium" skin, and had completed one year of college under a theater arts major. Unique Ancestry donor 757 is of African-American, French-Canadian, Irish, and American Indian descent, has dark brown curly hair, brown eyes, stands 5'6", weighs 130 pounds, has B-positive blood, "dark" skin, and had completed three years of college with a major in international relations and political science.

Without in any way diminishing its importance to those contemplating AID, one can say that the act of poring over the Donor Catalog is the ultimate extension of the consumer society: shopping for the future progenitor of one's children. Forever young, locked in an eternal present, the donors are putative exemplars of their gender. Women and couples comb through the catalogs assiduously, searching for the "best" donor, weighing the imponderables. To satisfy clients' desire to learn as much as possible, California Cryobank also will, for five dollars and nine dollars respectively, provide short (two-page) and long (ten-to-twenty-page) profiles of candidates which may include job and salary histories, as well as other personal details. In fact, the company recommends that people consult the long profiles before finally settling on a donor.

California Cryobank also offers a special service. They will assign a "donor matching counselor" to help with the selection process for a fee (in 1994 the charge was thirty-five dollars per half hour). Clients fill out a "Donor Matching Questionnaire," rating on a scale of one to three (one for "essential," two for "nonessential, desirable," and three for "least importance") assorted qualities, including race, ethnicity/country of origin, religious heritage, blood type and Rh factor, hair color, hair texture, eye color, height and weight, bone structure (small, medium, or large), occupation (the questionnaire notes that "most of our donors are students, so their major would be considered their occupation"), interests, and skin characteristics—the five types itemized are very fair ("little to no ability to tan"), fair ("skin will tan lightly"), medium ("light color, will tan moderate to dark"), olive ("pigmentation of unexposed skin"), and dark ("unexposed skin"). Clients are asked to list "other characteristics which you feel are important" on the reverse of the form. Once the questionnaire is in hand, the donor matching counselor narrows the field of possibles to three, from which clients select.

About 60 percent of clients pay for another service. According to the company's explanatory packet, "Many of our married clients send photographs of their husbands with their initial application. Single clients often send photographs of themselves, male relatives, or other individuals of their choosing." Provided with such a photo, a donor matching counselor will compare it to photos of donors in order to identify the one whose features most closely match. (Xytex, an Atlanta sperm bank, advertises that it provides "a clear picture of your choices," that is, head-and-shoulders photographs of donors.) For couples who intend to keep their use of AID secret, this careful matching is geared to producing children whose appearances will not raise eyebrows by clashing with that of their "father." Other sperm purchasers clearly hold eugenic beliefs, however naively, assuming that if they pick a donor with certain qualities, their children will have the same qualities—or stand a higher chance of having them.

Clients discovering a donor who appeals to them are urged by California Cryobank to place their order "*at the earliest possible time.* Many donor specimens are sold daily and we CANNOT guarantee donor availability." (California Cryobank follows the American Society for Reproductive Medicine guidelines and limits the number of offspring for any donor to ten.) Prepared for intracervical insemination, specimens cost $142 per vial (enough for one insemination); those for intrauterine insemination, $175 per vial. Shipping by Federal Express, in a special double-walled container holding liquid nitrogen, runs from $85 for round-trip standard two-day delivery within the United States to $341 for round-trip overnight delivery to Canada (round-trip because California Cryobank retrieves the valuable double-walled containers). The cost of transporting frozen sperm overseas varies by location, but suffice it to say that substantial numbers of clients worldwide consider this American product worth importing. As Jerome Sherman remarked in 1973, the air delivery of sperm to distant places around the globe "should not be surprising, as international shipment is a routine practice in the animal breeding industry."[19]

T he Office of Technology Assessment estimated that during twelve months spanning 1986 and 1987, there were some 172,000 artificial inseminations in the United States, of which about half were with donor sperm. Partly, people had become convinced that adoptions entailed too long a wait and too much red tape. Too, they seemed to prefer having a child who was related to someone in the family. Experts put the total number of AID babies born in the country since the 1940s at anywhere from 500,000 to 1 million.

Although there is no way of knowing for sure, it is generally agreed that most of those born as a result of AID have been kept in the dark about their origins. Upon reaching adulthood, a few of those who were told or inadvertently found out have written or spoken about their deep sense of betrayal and anger. Those who have sought information on their biological fathers sometimes have encountered outright hostility from clinics. Others merely run into a blank wall: no leads, no names, no photographs. The machinery for preserving anonymity has generally done its job.

When southern Californian Suzanne Rubin found out after her mother's death in 1980 that she had been an AID baby, she set out to discover who her biological father was. Rubin had herself at age fifteen given up for adoption a baby girl, whom she had eventually tracked down, so perhaps she felt the uncertainty of her own origins more acutely for that. Her widowed father had told her the name of the doctor who had inseminated her mother, reputedly with the sperm of a medical student. Given the location of the doctor's office, Rubin surmised that the young man must have been enrolled at the University of Southern California. Probing the USC archives, she was able to narrow the field only to fifty young men who had attended the medical school in 1948 and 1949.

Although frustrated, Rubin continued the hunt, telling a London *Sunday Times* reporter in 1982, "It's an obsession with me. I must find my father even if it's only to discover what kind of man sells his sperm and ultimately his own flesh and blood for $25 and then walks away whistling a happy tune without any thought of the life he may have created." Rubin eventually concluded, based on the remarkable physical similarities between herself and the man who had been her mother's physician, that he, not a medical student, had provided the sperm and lied to her parents about its source.

Rubin subsequently waged a public campaign against the practice of AID. She recounted how out of place she had felt as a "tall, blue-eyed redhead in a family of shortish brunettes," how she had believed she must have been adopted. She talked about her anguish upon learning about her origins; she felt her parents had "cruelly deceived" her. Her own emotional responses led her to the conclusion that AID in general was inherently "immoral, unethical, and adulterous." Rubin, who has since taken the last name Ariel, which belonged to her maternal grandmother, asserted, "In my family, the lies warped the relationships and poisoned them beyond repair."[20]

Ethicists and mental health professionals, long bothered by the dark side of AID, have endeavored to carry out formal studies on its psychological impacts, but have often found themselves stymied by a lack of interest among both physicians and the organizations, public or private, which fund scientific research. British investigators Robert Snowden, Duncan Mitchell, and E. M. Snowden of the Institute of Population Studies at the University of Exeter were among those who succeeded in winning support for their research into AID's impact upon the family. In the late 1970s, they undertook a comprehensive study of couples who had undergone the procedure, as well as solicited testimony from AID children.

The team had long been associated with a gynecologist in Devon who ran a large infertility practice and had kept in contact with many of her patients over the years. She had inseminated 986 women from 1940 through 1982, 899 strictly with donor sperm. Out of the 899, 480 had given birth (8 of the babies died of congenital abnormalities or birth trauma). Concerned that so little was known about the impact of AID on the family, the gynecologist approached Mitchell and the Snowdens.

The Exeter researchers decided to contact 111 of the gynecologist's successful couples, testing the waters by sending an initial letter asking if they would be willing to be interviewed for a medical survey—nature unspecified. Only those who responded positively were informed that AID was the topic. Left with a pool of 57 couples, the researchers then visited the women and their husbands (if alive) at home, and carried out lengthy interviews concerning their experience with infertility. The couples were asked how they had settled on AID, whether they had told anyone about it at the time, and whether they had told or intended to tell their children.

The majority of couples told the researchers that they had been "immensely grateful that their involuntary childless state had been resolved" and had no regrets about having chosen AID. The husbands had, on the whole, readily accepted the children as their own, and showed at worst a mild sense of confusion about who was the "real" father. As the man designated Husband 877 for purposes of the study remarked, "My wife said was I at all upset about it being another man's, a donor. And I also had to ask her a straight question—did *she* mind it being another man's baby inside her?"[21]

Indeed, wives expressed more anxiety on this score, especially before giving birth. Some feared that the eye, hair, or skin color of the child would be a giveaway, triggering suspicions about its paternity. "Some wives confessed to spending the last few days of pregnancy worrying if the child might have red hair, or a long nose, or even, by some awful mistake, be of a different racial origin," reported the researchers.[22] After the delivery, some wives felt anxious when their children fussed or made trouble, and attempted to shield their husbands from any of the unpleasantries of child raising. They seemed to operate from the unconscious conviction that a man lacking a genetic tie to a child will have only a tenuous sense of responsibility or affection for it, and that he will reject that child when presented with something as inevitable as midnight wailing or the tantrum of a two-year-old. (It goes without saying that such an attitude is in large measure culturally constructed, not somehow part of the basic package of human responses. In matrilineal societies, or those in which uncles or other men take an active role in the upbringing of children, people place far less emphasis on biological fatherhood, and, in addition, are often not so concerned to curtail women's sexual behavior by censuring premarital or extramarital encounters.)

Notwithstanding their pleasure at having children, thirty-three out of fifty-seven couples had never breathed a word to anyone about having used donor sperm. They hoped by their concealment, which was often accompanied by guilt or uneasiness, to protect the public image of the husband, whom they feared would take flak in the workplace or community were his infertility to be discovered. They said they also wanted to prevent any possible embarrassment to the child's grandparents, who might be approached in public by curious or moralistic types. Finally, they wanted to avoid having the child held up to shame by a judgmental community. These impulses clearly were bred out of a strong sense of the British *Zeitgeist*. But insofar as they also stemmed from psychological denial, the researchers considered them insupportable. In a volume summarizing their findings, they asserted that "secrecy, while ostensibly being maintained for the sake of the child, is closely bound up with the concept of stigma, particularly the stigma of male infertility." In the long run, attempting to deny the husband's infertility was counterproductive. Instead, couples and physicians should work toward "an enlightened understanding of the concept of male infertility, through education and counselling."[23]

The researchers found that even those couples who had opted for a more open policy rarely chose to discuss what they had done with just anyone. Four couples had spread the news to everyone in their immediate family, while ten had informed both sets of grandparents only. Seven couples hid their actions from one set of grandparents and not the other. Two couples told only the wife's mother, and two others informed no relatives. Eleven couples told select friends. All claimed to have received universal support from those in whom they confided, yet this did not encourage them to make additional disclosures.

Regarding the children, 48 couples had determined never to tell them that they had been fathered by proxy, and five others were undecided. Only three couples resolved to reveal the truth to their children when they had reached an age at which they might comprehend. In a separate study carried out by physicians at the Queen's Medical Centre of the University Hospital in Nottingham, whose results were published in 1982, 257 out of 366 couples seeking AID declared that they would not inform their children, while 22 said they would do so only if the social climate had become more accepting.[24]

Snowden and his colleagues also spoke with eleven couples with AID offspring over eighteen years old. These pioneering couples had availed themselves of the Devon gynecologist's services in the days just after World War

II. Of these, three had, under circumstances that appear in each case rather unfortunate, disclosed the family secret. One woman, seeing her son's distress over a hereditary ailment which had manifested itself in his supposed father, explained to him that he needn't worry about developing the ailment himself since he wasn't genetically related. The researchers don't record the son's reaction to this information; however, one can imagine that any relief he experienced on his own account might have been tempered by other feelings of guilt, along with sorrow and perhaps even anger at having been misled over the years. A second woman reported to the researchers that her husband's habitual extramarital affairs had prompted her to tell her grown, married son about his origins. She said her motive had been to disabuse her son's wife of any doubts that he, too, was doomed to be a philanderer. Finally, a working-class couple with two AID sons noticed that the younger of the two was behaving rebelliously and falling behind in school. Aware that the sperm donor had been a professional man, the parents imparted this knowledge to the fractious son "so that he should know that he had the potential of a high intellectual capacity in his inherited characteristics." They also let the elder son know that he, too, was an AID baby.[25]

A number of couples seemed to have opted for AID on the basis of eugenic considerations. Some believed it offered advantages over adoption. Said Wife 906,

> Some adopted children are from teenage mothers, perhaps with not very good backgrounds, and you don't know how that child might turn out. Because I know a lot of it depends on environment, but it is partly heredity, and I think in a way you can feel more sure with AID than with adoption that you are likely to end up with a reasonable outcome.[26]

For reasons of predictability, too, several couples who had returned to the gynecologist for a second pregnancy requested the same donor as they had the first time, presuming that the "blood" relation would make for closer bonding between the siblings.

The Exeter team interviewed seven AID children who had been told of their history. Several claimed that they had suspected all along that they were somehow odd. However, most said they had not been extremely bothered to learn of their unusual paternity, nor felt inclined to hold their parents' silence against them. One young man felt only relief: "It was as if I'd always known there was something wrong, I'd always known there was something amiss—and suddenly, being told that [I was an AID baby], it was as if a huge great weight had been lifted off my shoulders."[27]

In hopes of gaining further insight into the feelings of those who had been conceived by AID, the investigators ran an ad in a London newspaper solicit-

ing testimonies. Of the handful of people who responded, one had suffered
an extreme reaction upon learning she was an AID baby. She wrote, "I felt
like escaping, carrying this great shattering boulder of information away with
me. I had a picture of grunting farm animals, test tubes, sperm and me. God
the father had deserted me. I was the child of the devil: a pubescent melo-
drama that I acted out in hate and revenge." She explained that her mother
had attempted to soothe her by supplying the "only information that she had.
To make sure that his sperm was 'all right' every donor had to have three children. (Somewhere I had more family.) He had to be of good moral character.
(I should be reassured that there was nothing of the criminal in my blood.)"
However, the woman perceived that her grandmother had grave reservations
about her origins. She wrote, "My grandmother always finds occasion to
speak darkly about 'blood' in my company—blood and its mysterious capacity to carry talents and traits. She always knew just how my mother had conceived me, for she had paid the necessary fee for my conception."[28]

In recent years, psychologists, psychiatrists, and social workers in the
United States and abroad have expanded their investigations into the impacts
of AID. While few people seem to undergo such devastating reactions as this
anonymous woman did, counselors have found that profound guilt often suf-
fuses families who keep AID secret, and that anger and rage at parental
betrayal often grip those who find out what has long been hidden. At the
least, people seem to experience a feeling of incompleteness, as if a part of
them exists out there in the world and they have not yet found it.

The Ethics Committee of the American Society for Reproductive Medicine, which issues advice that is not binding to those in the field, urges clinics
and sperm banks to keep nonidentifying information about donors on file,
complete with results of thorough genetic workups, to be provided to couples
and resulting offspring. Surprisingly, not until 1991 did anyone bother systematically to poll sperm donors with the aim of determining what their general attitudes might be toward the gathering and disclosure of such
information. Patricia Mahlstedt and Kris Probasco surveyed a group of sev-
enty-nine men who donated sperm at the Baylor College of Medicine in
Houston and Reproductive Resources in Metairie, Louisiana. Only 2 percent
of the men said that they would feel negatively about the release of nonidentifying information about themselves; 96 percent happily provided background information about themselves and their families (2 percent couldn't
decide how they felt). Said one donor, "This would seem essential, especially
since behavior and character traits appear to be highly correlated with genetics." Said another, "I think it is very reasonable. Recipient families should be
able to choose from similar looking donors, as well as pick from personality
traits/talents as they see fit."[29]

When Mahlstedt and Probasco asked the men how they felt about openness in AID, their reactions were mixed. Some would agree to meet the cou-

ple, not the child; others thought the child had the "right to know where he comes from"; others found the idea "not that excit[ing]." Asked whether they would want to meet any children conceived with their sperm, 37 percent said no, 41 percent said yes, if the child took the initiative, and 19 percent said they would "actively want to know" the child.

Mahlstedt and Probasco also gave the men a chance to toss a bottle out into the ocean, as it were. "What message would you want to give to a child conceived with your sperm?" they asked. The answers ranged from the pithy, "Be the best you can be and make the world better because of your presence," which managed improbably to combine an Army recruitment slogan with a lyric from a Michael Jackson song; to the didactic, "Be aware of the small sacrifices that your parents have made on your behalf. Artificial insemination is a difficult choice, and to have made it means your parents considered having you to be worth the struggle. Appreciate them. Also, pick a talent, focus on it. Stay away from rugby (it kills the shoulders) and strive for excellence"; to the reassuring, "My love and thoughts are with you. Also, don't worry if you're very tall and thin as a youth, because by the time you are about 20 years old, you'll have to knock the women off with a stick" (this donor seemed to assume he would father only males); to the heartfelt, "I'm glad I could help in your conception, but your real parents are the ones who raised you and took care of you. I hope you find life beautiful. Be happy!"

In considering the answers they had heard, Mahlstedt and Probasco were reminded of the "truths" that in earlier years were perpetrated concerning adoption, to wit:

> Raising an adopted child is the same as raising a biological child; birth mothers will relinquish custody of their children and simply go on with their lives; if adoptees have positive childhood experiences, they will have no interest in knowing about their birth mothers and fathers. These "truths" are now seen as myths by the experts in the field of adoption.[30]

Couples obviously have the right to privacy, to decide whom they feel comfortable telling about AID. However, as Mahlstedt pointed out at a 1994 workshop on infertility counseling held in San Antonio, Texas, in conjunction with the annual meeting of the American Society for Reproductive Medicine, there is a difference between privacy and secrecy. A couple's "decision not to tell the child," she said, "intrudes on the child's right to know his genetic origin. Withholding this fact involves a negative process . . . a pretending that the child is the biological offspring of them both."[31] Everyone involved would profit, Mahlstedt contended, if the psychological realities of AID were acknowledged: the feelings of loss that men who desire to father children experience when they find out they are physically incapable of doing so; the sadness women may feel if the men they have married cannot father children; the confusion and anger that children may feel when they discover that they have a biological father, in addition to the father who has raised them.

These notions appear sensible and well founded, yet many in the infertility field still equivocate when it comes to secrecy. At the October 1994 workshop in San Antonio, Mahlstedt presented the case for bringing AID into the light. Addressing a roomful of mostly female psychiatrists, psychologists, social workers, and family and marriage therapists who counsel infertile couples, she ended her remarks by lobbing a bomb into the audience, saying, "In a more open environment, Dr. Cecil Jacobson could not have 'cared' for his patients in the manner he did." During the question-and-answer period following, several women rose to take issue with her and insist that drawing a veil over AID was not necessarily harmful. Indeed, with the exception of one dissenting member, who argues that both sperm and egg donation are "ethically inappropriate," this is also the view of the physicians, scientists, lawyers, and ethicists who sit on the Ethics Committee of the American Society for Reproductive Medicine, which asserts that "there is a lack of information about whether secrecy is better for a child." Unfortunately, no one can say for sure because the policy of disguise and denial that has been resolutely pursued for almost a century has obviated any attempt to assess AID's impact on the children born as a result of it, since those children are, for the most part, unidentifiable, even to themselves.

11
Life
in the Lab

From the earliest days of embryology, neither basic researchers nor medical specialists who attempted to understand the mechanisms of reproduction perceived much of a distinction between animals and humans. To their minds, species were readily interchangeable, so that an investigator might refer to results in cattle and women in the same breath, and base expectations for what should be seen in the clinic upon what had already been observed in the laboratory or on the farm. If they could boost fertility in cows or "improve" birth outcomes among pigs, reproductive physiologists figured they could do the same for their own kind. As F. H. A. Marshall wrote in *The Physiology of Reproduction*, "Generative physiology forms the basis of gynaecological science, and must ever bear a close relation to the study of animal breeding."[1]

Just as with artificial insemination, the technology which would ultimately enable embryologists to concoct life in the lab was developed through years of experimentation with domesticated species. Initially, the work was undertaken for the sake of acquiring pure knowledge, out of a desire to elucidate the manifold aspects of the biological wonder that is sexual reproduction. Later, the interest became pragmatic: Animal scientists hoped to amplify the reproductive powers of "superior" females, as they had done for "superior" males. Investigators had solved one side of the quality equation by gaining the ability to endow select sperm upon thousands of females. Now they sought methods for exploiting the potential of well-bred females.

There were several possible tacks. One idea was to inseminate chosen females, then remove any resulting embryos and implant them in another animal for gestation. This way, an exceptional female would not have her uterus tied up for months and could be inseminated as many times as she came into heat. In the end, more offspring carrying her admirable heritage would be produced than if she gestated them herself, seriatim.

Another idea was to retrieve eggs from the ovaries of first-rate females, to fertilize these eggs, and then, once again, to bring the resulting embryos to term in lesser females. Further, it might be possible to take an egg from lower-quality stock and replace some or all of the material in its nucleus to obtain a better-bred animal.

Either of these schemes required a number of skills which, at the turn of the century, scientists possessed only in half measure or not at all. To accomplish the first, that is, to transfer embryos from one female to another, it was necessary to develop a method for removing fertilized eggs from the fallopian tubes or uterus. After that, there was the tricky business of reinserting the embryos in another receptacle: Would they take?

The second scheme, which came to be known as in vitro fertilization (literally fertilization in glass) after the laboratory vessels embryologists used, required that eggs be obtained at the point just prior to their release from the ovaries, when they are mature enough to undergo fertilization. Then the eggs would have to be kept alive outside the body for some period of time, hours to days. Next, the eggs would have to be fertilized while bathed in culture medium sitting in a dish. Finally, the resulting embryos would have to be kept alive until they could be returned to a receptive uterus.

Complicating matters was biology's almost utter ignorance of what made for fertility. No one had the slightest clue how ovulation was controlled—maybe, Walter Heape suggested, a "generative ferment," a circulating substance in the blood?—and well into the twentieth century, reproductive physiologists wrongly believed that females bled and expelled eggs from the ovary at the same time, and therefore entered their most fertile phase directly after estrus or menstruation. Piecing together the real picture took several decades of work by scores of scientists coming at the puzzle from many different directions.

The in vitro fertilization story line, whose starting point can be taken as 1878, with the work of S. L. Schenk, and whose logical climax comes precisely a century later, with the birth of Louise Brown, the premier IVF baby, is punctuated with "firsts." Somewhat like Ronald Reagan's war on Grenada, IVF seems to have afforded just about everyone involved the opportunity to come away decorated: Overviews of the field are sometimes little more than lengthy lists crediting researchers recognized as the first to obtain eggs from rabbits, mice, guinea pigs, hamsters, opossums, rats, cows, horses, goats, sheep, pigs, dogs, monkeys, chimpanzees, humans, and so on (not necessarily in that order). In addition, there are the researchers—sometimes the same people, sometimes not—who were the first to successfully fertilize the eggs of each of these and other species; then those who were the first to transfer the embryos of same. There are also the first researchers to produce a chemically detectable pregnancy; a pregnancy which persisted for some discernible period of time before miscarriage; and a full-term pregnancy in each of many species, culminating in humans.

Because assisted reproduction is an international enterprise, the opportunity for a few additional winners arises: One can be the first to have successfully applied a certain technique in a certain country, a parochial honor but nonetheless coveted.

A male rabbit held down and pierced with a needle, a female cut open and robbed of a cluster of eggs—IVF effectively began with these actions, in the Vienna laboratory of embryologist S. L. Schenk. Drawing sperm from the epididymis of a male rabbit, or buck, adding them to a flat glass dish containing unfertilized eggs and snippets of tissue from the female's uterine wall, Schenk made the earliest-known attempt to produce a mammalian embryo outside the warm, moist confines of the reproductive system. Although his initial efforts failed, Schenk ultimately succeeded in fertilizing both rabbit and guinea pig eggs and watched as they spontaneously divided under his gaze—life proliferating under alien conditions, at least until the cellular engine inexplicably ceased.

Oddly enough, female mammals are born with all the eggs they will ever have. Human fetuses generate some 6 to 8 million primordial germ cells, or oogonia, which are stored in the ovaries. These cells begin the chromosomal shuffling known as meiosis but spontaneously arrest themselves partway through the process and enter a state of suspension in which they sit until just prior to ovulation.

Even before gestation has been completed, female germ cells begin dying off, so that at birth, the ovaries of baby girls contain only about 1.5 million oocytes, which continue to decay and die, perhaps owing to fatal errors that crept in during their manufacture. At the time of first menstruation, about four hundred thousand remain. As a woman enters her cycle each month, a certain number of oocytes regain momentum, completing a few more steps in meiosis and undergoing an unequal division. The larger of the two resulting parts is called a secondary oocyte; the smaller, the first polar body. At the same time, an oocyte readying for ovulation acquires a carapace of cells and forms what is known as a primary follicle. As the cells surrounding the oocyte multiply, the follicle enlarges and fills with fluid. Gradually, the speck of egg migrates toward the surface of the ovary. Around day fourteen, the follicular bubble bursts, ejecting the egg clad in its corona of cells. Hairlike cilia lining the fallopian tube propel the egg toward the uterus as the final chromosomal regroupings of meiosis take place and the oocyte divides once again, producing a second polar body.

If a sperm penetrates the egg and the gametic pronuclei fuse, yielding a solo nucleus with the full chromosomal endowment, the resulting zygote undergoes cleavage, its single cell dividing into two, four, eight cells and so on. Biologists refer to these cells as blastomeres, and to the solid mass they form after about five cleavages as the morula. Until this stage, the embryo remains roughly the same size, about as large as a pinprick. With the cleavages that follow, it inflates, the sphere of accumulating blastomeres hollowing out. Some blastomeres cover the interior of the zona pellucida like tiles, in a one-cell-thick layer called the trophoblast. The rest cluster together at one pole. From this so-called inner cell mass, the myriad types of cells composing the fetus will differentiate.

By day six or seven after conception, the embryo has reached the uterus, where it may or may not attach itself to the uterine lining, or endometrium. Meanwhile in the ovary, the empty follicle hardens into the corpus luteum, a bright yellow fibrous body webbed with blood vessels which will persist through month five or six of a pregnancy. Where implantation has failed to occur, the corpus luteum quickly lapses into an amorphous scar known as the corpus albicans.

Schenk's rabbit and guinea pig embryos underwent cleavage to the two-cell stage, and his 1880 report describing his accomplishment provided tantalizing evidence that embryos might be contrived outside the body. Word spread, and other investigators attempted to replicate his results. One, M. J. Onanoff, apparently succeeded in brewing up a medium so conducive to the growth of rabbit and guinea pig embryos that they divided until reaching the eight-cell stage before dying.

Onanoff also claimed to have inserted such embryos into the abdominal cavities of both male and female rabbits and guinea pigs. Bizarrely, the embryos behaved as if housed within the protective uterus: When Onanoff dissected the host animals a few days later and examined the embryos, they displayed what is known as the "primitive streak," the ghostly seam of cells which becomes the head-to-tail axis of a forming creature. (In humans, the primitive streak appears on the fifteenth day after fertilization.)

Onanoff had discovered an extremely strange aspect of embryos, revealing their enormous integrity and relative independence. Yet the account of his findings, published posthumously in 1893, stirred only minimal curiosity, in the main because it lacked rigorous documentation. Not until the 1930s would researchers revisit Schenk's and Onanoff's work, and by then, enormous headway had been made in the area of embryo transfer.

The acknowledged father of embryo transfer is Walter Heape of Cambridge University, the very same biologist who had dabbled with artificial insemination. Heape's primary interest lay in the maternal side of reproduction. He wanted to elucidate how uterine structure and biochemistry shaped the phenotype of offspring. Mainly, he wondered what accounted for telegony. Could it be that a male, in mating with a young female, altered her uterine environment in a way that affected all subsequent offspring? (Heape was unaware of Onanoff's peculiar work gestating embryos within the abdominal cavity.)

One way of getting at these core questions, Heape decided, was to incubate embryos from one breed of animal in a female of a different type. If the offspring emerged substantially altered, it would be a sign that the uterine environment conveyed something essential. For his experiment, he settled on Angora rabbits and Belgian hares. In the spring of 1890, after mating two Angoras, he anesthetized the female, or doe, with chloroform and, flushing her ovaries out, probably with a saline solution, retrieved a clutch of fertilized eggs. Next, he anesthetized a Belgian doe, and, using a spear-tipped needle, conveyed the embryos into her oviduct. Four weeks later, the doe produced a litter, of which two out of six were Angoras, their phenotype gloriously unaltered by the foreign uterus in which they had grown.

Heape sent off a missive to the Royal Society of London. With typical scientific understatement, he wrote: "In this preliminary note I wish merely to record an experiment by which it is shown that it is possible to make use of the uterus of one variety of rabbit as a medium for the growth and complete fetal development of fertilized ova of another variety of rabbit."[2]

To embryologists, Heape's work was less interesting for what it revealed about the maternal environment than for its sheer technical virtuosity. Many researchers mimicked his approach, but not until the 1930s did anyone consistently match his dexterity. The man most talented at manipulating ova was the American Gregory Pincus.

Pincus had grown up in a family of agricultural scientists. His father had a degree from a Connecticut agricultural school and edited a farm journal. His maternal uncle was dean of the agricultural college at Rutgers University and director of New Jersey's Agricultural Experiment Station. As a graduate student at Harvard, Pincus studied both genetics and animal physiology, and after graduating in 1927 he did postdoctoral work with Marshall and Hammond at Cambridge and with geneticist Richard Goldschmidt at the Kaiser Wilhelm Institute. Back in the United States, he took a position at his alma mater and published *The Eggs of Mammals,* a tour de force which covered not just ova but virtually every aspect of mammalian reproduction. In 1944, Pincus cofounded the Worcester Foundation for Experimental Biology in Shrewsbury, Massachusetts, with Hudson Hoagland, their intention being to pursue basic research and also to develop technologies that might benefit the animal industry. That same year, Pincus also launched the Laurentian Hormone Conference, for years the meeting where scientists seriously interested in hormones went to see and be seen. Pincus would win worldwide acclaim for his contributions to the development of the birth control pill, but among embryologists, he is still revered for his studies of gametes and zygotes. In 1988, one reproductive scientist characterized Pincus's lab during its heyday as "the only place where the mammalian egg was studied and honored."[3]

During the 1930s, Pincus got interested in embryo transfer and began a series of experiments. In his experience, mouse, rat, and guinea pig eggs were so fragile that they fell apart after being retrieved. Rabbit eggs showed more durability, withstanding a fair amount of handling. With ease, Pincus was able to sluice fertilized eggs from rabbit tubes with a saline solution, channel them into pipettes, and capture them in watch glasses, shallow dishes often used for evaporating liquids. The fertilized eggs might sit for several hours at room temperature before he placed them in another animal. During this period they might cleave to the blastula stage.

However, Pincus was unable to obtain any baby rabbits in this manner. The fertilized eggs appeared normal while sitting in culture medium before transfer, but Pincus suspected they had not been properly activated, so he performed various tests to chart the early stages of embryonic growth and assess the viability of embryos. Much to his surprise, he found that even dead sperm could cause mammalian eggs to behave as if they had been fertilized. Exposed to dead sperm, eggs showed a telltale shrinkage of their outer covering, or zona pellucida, and formed the second polar body even though they had not been penetrated. After this they inexorably degraded. This type of development without sex is dubbed parthenogenesis.

Pincus persisted in attempting to transfer embryos, and in 1934 hit upon a protocol that worked. With coinvestigator E. V. Enzmann, he recovered eggs from an anesthetized agouti doe, then placed the eggs in a dish with sperm from a black-furred male for about twenty minutes and watched as the cladding of cumulus cells fell away and the zona shrank, signaling that a sperm had done its work. Scooping the fertilized eggs up in a pipette, the researchers inserted them in the right fallopian tube of a New Zealand red doe. Thirty-three days later, this animal delivered seven dark gray young. In a second experiment, the pair gleaned fertilized eggs from an English-spotted doe, which they cultured in flasks for twenty hours before implantation in an albino doe. The albino yielded two English-spotted young.

Pincus and Enzmann's accomplishment, officially aired in the *Proceedings of the National Academy of Science,* was still being talked about three years later when Harvard Medical School researchers discovered that by attaching electrodes to a woman's vagina and abdomen, it was possible to monitor ovulation: A slight electrical pulse marked the moment when the bursting follicle launched its tiny missile toward the anemone-like fringe surrounding the opening of the overhanging fallopian tube. The October 21, 1937, issue of the *New England Journal of Medicine,* in which the Boston researchers announced that they had found a way to track ovum release, ran an accompanying editorial musing upon this advance. Electronic monitoring of ovulation, the editorial explained, provided a "direct objective yardstick" which would appear to "enable women to know when they are fertile and when infertile from a record of their menstrual dates." Obviously, this would be highly advantageous to both those wishing to limit births and those hoping to conceive. But, the editorial writer went on,

> contemplating this new discovery, one's mind travels much farther. [Warren] Lewis and [Carl] Hartman [at Philadelphia's Wistar Institute of Anatomy and Biology] have isolated a fertilized monkey egg and photographed its early cleavage in vitro. Pincus and Enzmann have started one step earlier with the rabbit. . . . If such an accomplishment . . . were to be duplicated in human beings, we should, in the words of "flaming youth," be "going places." The difficulty with human ova has been that those recovered from tubes have regressed beyond the possibility of fertilization in vitro. But by utilizing the electrical sign we may be able to obtain them from the follicle at the peak of their maturity. If the new peritoneoscope can be developed along the lines of the operating cystoscope, laparotomy [in which the abdomen is incised from ribcage to pubic bone] may even be dispensed with. What a boon for the barren woman with closed tubes! [Cambridge University sperm specialist Arthur] Walton [who works with John Hammond] is quoted as saying that it is theoretically possible to separate male-determining from female-determining spermatozoa. Will it be possible to obtain son or daughter, according to specifications, and even deliver them of women who are not their mothers? Truly it seems as if the forge were being warmed, and another link may be welded in the chain by which mankind strives to hold nature under control.[4]

Palpably excited, this editorialist saw the future—all but anticipating egg retrieval by laparoscopy, egg donation, gender selection, and surrogate motherhood—and pronounced it marvelous.

Other prognosticators liked less well what they saw. From the turn of the century on, as the news of scientists' machinations with sperm and eggs filtered out of laboratories and clinics, artists took the liberty of imagining what would happen if such techniques were systematically applied to humans. Among the most notable literary works from the early part of this century which deal with assisted reproduction are *We*, written in postrevolutionary Russia; *Brave New World*, which has come to be considered a classic of the dystopian genre; and the prophetic *1984*.

Composing the elliptical *We* in 1920 and 1921, Yevgeny Zamyatin, a naval-engineer-turned-novelist and leader of a band of visionary writers in St. Petersburg, modeled a totalitarian fictional state upon the "unanimous" philosophy of Soviet Communist ideologues. Relations between the sexes, in Zamyatin's glass-enclosed city of tomorrow, have been reduced to a dry, government-supervised routine: Lab technicians in the Sexual Department determine "the exact content of sexual hormones in your blood" and provide you with a table of days on which intercourse is allowable. "After that, you declare that on your sexual days you wish to use number so-and-so, and you receive your book of coupons (pink). And that is all."[5] Sex is purely recreational: Reproduction is too important for the numbered citizenry to be permitted to breed at will. Only those deemed eugenically fit by the Benefactor, ruler of the One State, can procreate. The narrator, D-503, a mathematician, debunks the permissiveness of earlier days, when "the state (it dared to call itself a state!) could leave sexual life without any semblance of control."[6]

Aldous Huxley, torn between technophilia and technophobia, alternately fascinated by scientists' legerdemain and repelled by the possibility that their methods would be ruthlessly exploited by politicians, constructed a whole society based on control of reproduction. In his 1932 novel *Brave New World*, people enjoy complete sexual freedom. However, they are banned from procreating. Instead, there is the Central London Hatchery, a baby factory where embryos are put together in assembly-line fashion and brought to maturity in artificial wombs—large bottles filled with a culture medium whose recipe Huxley cribbed from scientific papers of the time, a bit of cow peritoneum here, a dash of hormone there. Stacked in tiers three high, embryos are trundled along conveyor belts under red light (embryos, says the hatchery director, only like red light), completing their transit in 267 days, whereupon they are not born, but rather "decanted."

The Central London Hatchery does more than merely gestate embryos. It has stepped "out of the realm of mere slavish imitation of nature into the much more interesting world of human intervention." Depriving the embryos of specific amounts of oxygen, the hatchery produces an array of infants with differing mental abilities. Those who receive the least amount are so profoundly damaged that they possess no "human intelligence"; they will join the ranks of the lowest caste, the Epsilons, the brute laborers. Hatchery technicians also (and here Huxley swerved into pure conjecture) expose some embryos to excessive heat, conditioning them to despise cold and preparing them for a life in tropical mines. Similarly, those destined to toil in chemical plants are dosed with lead, caustic soda, tar, and chlorine so that they will develop a tolerance to these toxins. At the top of the social ladder, Alpha Plus intellectuals, permitted a full complement of oxygen, undergo rigorous brainwashing after decanting. In their nurseries, they receive Pavlovian conditioning, and are taught to think of the historical existence of conventional sexual reproduction as an obscenity, and of the words "mother" and "father" as embarrassing smut.

George Orwell's *1984* offers the bleakest literary take on assisted reproduction. Written in 1949, the novel draws heavily on real-life eugenics, distorting its beliefs and campaigns only slightly to arrive at a future in which children are begotten for the sole sake of the Party. The narrator, the hapless Winston Smith, reports that "sexual intercourse was to be looked on as a slightly disgusting minor operation, like having an enema." Some citizens, including Winston for a time, commit the "thoughtcrime" of desire, and find virtue in old-fashioned coupling. But the Junior Anti-Sex League propagandizes incessantly for complete celibacy: "All children were to be begotten by artificial insemination (*artsem*, it was called in Newspeak) and brought up in public institutions."[7] Except among the proles, the dregs of society, technological reproduction is Party policy.

If such dark fantasies touched a chord with the public, the criticisms implicit in them hardly disturbed scientists. The fertilization of human eggs was virtually a reality, and researchers were not about to rein themselves in after coming so close.

Pincus went on in 1935 to show without doubt that rabbit eggs would function entirely normally after being taken from the ovaries, fertilized in vitro, and reimplanted. Other researchers actually filmed rabbit eggs dividing happily in culture to the eight-cell stage. Guinea pigs, mice, rats—eggs from these species also proved capable of being fertilized outside the reproductive tract if properly handled, although cleavage stalled early on or failed to take place altogether. Thanks to Pincus, who had in 1940 devised a way of harvesting many more eggs than usual from animals by injecting them with hormones, researchers saw a big jump in the amount of material they had to work with.

Next on the list were eggs of higher mammals. Warren Lewis and Carl Hartman of the Wistar Institute had had fairly good luck fertilizing monkey eggs in vitro, but Pincus eschewed monkeys and went for the big score. With B. Saunders, he carried out preliminary trials in which he placed immature human ova in blood serum, the yellowish liquid which separates out from coagulated blood. Sitting in serum for eight to twenty-four hours, Pincus's eggs acted as if they were still inside their follicles, leaving their suspended state and resuming meiosis. The eggs matured to the time of first polar body formation. If they could be urged to proceed a bit further, they would be fertilizable. Maybe it wasn't necessary to glean eggs at precisely the right moment in the menstrual cycle, merely to hit upon the correct formula for bringing them up to speed in vitro.

Encouraged by Pincus and others to extend this work, the physician John Rock and his colleague Miriam Menkin, both then at the Harvard Medical School, attempted to mature some eight hundred follicular oocytes, a huge trove obtained thanks to obliging patients undergoing operations at the Boston Free Hospital for Women. In addition to "surgical material," the women donated blood so that their eggs could be incubated in their own serum.

Of the eggs they were given, Rock and Menkin exposed 138 to sperm. They varied the culture medium, the length of time that egg and sperm had contact, and the concentration of sperm. After all their jiggering, cleavage ensued in just three embryos: but, to the scientists, what an amazing trio! Rock and Menkin eyed the embryos under the microscope, sketched them, fixed them in wax, sliced them into paper-thin sections, stained them, photographed them, measured them, and described them in exquisite detail in a March 1948 paper published in the *American Journal of Obstetrics and Gynecology*. Scrutinized and ballyhooed, these were for a time the most famous embryos in history.

In vitro fertilization in humans, however, was by no means a done deal. Eggs and sperm turned out to be notoriously temperamental, performing differently in the hands of every researcher. Over the next decade, scientists engaged in repeats, reruns, redos. They muddled, goofed, and bollixed things up more often than they attained the outcome they wanted. Perplexed by every failure, urged onward by every success, researchers around the world raced to reach the goal—a combined in vitro fertilization and embryo transfer resulting in a live birth—before anyone else.

A key discovery came in 1951, when Colin Austin of Cambridge University and Min Chueh Chang, a Chinese-born Pincus protégé at the Worcestor Foundation, independently demonstrated that sperm exposed to the female reproductive tract gained the ability to penetrate an egg because it underwent a kind of chemical undressing, shedding part of itself in a process that Austin dubbed capacitation. During capacitation, the protein coat over the head of the sperm is removed, freeing enzymes that enable it to penetrate the zona pellucida. Under natural conditions, this reaction takes place as sperm migrate through the mucus surrounding the cervix and through secretions lining the uterus and fallopian tubes. After capacitation, the straight-line motion of the sperm becomes wavelike, and its long tail, which acts as a motor, revs up and whips back and forth more exaggeratedly. If capacitation fails to occur, sperm encountering an egg may fail to navigate the surrounding cloud of cumulus cells, which have been expelled along with it from the bursting chamber, or follicle, in the ovary. Or the sperm may slip through the cloud and succeed in attaching themselves to the egg surface yet wriggle there uselessly, lacking the ability to push on through the zona to complete fertilization. Landrum Shettles of Columbia University repeated Chang's and Austin's results in humans, showing that sperm incubated with mucosa from the fallopian tubes gained the ability to tunnel through the zona pellucida in vitro.

For all the strides experimentalists had made, IVF still struck many scientists as dubious, and they debated among themselves about how actually to determine whether an egg had been fertilized. In principle, the genetic material of the sperm and egg must fuse, and the resulting zygote must divide and then differentiate. But what did that look like under a microscope? Sometimes, an egg in the process of fragmenting looked uncannily like one that had cleaved, and as Pincus had shown with dead sperm, cleavage could occur spontaneously. So what constituted surefire proof of fertilization? Generally, people in the field agreed it was fair to say fertilization had taken place if you spotted a sperm burrowing through the zona pellucida or lodged in the perivitelline space, a small gap between the zona and the membrane enclosing the egg. The presence in the egg cytoplasm of sperm tails snapped off like those of lizards, of two pronuclei, or of an extruded polar body were also good signs. Researchers would quibble over this issue into the 1980s.

Another ongoing topic of controversy was the quality of embryos produced in vitro. Pincus's litters of rabbits would seem to have amply demonstrated the viability of embryos made in the petri dish, yet many remained unconvinced that such embryos would invariably be normal. This was a plaguesome uncertainty which weighed against any widespread application of IVF and embryo transfer in cattle, much less in humans. No one was interested in hazarding the procedure commercially unless it could be shown unassailably that IVF yielded genetically normal embryos which would develop into normal young. Pincus himself, who envisioned hordes of improved beef cattle bred in glass, wanted further proof, and it would come in 1959 from experiments carried out at his own foundation by M. C. Chang.

At nine o'clock one evening, Chang bred fertile bucks to a rabbit in estrus. The same evening, he injected two other rabbits with hormones extracted from sheep pituitary glands, to bring about ovulation. At nine the next morning, he collected blood straight from the carotid artery of a rabbit before killing it. From this blood, serum was prepared. Next, Chang fetched the mated doe, inserted a needle filled with bicarbonate solution into her uterus, emptied the syringe, and then immediately withdrew the plunger. In this way he captured capacitated sperm, which he placed in flasks. Taking the two does he had dosed with hormones, Chang flushed their fallopian tubes with bicarbonate solution and obtained ova, which went into the same stoppered flask containing the sperm. The flask he set in a gentle rocking device inside an incubator set at 38°C (100.4°F). Three hours later, he transported the ova with a pipette to a second flask of rabbit serum and saline solution. After letting the eggs sit another eighteen hours in culture, he mounted each one on a slide, examined them under the microscope for signs of fertilization, then carried out a thorough analysis of their state.

Chang counted eggs that were clearly unfertilized and eggs that may have been fertilized or parthenogenetically activated. He counted eggs that had fertilized and sometimes cleaved but died, and eggs that had cleaved normally. Out of a total of 266 eggs, 62 percent remained unfertilized, 23 percent might or might not have been fertilized, 22 percent had fragmented after cleaving, and 21 percent were in the process of cleaving. Out of this last group of vigorous dividers, Chang transplanted 36 into the oviducts of six does. Four of the does ultimately delivered fifteen baby rabbits. Chang, in presenting his results in *Nature*, made a lofty claim—that finally, science had "a repeatable procedure for fertilizing mammalian ova" in vitro—to which his peers by and large acceded.[8] The regimen worked so well that Chang shortly topped his rabbit work with the fertilization of golden hamster eggs, and other researchers racked up similar successes with Chinese hamster, mouse, and rat eggs.

Still, scientists continued to pooh-pooh one another's reports. Certainly, almost no one but the popular press bought the stories of Bolognese physician Daniele Petrucci, who, while carrying out basic research on how to facilitate organ transplants, claimed in 1959 to have fertilized a human egg in vitro and to have kept the embryo alive for twenty-nine days (at which time it would have resembled in size and shape a small shrimp). Petrucci reported that the embryo had become "deformed and enlarged—a monstrosity," and he had terminated the experiment.[9] *L'Osservatore Romano*, the Vatican daily, called upon Petrucci to desist from his diabolical experimentation, while many in the international scientific community dismissed him as a crank. Petrucci next divulged that he had maintained an embryo for nearly two months in the lab (at which time it would have been close to three inches long and had recognizable limbs and head). Feted in the Soviet Union, awarded a medal for his achievements in Moscow in 1961, the Italian commenced a collaboration with two physicians there, Pyotry Anokhin and Ivan Maiscki, who announced in 1966 that they had kept 250 human embryos alive for at least two months. One fetus supposedly reached six months, and weighed one pound, two ounces.

Profound skepticism greeted these disclosures. At issue was the believability of the communications, not the nature of the work—European scientists had themselves made attempts to assemble an artificial placenta, both for the sake of basic research and to salvage premature infants who would otherwise die. Teams in Germany and Britain had kept pig fetuses and aborted human fetuses alive by hooking them up to apparatuses which kept blood circulating, oxygenating it and removing impurities. Some such research was supported for eugenic reasons, on the presumption that it might help spare premature infants from disabilities. "Human intervention" was deemed not only much more interesting than "mere slavish imitation of nature," but also more humane.

The formidable complexity of the human female reproductive system, involving the brain, various glands and tissues, a battery of hormones, and intricate feedback loops, can confound researchers even today, after eight or nine decades of highly productive probing. Some 30 percent of couples who visit physicians complaining of an inability to conceive are poked, scanned, and tested, only to be told that their infertility is idiopathic; that is, its cause cannot be identified. Moreover, the fine points of conception and implantation remain unexplicated.

Back in the late 1800s, Heape had suspected that some "generative ferment" drives the round of physiological changes which constitute the monthly female cycle. This notion had been supported by experiments which showed that the ovaries had a glandular function. However, the endocrine system was terra incognita. Not until 1908 did British scientist William Bayliss discover the first of the hormones, those multipurpose proteins which serve as bodily messengers and provocateurs. Two years later, Marshall at Cambridge collected evidence that the ovaries are organs of "internal secretion," and that substances issuing from developing follicles are responsible for estrus.

Soon, it became clear that reproduction involved not one circulating substance, as Heape had thought, but several. Now scientists had a whole new class of hormones, the sex hormones, also known as gonadotropic hormones or gonadotropins, to explore. By 1929, researchers in the United States and Germany had identified estrone, a hormone manufactured and released by the ovaries. Within a few years two similar steroids, estriol and estradiol, had been identified and synthesized. Only minuscule amounts of these circulate in the bloodstream: The American researcher who isolated estradiol mined one ton of pig ovaries to obtain just 6 milligrams (.0002 ounce) of the hormone.

Estrone, estriol, and estradiol, classed as estrogens, perform a variety of functions. In fetuses, estrogens trigger the growth of the uterus and vagina, while in mature females they monthly prime the uterus and mammary glands for a possible pregnancy. Through a feedback loop, estrogens also signal the hypothalamus and pituitary gland, helping govern the production of two other key gonadotropins, follicle-stimulating hormone (FSH) and luteinizing hormone (LH).

It is now known that at the beginning of a menstrual cycle in a normally ovulating woman, the hypothalamus sends a chemical signal to the pituitary in the form of gonadotropin-releasing hormone (GnRH). The pituitary in turn disperses FSH. At this point, the ovaries together will usually have up to thirty follicles poised to mature, the largest of which will measure between 5 to 8 millimeters (about .2 to .3 inch). In addition to an egg, every fluid-filled follicle contains estradiol, which leaks into the bloodstream and also fuels the division of so-called granulosa cells making up the follicular wall. Granulosa cells are equipped on their surface with molecular docking sites for FSH, which is taken up as it arrives from the pituitary and stimulates further cell division. More granulosa cells translates into additional docking sites and a greater overall uptake of FSH. (With the onset of menopause, which is caused by a depletion of eggs, FSH levels skyrocket. As the egg supply dwindles, the number of FSH receptors falls, leaving FSH to wander the bloodstream without a port.)

As the menstrual cycle proceeds, the amount of estradiol entering the blood during a twenty-four-hour period climbs steadily from .03 milligram (.000001 ounce) to .40 milligram (.000014 ounce). Follicles swell to 12 millimeters (almost half an inch) or larger. In the meantime, the granulosa cells construct docking sites for LH. Deeper in the follicle, specialized theca cells have already been fielding this chemical. After fourteen days, estrogen levels reach their peak. Cued by the high estrogen levels, the pituitary releases LH in a surge, and more molecules of the steroid swarm the granulosa and theca cell docking sites. Finally, the pressure of the teeming cells within a follicle becomes too great and it bursts, jettisoning its entire contents. Occasionally, more than one follicle will reach this stage simultaneously, paving the way for a multiple conception. When one or more eggs have been released, the remaining follicles regress, having fulfilled their lifetime function. In the absence of a pregnancy, a whole new cohort of follicles will be recruited for the next cycle.

After ovulation, the menstrual cycle enters what is known as the luteal phase. The corpus luteum arises and pumps out progesterone, which incites changes in the uterine lining, or endometrium, that prepare it for the possible arrival of a fertilized egg. Progesterone also inhibits the secretion of LH. If an embryo succeeds in implanting itself in the endometrium, a multilayered sac called the chorion forms around it. This intermediary between the uterus and the embryo consists of three surfaces: the villi, a tufted layer bathed in maternal blood; the trophoblast, which passes nourishment into the embryo; and the extraembryonic somatic mesoderm, which will become the amnion, or membrane holding the amniotic fluid. The chorion emits yet another hormone, human chorionic gonadotropin (HCG), which signals the ovaries to continue secreting the estrogen and progesterone which are vital to the maintenance of the conceptus during the first trimester of pregnancy.

As scientists were puzzling out this system, they noticed that gonadotropic hormones, when injected into lab animals, had the power to modulate the progression of estrus. Administered at given points in the cycle, not necessarily in sync with the normal bodily rhythms, they could inhibit ovulation or promote it. The realization that estrus could be controlled from the outside prompted an exploration, from the late 1940s on, of ways in which hormones might be used as contraceptives. After being visited in 1951 by the aging firebrand Margaret Sanger, whose International Planned Parenthood Federation of America was leading the charge against global overpopulation, Gregory Pincus and M. C. Chang intensified their efforts in this vein.

Although shying from the alarmism which characterized many contemporary discussions of the population explosion, Pincus considered it incumbent upon the "physiological research worker" to address the "deficiencies of conventional contraceptive methods," while at the same time continuing to accumulate knowledge concerning "the delicately balanced set of sequential events" involved in reproduction, from the production of egg and sperm to delivery.[10] Finding that certain enzymes made in mammalian testes prevented ovulation in rabbits and rats, Pincus went on to test a range of steroids for similar effects. In 1953, he and John Rock collaborated to screen a battery of compounds, some of which had been synthesized by the pharmaceutical company G. D. Searle, for their power to inhibit ovulation. The two most promising candidates then underwent trials with human volunteers in Puerto Rico. Approval by the U.S. Food and Drug Administration followed, and in 1960 Searle began marketing its oral contraceptive. Ortho and Wyeth Laboratories, having undertaken their own intensive research, soon followed suit.

During the 1950s, researchers tested another drug, clomiphene citrate, for its contraceptive effects. A chemical analogue of one type of estrogen, clomiphene looked at first glance as if it would inhibit ovulation, and thereby forestall conception. But to everyone's surprise, it had precisely the opposite effect when injected into animals. Instead of blocking egg release, it promoted the maturation of multiple follicles. Soon, clomiphene was being given to women with recalcitrant ovaries, as was another hormone called human menopausal gonadotropin (HMG).

Definitely one of the odder drugs on record, HMG was refined by Italian scientists who reasoned that women who were past childbearing age would have high levels of FSH and LH in their bloodstream. And what better place to find a ready supply of menopausal women than a nunnery? The scientists approached a group of nuns, who obligingly supplied them with urine, from which FSH- and LH-rich HMG was extracted. Marketed under the name Pergonal, HMG was an extremely expensive (today it runs about forty-four dollars, per vial) but highly effective fertility drug—almost overly so, since it frequently resulted in multiple births among women to whom it was given. By the late 1960s, both clomiphene and HMG had become the most potent treatment physicians could offer women whose infertility was related to ovulatory problems. Soon, they would also be found in the kit bags of researchers pursuing embryo transfer.

long with the advent of the Pill and synthetic hormones, there was
another advance which in the 1950s galvanized those involved in reproductive physiology—not to mention the world intellectual community. This,
of course, was the elucidation of the molecular structure of DNA by English
physicist Francis Crick and American biologist James Watson at the
Cavendish Physics Laboratory in Cambridge. Intuiting DNA's peculiar, spi-
ral-staircase structure from fuzzy X-ray images produced by Maurice Wilkins
and Rosalind Franklin at a laboratory in London, Crick and Watson ushered
in the age of molecular biology.

In a very short time, biochemists had developed methods for deciphering
the code carried within DNA molecules. They thus in principle gained the
ability to obtain the genetic blueprint of any protein that exists. But there was
more. Biochemists would also devise enzymatic scissors and paste for cutting
up and recombining segments of DNA. This meant that, blueprint in hand,
they could in theory build any protein. And they could build it exactly as evolution had ordained, or they could build it to spec. They could, in short, engineer everything that lives, tinker with it at the most fundamental level. This
power to rearrange far exceeded any that animal breeders or plant hybridizers
had possessed. It was now possible to splice a gene from any organism into
the DNA of any other: a human gene in a bacterium or a mouse, a bacterial
gene in a strawberry plant, a pig gene in a human.

Many of those closest to the ferment in biology found it exhilarating. For
Julian Huxley, the discovery of the molecular basis for heredity brought to
mind nothing so much as the vast panorama "of [man's] possible future
improvement." To bring about this improvement, he told members of the
Eugenics Society at their annual Galton Lecture in London on June 6, 1962,
"we shall need . . . new ideas and attitudes about reproduction and parenthood in particular and human destiny in general. One of those new ideas will
be the moral imperative of eugenics."[11]

Imperfect man, said Huxley, in order to achieve "full evolutionary success,"
must eliminate from his line "severe and primarily genetic disabilities like
hemophilia, color-blindness, mongolism, some kinds of sexual deviation,
much mental defect, sickle-cell anemia, some forms of dwarfism, and Huntington's chorea," among other things.[12] He must raise the overall IQ of the
world's population, since "the great and striking advances in human affairs,"
whether in art, science, politics, or war, "are primarily due to a few exception-
ally gifted individuals." This might be accomplished fairly quickly, because
hormonal birth control, embryo transfer, artificial insemination (Huxley
favored Muller's VCOG scheme), sperm cryopreservation, and other techniques "raise the effective speed of the selective process to a new order of
magnitude."[13]

Huxley argued that while both nature and nurture shape the phenotype,
the most pressing problem faced by humanity, to wit "genetic deterioration,"

could not be "cured, or prevented by improving social environment or social organization." Only by directly manipulating the germ plasm through AID (which Huxley proposed should be renamed EID, the E for *eugenic*), through ova donation, and through other advances could man become the "senior partner, the directing partner" in his own evolution.[14]

Huxley's "new idea" was in fact rather long in the tooth, and his invocation of that shibboleth, the germ plasm, had an anachronistic ring even in 1962, when the word "DNA" was already well ensconced in the lexicon, if not yet a household word. Nonetheless, his talk of "purposeful planning" resonated with many scientists.

It made, for example, supreme sense to Joshua Lederberg, a geneticist then at Stanford University, winner of the 1958 Nobel Prize for Physiology or Medicine for his work on bacteria. Lederberg was attracted to the possibilities that advances in biology presented for "the prediction and modification of human nature." He envisioned a bold future for the aborning field of "genetic alchemy"—that is, molecular biology—which provided the means to alter the "substratum of posterity."[15] Lederberg foresaw that through applied biology, researchers might make epochal revisions to human evolution.

In a 1966 article for the *Bulletin of the Atomic Scientists,* he offered a replacement for the old, inefficacious eugenics: What would take its place was euphenics, or the eradication of "genotypic maladjustment" of individuals. Negative eugenics was futile, Lederberg said, because the bulk of genes which cause deleterious effects are recessive, and one or another of them is carried by virtually every human on the planet. To eliminate all carriers of harmful alleles, or variants of a particular gene, would be effectively to eliminate the species. And, positive eugenics was useless except in enabling high-risk couples to avoid passing on genetic diseases. Moreover, previous eugenicists had established overly restrictive criteria for who might be considered wellborn:

> Western culture and its limited population is being succeeded by a much broader world culture. Is there much point in setting eugenic standards relevant only to a small minority of the world's population even as we watch the unprecedented breakdown of intercultural barriers? The jet airplane has already had an incalculably greater effect on human population genetics than any conceivable program of calculated eugenics would have.[16]

Supplanting the outworn methodology of eugenics, Lederberg said, would be an array of molecular technologies with the power to effect a "major evolutionary perturbation."[17] Recombinant DNA techniques, whereby pieces of DNA can be assembled, cut, fused together, inserted into other lengths of DNA, and so on, had not yet been invented, but Lederberg foresaw that they would be and would have massive impact. He predicted that they would deliver the means of treating any genetic disease by correcting faulty genes before birth, or perhaps in childhood. In other words, they would, as Alfred Ploetz had desired, create the possibility of pushing selection back to the level of the embryo. They would also enable researchers to discover the cellular mechanics of aging and cancer, and to halt or reverse these processes. They would be used to create "chimeras," animals incorporating human genes, which might provide insights into the function and development of characteristics unique to humans. But most astonishing, they would open the way for vegetative propagation of animals, that is, the creation from a set of chromosomes of an identical replica, or a clone, of an already existing adult. And why not?

> If a "superior" individual—and presumably, then, genotype—is identified, why not copy it directly, rather than suffer all the risks, including those of sex determination, involved in the disruptions of recombination. The same solace is accorded the carrier of genetic disease: why not be sure of an exact copy of yourself, another healthy carrier, rather than risk an overtly diseased offspring; at worst, copy your spouse and allow some degree of biological parenthood.[18]

Lederberg thought the advantages would be many: Parents, especially those desiring sons, would not be surprised or disappointed by an unpredictable outcome. Organs and tissues could be readily transplanted among members of a clonal population because they would be recognized by a recipient's immune system as "self" not "other" and therefore would not be rejected. Because one might assume that "genetic identity confers neurological similarity," communication, especially between generations, would be greatly improved. Talking to one's clone would be akin to talking to oneself.

Lederberg's only real concerns were that cloning might lead to the reinforcement of racial, class, or gender discrimination. If scientists weren't careful, the nationality, batting average, famousness, or beauty of the first successful clone might set some long-term standard of what was considered most desirable in a clone.

Although, given the standing of the national pastime in the national consciousness these days, the mention of batting average rather dates Lederberg's piece, one is struck thirty years later by its overall prescience. Every prediction Lederberg made about molecular manipulations has come true, except one. No one has yet cloned a whole animal from adult DNA, the movie *Jurassic Park* notwithstanding. However, many scientists have succeeded in performing a sort of cloning, the production of artificial twins by splitting a single embryo in half, with a range of animals, including humans.

But in another respect, Lederberg missed the story altogether, as had so many eugenicists before him. British biologist Peter Medawar, who had won

the Nobel Prize in 1960 for his work elucidating the immune system's rejection of transplanted tissues, pointed out in a lecture several years later the
bankruptcy of schemes which aimed at a genetic improvement of humans.
Population geneticists had realized by that time, 1969, that stock breeding did
not, in fact, bear much in common with natural selection, although everyone—stockbreeders and evolutionists alike—had thought since Darwin's day
that they did.

Whereas breeders devote their efforts "towards the establishment or fixation
of a particular genotype or genetic makeup," Medawar explained, it turned out
that "natural populations of outbreeding organisms, including human beings,
are persistently and obstinately diverse in genetic makeup."[19] Breeders produce
consistency in their animals only by resorting to inbreeding, or line breeding,
mating fathers to daughters or granddaughters, and so forth, for generations,
whereas natural populations exhibit "a huge profusion of variant forms" of
proteins, such as those in the blood, which is to say they possess a corresponding abundance of alleles, or variations of a particular gene.

There is no one genetic makeup that qualifies as best; instead, populations
contain a range of genotypes arranged in a relatively stable pattern. These
patterns of genotypic variability account for the fact that populations "breed
true." But they also allow for the incredible variety of individuals in any given
generation. In the end, Medawar said, creating "a population of supermen"
would be scientifically infeasible. Moreover, it would be morally and politically indefensible. This was the problem with all utopian speculation—"the
inadequacy of man, the extreme unlikelihood that man can live up to his own
ambitions," for whatever reason.

The year Medawar spoke these words was the same year that British
researcher Robert Edwards announced that he and his colleagues had fertilized thirty-four human eggs in a petri dish. If no one else was, reproductive
physiologists were living up to their ambitions.

12

"Who Loves Ya, Baby?"

During World War II, when thousands of British women and children were evacuated from cities to safe haven in the countryside, the young Robert Edwards spent time in the Yorkshire Dales. There, he received a rudimentary sex education watching farm animals, and set to musing on the mysteries of fertilization and birth. Edwards would remember this experience partway through his college career. Dissatisfied with his course work in agriculture, he would decide to switch to a zoology major. After graduating in 1951, in debt and unsure of his prospects, he grabbed at a postdoctoral opening in the Institute of Animal Genetics at the University of Edinburgh.

Soon after arriving, Edwards found himself mesmerized by a film made by Alan Beatty, a researcher at the institute. The film, titled *Inovulation*, showed a scientist retrieving a fertilized egg from a mouse, then carefully transferring it through the cervix into a second female's uterus. Edwards promptly indentured himself to Beatty, and was soon nightly haunting his new adviser's lab, a run-down outbuilding known to all as the "Mouse House," where the rodents were raised in cages stacked five high. The institute maintained dozens of inbred experimental strains which, generation after generation, produced nearly identical mice with particular qualities.

One of Beatty's interests was chromosomal anomalies. Like sperm, eggs are specialized cells which carry only half the chromosomes of normal cells, a condition termed haploid. Thus, when the pronuclei of sperm and egg fuse after fertilization, the resulting embryo contains forty-six chromosomes, as do all normal cells, rather than twice that many. Sometimes, however, mistakes occur, and germ cells with more than twenty-three chromosomes arise. These botched sperm or eggs yield embryos crowded with redundant genetic material. Known as polyploid, they may contain multiple sets of chromosomes, or some portion of an extra set. Such abnormalities are responsible for numerous diseases and disabilities, including trisomy 21, or Down syndrome, wherein an embryo receives three copies of chromosome 21.

Beatty set Edwards to gathering eggs from mice and to dosing sperm with X rays, ultraviolet light, and chemicals to upset their chromosomal material. The aim was to cook up and examine embryos that were artificially polyploid. Almost invariably, those Edwards produced lived only a very short time. It is now known that polyploidy accounts for a large proportion of fetal deaths and spontaneous abortions: Nature is quite efficient at weeding out flawed embryos.

Beatty had in mind applications for this work. The idea was to become adroit enough at rearranging the nuclear contents to be able to effect a rough-hewn sort of genetic engineering. Edwards, making a foray into organized religion, briefly worried about the ethicality of doing this, but decided there was nothing terribly wrong about it. Soon thereafter, he gave up on churches: "I felt eventually that the numinous and mysterious could be found rather in the laboratory where each night I peered through a microscope at primitive sex cells. Truth to tell, I found scientific concepts more to my taste than theological fumblings."[1]

For that matter, Edwards had little patience for fumblings of any kind. Accordingly, his next goal after having produced a few examples of polyploidy was to speed up the production of eggs—and therefore embryos—by injecting gonadotropic hormones into his immature mice to induce superovulation, or the simultaneous ripening and release of many more eggs than usual. This tactic enabled him to strip over one hundred eggs from each mouse in a fell swoop. Although accepted wisdom held that gonadotropic hormones would not promote ovulation in adult females, Edwards gave it a go anyway and injected a bunch of mice as he was about to leave on a two-week vacation. Upon returning, he peered into a set of cages occupied by "superbly pregnant, superlatively pregnant" mice. Pumped full of gonadotropins, the mice had mated and, as it were, superconceived. Edwards killed and autopsied several animals immediately, and found their abdominal cavities jammed with fetuses—behind the liver and kidneys and in the folds of the intestines. One female carried thirty-seven fetuses, all alive and normal.[2]

If one could superovulate farm animals, there would be enormous profit in it: This was what leaped to Edwards's mind immediately. Instead of gathering only a handful of eggs from a prime female for in vitro fertilization and embryo transfer, one might collect hundreds at a time. Just as artificial insemination enabled breeders to endow multiple offspring with a prime male's genetic material, so would superovulation do the same with that of prime females. But Edwards foresaw that superovulation might serve another, more important purpose as a key component in the treatment of infertility in humans.

He would not, however, take action on these notions right away. Soon he was winging his way to the West Coast of the United States, having won temporary berth at the California Institute of Technology. There, his attention was diverted by a project having to do with sperm. Since the early part of the century, researchers had attempted to devise a means of determining the sex of a child before birth. Often, they had concentrated on identifying those hormonal changes which characterized pregnancy. They had looked for steroids excreted in urine, like the androgens responsible for male sex characteristics, or had analyzed maternal blood for chemical giveaways of fetal gender. In the 1930s, researchers pursued the peculiar expedient of injecting the urine of pregnant women into male rabbits and adding it to tanks holding female fish. They claimed, as a result, to have observed alterations in the sex organs of these animals that could be related to the gender of the fetus. A few investigators tried to determine whether women mount an immune reaction against the alien tissue of male fetuses; however, they obtained only ambiguous answers.

Finally, in 1955, four separate teams around the world hit upon a reliable test involving the sex chromatin, or Barr body. This small structure, first identified in the cells of cats, is a compressed, inactivated X chromosome that appears as a tiny nodule inside the nuclear membrane of female cells. (They also sometimes appear in cells of those rare, apparently male individuals who have, through meiotic errors, been endowed with one or more stray X chromosomes in addition to the customary X and Y.) When certain standard cellular dyes are applied, the sex chromatin colors intensely. Not all female cells sport this feature, but enough do that researchers seized upon it as an ideal way to discriminate between female and male fetuses. All that was required was to obtain cells from the fetus, which was accomplished by amniocentesis, whereby a needle is inserted either through the abdominal wall or vagina into the amniotic sac around week sixteen or seventeen of a pregnancy and a bit of fluid is withdrawn. These amniotic cells were then examined for sex chromatin.

One of the scientists who hit upon this test, Columbia University's Landrum Shettles, also tried another, proactive approach to sexing the fetus: His goal was to perform prenatal gender *selection,* as opposed to gender determination. Initially, Shettles asserted that it was possible to distinguish X and Y sperm by sight. According to him, X sperm had large oval heads, while the Y head was smaller and rounder. Several scientists immediately labeled this assertion fallacious. Shettles, they said, had completely misread the situation. The differences between sperm heads were not real, but distortions caused by the microscope he was using.

Shettles persisted in his studies. He subjected sperm to a kind of chemical obstacle course and concluded that those carrying X and Y chromosomes migrated through both acidic and basic media at different rates. Since vaginal pH slides up and down the acid-to-basic scale over the course of the monthly cycle, Shettles suggested that it was possible to schedule coitus at certain times of the month to give either the X or the Y sperm an edge. (Some breeders of dogs and other animals still swear by this tactic.)

Shettles, of course, did not have the last word. Many other investigators remained in the hunt, staining, centrifuging, and filtering the tiny swimmers with the intent of devising a foolproof mode for sorting them and thereby guaranteeing the gender of offspring. As for Edwards, nothing much came of his Cal Tech project, but the experience cemented in his mind a conviction that separating sperm would be a valuable medical tool. It would afford substantial eugenic benefits by enabling physicians to prevent the perpetuation of sex-linked hereditary ailments, that is, those caused by genes carried on the X or Y chromosome.

Back in Britain again, Edwards joined a National Institute for Medical Research team headed by Alan Parkes, which was exploring ways of employing the body's immune reaction as a form of contraception. Still, Edwards couldn't get embryos out of his mind, and around 1960 began to "wonder for the first time about the practicability of replacing [laboratory-produced] human embryos in the womb of a woman."[3]

Over the next eighteen years, Edwards concentrated on achieving that goal. Immediately, at the institute, he set himself the task of contriving the best culture medium for ripening eggs. He tried a soup of salt, potassium chloride, glucose, protein, and human chorionic gonadotropin, which he considered a key ingredient, vital to proper maturation. Much to his surprise, a group of eggs sitting in a control solution lacking the HCG divided vigorously, exactly on schedule. Certain that he had made a monumental discovery, Edwards hurried to the institute library to comb through the embryological literature. At one point he looked up and punctured the room's hush with an expletive. "Sod it!" he said. He had been beaten to the punch. Twenty-five years before, Pincus had noted this phenomenon in his work with rabbits. Eggs in culture sometimes cleaved parthenogenetically, without benefit of the male contribution; however, these eggs soon imploded, their development nothing more than a cellular mockery.

Edwards soldiered on. His next project was to obtain as many immature human eggs as possible, get them to ripen, and then attempt fertilization in a petri dish. Again suppressing a few ethical qualms, he sought the cooperation of two gynecologists who routinely performed oophorectomies, in which one or both ovaries are removed. The gynecologists agreed to provide Edwards with excised ovarian tissue for his investigations. Six months later, unable to force maturation of any eggs gained from the surgical patients, he switched to rhesus monkeys. While culturing rhesus eggs, he realized that he had perhaps rushed the whole process. Maybe primate eggs ripened more slowly than those from rabbits. Indeed, when he let rhesus eggs sit for twenty-eight hours in his broth, he was able to glimpse them reviving from their suspended state and recommencing meiosis. Again, Edwards was led to contemplate the possibility of performing in vitro fertilization in a medical setting in order to overcome infertility.

Eager for more human ovarian tissue, he began pitching this prospect to other physicians, despite a suspicion that in vitro fertilization would encounter a good deal of censure, just as artificial insemination continued to do. "Perhaps I explained myself badly, or was too enthusiastic," Edwards wrote in 1980 of the in vitro breakthrough. "Or the issues of fertilization carried a more emotional charge than I realized. Whatever the cause, the door closed behind me and I came away empty-handed."[4] The press got wind of his activities, and a British television producer rang him, fishing for a story. Alan Parkes, more savvy in media relations than he, overheard the conversation and handed him a note which read, "Don't let yourself be interviewed over the telephone. Take care." Worse, the new head of his department ardently opposed in vitro fertilization, and, Edwards speculated, engineered a fateful meeting with Sir Charles Harrington, director of the institute, who commanded, "I don't want any human eggs fertilized here." That looked like the end of it until some months later, when Harrington stepped down and his replacement, Nobel biologist Peter Medawar, okayed the project.

By then, the peripatetic Edwards was about to leave anyway. He did a short stint at Glasgow University, then rejoined Parkes, who had moved to the Physiological Laboratory at Cambridge University. That was 1963. At Cambridge, Edwards received a steady supply of eggs—from cows and sheep felled at the local slaughterhouse, from laboratory monkeys, from gynecologists at nearby Addenbrooke's Hospital.

By 1965, Edwards decided to attempt fertilization of a human egg. The twist was, he wanted to use his own sperm, gathered fresh, in flagrant disregard of the paradigm, accepted by virtually all embryologists, that only sperm which had been exposed to the uterus (and some as yet unknown chemical priming) would do the trick. Before attempting this, he solicited the opinion of his wife, Ruth Fowler, herself a geneticist, and granddaughter of two of Edwards's scientific heroes, the revered Cambridge physicist Ernest Rutherford and the meteorologist Luke Howard. (When Edwards had first learned of Ruth's heritage, he had thought to himself, "The same genes as Rutherford, my God!") Regarding her husband's plan, Ruth said, "Best of luck." So one evening around eight, Edwards pedaled over to his lab. To the accompaniment of a dripping faucet, he masturbated into a sterile vessel, then added his sperm to the eggs in culture.[5]

Checking his experiment the next morning, Edwards disappointedly observed that no fusion had taken place. So maybe it was true. Maybe only sperm that had navigated the uterus had the wherewithal to penetrate the egg. Try again, Fowler suggested. Enlist outside help. Edwards took her advice and arranged, through Victor McKusick, the eminent Johns Hopkins University Medical

School geneticist, to spend a six-week sojourn in Baltimore. Well supplied with eggs for the next month and a half from the busy Hopkins obstetrics and gynecology department, headed by Howard Jones, Jr., Edwards tried one approach to fertilization after another, to no avail. Finally, toward the end of his stay, he decided to shift focus slightly and concentrate simply on prodding eggs into resuming meiosis in vitro. Here, he hit upon the right combination. When one of the surgeons would hand him an ovary or a piece of an ovary, he would wash it in saline solution containing an anticlotting factor, heparin. With a scalpel, he would cut the largest follicles from the surrounding tissue and puncture them to release their oocytes. These he cultured in a standard medium fortified with fetal calf serum and laced with penicillin and streptomycin, since any bacterial growth would be fatal.

In all, Edwards prepared 250 oocytes in this fashion and was able to inspect 133 of these under the microscope. He began checking their progress twenty-five hours after culturing them, and looked in on them at frequent intervals until fifty-two hours had passed, making a final check between fifty-four and seventy hours. After twenty-five hours, 89 oocytes appeared to be resuming meiosis, and three hours later 71 had undeniably done so, a remarkable 79 percent of them (Edwards rounded up to 80 when doing his calculation). Edwards was surprised to note that no matter what phase of the monthly cycle the ovary had been in when removed from the patient, the eggs all proceeded at the same rate through meiosis. This struck him as extremely advantageous for in vitro efforts, since most females had generous supplies of immature eggs, but mature eggs were only yielded up parsimoniously. Edwards thought that his procedure for maturing eggs in vitro would also aid embryologists who wished to probe genetic abnormalities. They could culture immature eggs and carefully monitor them to determine whether an error called nondisjunction—wherein one or more chromosomes cling together rather than break apart during one round of the meiotic square dance—causes polyploidy, as had been theorized.

Edwards rushed an account of his Hopkins work to the British journal *Lancet* and was taken aback when the editor rejected the first draft: But this was groundbreaking stuff! A dogged Edwards phoned the editor, who suggested a number of changes, then printed the piece in the November 6, 1965, issue. Edwards got his first fifteen-minute increment of fame (many more would follow). The *Times* of London trumpeted his success in an article which was headlined "Births May Be by Proxy" and invoked Huxley's *Brave New World*.

Over the next several years, Edwards honed his methods—mostly by jiggering culture media for sperm, eggs, and embryos—and published a number of reports in the *American Journal of Obstetrics and Gynecology, Science, Nature,* and *Scientific American*. Exposure in such high-profile periodicals assured that advances in in vitro fertilization would be marked closely by the general press, whose reporters were astute enough to perceive the stunning implications behind dry pronouncements like, "There may be certain clinical and scientific uses for human eggs fertilized by this procedure."

Various people claim to have introduced Edwards to his future partner, the gynecologist Patrick Steptoe, or to have been present at their first meeting, but according to both parties, their initial contact was a brief and rather inconclusive phone call placed by Edwards in 1967 regarding Steptoe's pioneering work with laparoscopy.

The laparoscope, an instrument consisting nowadays of fiber-optic tubing, a low-heat halogen bulb, and a probe which can be fitted with cutting and grasping tools, evolved from somewhat cruder endoscopes that physicians had used from the early 1800s on to scan the interior of the abdominal cavity. In the 1920s, a Bristol, England, surgeon devised a method of distending the abdomen with a gas to separate the organs and improve the view afforded the endoscope. Twenty years later, a New York gynecologist began threading an endoscope through the vagina to examine the pelvic viscera, a procedure called culdoscopy.

Steptoe became interested in learning this procedure in the early 1950s, after taking up a post at Oldham General Hospital outside Manchester. Often, after submitting patients to exploratory surgery, he had spied no sign of pathology. He began to wonder whether culdoscopy might offer a way of visualizing the pelvic viscera without the trauma of laparotomies, and in 1958 made a special swing through Boston and New York to observe colleagues performing the procedure. Unfortunately, the trip proved disillusioning. As far as he was concerned, culdoscopy was clumsy and provided an unacceptably murky view of the cavelike pelvic interior.

Through the professional grapevine, Steptoe heard that a French physician, Raoul Palmer, was trying out a far more serviceable procedure called laparoscopy. The abdominal cavity would be pumped full of an inert gas like carbon dioxide. Then the physician would thread the view piece of an endoscope through a slit above the navel and the knitting-needle-like operative instrument through another slit in the lower abdomen. Steptoe paid Palmer a professional visit, and in 1959 heard him speak at an international conference held in Amsterdam. There, a film on laparoscopy by Hans Frangenheim, an innovative West German physician, was also screened. Steptoe went back to Oldham and gained facility with the laparoscope on thirty corpses before attempting to insert it into a live patient, a nurse with unexplained abdominal pain, who patiently granted Steptoe another chance after his first attempt failed. The second try gave him a clear view of her ovaries, fallopian tubes, and uterus, all perfectly healthy. The nurse wouldn't need further surgery, and notwithstanding Steptoe's initial botched performance, that news had come at the price of only a few inch-long scars on her abdomen, rather than an eight- or nine-inch gash.

Palmer and Frangenheim further refined the laparoscope, miniaturizing the light source and substituting fiber optics for the conventional lens and tubing of the endoscope. In 1964, gynecologists gathered in Palermo, Sicily, for the first international symposium on laparoscopy, and Frangenheim unveiled the newly streamlined model. Also while there, Palmer and Frangenheim convinced Steptoe that he should try performing tubal ligations for sterilization by laparoscope. Requiring only a couple of cuts, one version of tubal surgery could be easily accomplished, they contended. Back at Oldham, Steptoe followed up on their suggestion and successfully operated on fifty women who had come in to have their tubes "tied."

Steptoe was on the verge of publishing a textbook on laparoscopy in 1967 when Edwards phoned, asking for a favor. Edwards had heard about Steptoe's work and realized that laparoscopy might allow him to obtain sperm that had been exposed to the female reproductive tract—sperm that might be capacitated and capable of fertilizing eggs in vitro. Would Steptoe be willing to let him have access to segments of fallopian tube that were excised during tubal surgery?

Steptoe didn't see why not.

Well, would he be willing to ask his patients to have intercourse the night before surgery, so that sperm might be recovered?

This might be somewhat delicate, but Steptoe was willing to see what he could manage.

Having negotiated a tentative agreement with the Lancashire physician, Edwards hung up; however, with one thing and another, he didn't get around to calling back. The following year, he traveled to London to attend a Royal Society of Medicine meeting. Among other things, the assembled physicians were interested in the possibility of using superovulation drugs to treat women whose infertility seemed to stem from ovarian malfunction. The physicians agreed that when attempting to chemically goose multiple follicles into ripening, it would be highly desirable to be able to see the ovaries, so that one could track their progress. Someone suggested that laparoscopy might be a relatively easy way of monitoring the situation. Edwards recollected that a "distinguished-looking gentleman" immediately pooh-poohed this idea, insisting that a laparoscope was useless for viewing the ovaries. But a "thick-set grey-haired man" at the back of the hall "leapt to his feet, evidently impatient with the speaker. He did not actually say, 'Rubbish,' but his remarks were pungent and direct."[6] He reported that in his practice he had found the laparoscope more than adequate to the task. In the right hands, it gave excellent views of the pelvic cavity and the organs therein. After the meeting, Edwards sought out the man and introduced himself: It was, of course, Steptoe. Thenceforward, the two men forged an intense partnership, which lasted until Steptoe's death in 1988.

Soon, Edwards made the 180-mile drive from Cambridge up to Oldham to plot a course of action. Edwards pointedly noted in *A Matter of Life,* a chronicle he cowrote with Steptoe about their collaboration, that "ethical concerns hardly entered our conversation." From their perspective, IVF would not only be a boon to the thousands of women in Britain with tubal blockage, which rendered them incapable of conceiving normally, but also usher in an age in which medicine might address disease at the level of the genes, prenatally. Fertilization in the lab would enable the researchers to "examine a microscopic human being—one in its very earliest stages of development—and as a result of our scrutiny we could well gain new knowledge about genetic disorders." Embryos fertilized in the lab could be discarded if flawed. At some future date, those that appeared likely to develop abnormally might be doctored before implantation. Edwards had already considered the possibilities and decided that embryonic and genetic engineering "are challenges which we should not fear, though we must be on our guard against abuses."[7]

At a December 1968 lecture given with National Law Center professor David Sharpe at George Washington University in Washington, D.C., Edwards argued that no laws then on the books could be construed as preventing embryologists from pursuing IVF. Antiabortion statutes did not apply, since lawmakers had "provided criminal penalties only for destroying or attempting to destroy implanted foetuses." The death of a blastocyst in a petri dish could not be equated with an induced miscarriage. Moreover, even if destroying a blastocyst should be deemed illegal, there seemed no impediment to eugenic sorting of blastocysts; in fact, "the opportunity to give the mother a healthy embryo seems fundamentally humane." To achieve this end, even direct interventions might be warranted, including modifying an embryo's genetic material to eliminate disease-causing genes, or combining genes from two individuals to yield a hybrid, or chimera, expressing traits of both.

Edwards and Sharpe did voice one major reservation. Theoretically, in vitro fertilization would facilitate the cloning of eggs, by which Edwards and Sharpe meant the removal of a nucleus from one egg and insertion of a nucleus from another (note that this is different from vegetative cloning, in which cells from an adult cactus, say, are induced to bud into genetically identical little cacti). Scientists had already carried out such nuclear replacement with frog and mice eggs, and in the process shown that blastomeres were all alike, not yet displaying any specialized character. In fact, it was possible to rob a blastocyst of a blastomere or two without fazing it in the least; it would go on robustly cleaving and could develop into a normal organism. But performing nuclear replacement with human embryos struck Edwards and Sharpe as not only unlikely but also suspect:

It is highly doubtful that scientists or doctors could know in advance of any social benefits arising from these methods when we have so little information about genetic and other factors that influence complex human characteristics. It is also questionable that the donor's characteristics would be totally inherited in the offspring, since even small mutations in the donor's cells or uterine effects might have subtle effects on the offspring.[8]

Moreover, it could be asserted that each child has "a right to be an individual," not a copy of some other being.

Cloning aside, though, "eugenic techniques providing healthy births" would, Edwards and Sharpe believed, "enhance rather than harm the sanctity of life," and both authors firmly opposed any attempts by government to prohibit embryological research. Without dissection of human corpses, which had been deemed criminal for centuries in medieval Europe, medicine would have remained at a rudimentary level. So, too, banning embryological research would impair scientific and medical progress. Better for both society and scientists would be a laissez-faire stance which permitted biologists to police themselves. Failing this, a nonpartisan oversight body which seated a range of interested professionals—"doctors, scientists, lawyers, authors, and other laymen"—could be assembled: "It would frame public debate, act as a watchdog, and yet interfere minimally with the independence of science."[9]

In fact, although Edwards and Sharpe failed to mention the U.S. National Institutes of Health, this body, through its process of evaluating applications for federal grant money, was at the time already performing similar functions in the medical research arena. It maintained, however, a rather lower profile than Edwards and Sharpe seemed to desire. They wanted more, not less, debate on the issues and urged scientists to embrace, rather than shrink from, conflicts, even if disclosing the nature of their work seemed to "run against their immediate interests." To prevail, "they may have to stir up public opinion, even lobby for laws before legislatures, in the hope that the attitudes of society as evidenced in its laws will mature at a rate not too far behind the transition of scientific discovery into technological achievement."[10] Reform might be hard-won, but surely biologists could succeed in convincing people that research on early embryos would provide benefits that far exceeded any potential hazards.

Over the next decade, Edwards and Steptoe, aided by graduate students and technicians in Cambridge and by nurses and staff at Oldham, struggled to systematize fertilization and to get the timing right on egg harvesting and embryo transfer. They tinkered with the laparoscope, fitting it out with a hollow metal tube the diameter of a fine-gauge knitting needle. With its beveled tip, Steptoe could pierce a swollen follicle and, applying a light suction, draw off the fluid within—and the sought-after egg.

So that each retrieval might garner more than one ripe egg, Steptoe initiated the practice of superovulating his volunteers. He dosed them with human menopausal gonadotropin three times between the third and ninth days of the menstrual cycle to promote the growth of follicles, then with human chorionic gonadotropin between days nine and eleven to trigger egg maturation. About thirty hours after the HCG injection, the volunteers were anesthetized so that Steptoe could insert the laparoscope and suck ripened eggs from the follicles, which were on the verge of bursting.

Edwards logged thousands of miles zipping up the M roads to Oldham and back, often accompanied by his lab assistant Jean Purdy. Frequently, he ferried eggs and embryos between the two places, Lancashire clinic and Cambridge lab.

In between road trips, Edwards doggedly kept at the capacitation problem, aided by a graduate student, Barry Bavister. Edwards had already undertaken to capacitate human sperm by passing them through the reproductive tract of a rabbit and two species of monkey, which turned out to be ineffective. (He had also transferred human embryos into rabbits in hopes of overriding their tendency to stall out in culture medium after a few cleavages: more evidence of the relative indifference of reproductive researchers to species boundaries.) But experiments by Min Chueh Chang and others had revealed that if sperm were taken from the epididymis of a hamster and cultured with follicular or tubal fluids, they would shed their protective caps, readying themselves for an assault on the zona pellucida. This suggested that the theoretical guideline might be wrong. Uterine factors might not be essential for capacitation. Accordingly, Bavister doctored a basic culturing medium of Tyrode's solution (a medical staple made up of chlorides of sodium, potassium, calcium, and magnesium; sodium phosphate; bicarbonate of soda; and glucose) with extra bicarbonate of soda and pyruvate, an energy source, added a dash of follicular fluid and a few other ingredients, and tried letting eggs and sperm sit in this mix both separately and together.

The mixture worked its magic. Of fifty-six eggs Bavister cultured and inseminated in it, thirty-four underwent fertilization within twelve hours and began developing—a 60 percent success rate, far and away the greatest anyone had ever seen. Dated the day after Valentine's, 1969, the four-page *Nature* article presenting his results, accompanied by remarkable photomicrographs in which sperm can be seen arcing through the zonae pellucidae of eggs like space probes heading into the soupy atmospheres of gaseous planetoids, caused a tremendous stir. It galvanized the scientific community, some of whose members questioned the claim that fertilization had taken place. A letter from the Cambridge biologist Lord Rothschild in the March 8 *Nature* asserted that it probably had not. His doubts, Lord Rothschild wrote, "in no sense imply that I disapprove of the work; nor that I do not admire it as a preliminary experiment. . . . Nevertheless, the claim to have fertilized a human egg outside the mother . . . is premature." The team fired back a testy reply, charging Rothschild with having misstated their case, among other things.[11]

But Edwards, Bavister, and Steptoe were less well equipped to fend off attacks from the likes of the archbishop of Liverpool, who condemned their work as "morally wrong."[12] Nor could they control the broader public response. An article in the London *Times* anticipated a test-tube baby race that would rival the space race, in which countries vied to improve their "wealth and influence" by adopting policies of selective breeding.[13]

Some critics worried about the potential dangers of vegetative cloning, of taking single cells from an adult organism and manipulating them in such a way that a whole army of genetically identical organisms would arise. Edwards had grown extremely weary of these speculations, considering them overly pessimistic and irrational. "The whole edifice of their argument is fragile," he wrote in *A Matter of Life*, "that nuclear physics led inevitably to the atom bomb, electricity to the electric chair, civil engineering to the gas chambers. Surely acceptance of the beginning does not necessitate embracing undesirable ends?"[14] Other defenders of IVF also belittled such fears, yet it was not, after all, journalists who had come up with such scenarios in the first place, but scientists like Joshua Lederberg, who gave every impression of being rather smitten by the prospect of cloning. Reporters and commentators weren't, at least for once, guilty of ginning up hysteria on the basis of their own ill-grounded fantasies.

Around this time Edwards turned down a request by a BBC producer to appear on a program about "the increasing control over early life." When it aired, it included footage of the building housing Edwards's lab, with a voice-over saying, "It's only a few hundred paces down the road from this laboratory that Rutherford split the atom—and we all know what that led to." Watching the show, Ruth Fowler commented, "Good job they don't know our relationship to Rutherford."[15]

A second program appeared in February 1970. In this instance, Steptoe had cooperated, but both he and Edwards felt ill-used by the filmmakers. Steptoe had mentioned that he had successfully fertilized an ovum taken from thirty-four-year-old Sylvia Allen. When the filmmakers tracked her down, Allen told them she expected to have the embryo transplanted in the next "two to six weeks"—an indication that she was either misinformed or confused. A précis of the show went out over the Reuters wire, sparking international interest, while in Britain the denizens of Fleet Street fanned a national debate in which representatives of the major scientific, religious, and political institutions urged a wholesale review of IVF research. Amid the roiling waters, Edwards stood firm: "I had no doubts about the morals and ethics of our work. I accepted the right of our patients to found their family, to have their own children."[16]

More than anything, Edwards and Steptoe wanted to nurse an embryo in culture to the eight- or sixteen-cell stage. Because embryos naturally reach the sixteen-cell stage immediately prior to implantation, this seemed the ideal point at which to make the transfer from petri dish into a waiting womb. Steptoe could wield his laparoscope like an artist; now Edwards and his crew needed to come through with blastocysts capable of sustained growth. Once again, they tried a range of fertilization media, Bavister's lucky brew included, adjusting them by fiddling with the pH and reducing sodium content. They also reviewed the very latest findings on capacitation and analyzed the composition of follicular fluid, measuring estradiol, LH, and progesterone levels with the ultimate aim of determining whether capacitation varied depending upon concentrations of these steroids.

Too, they harkened to reports coming out of American laboratories concerning the best medium for meeting the metabolic requirements of cleaving eggs. One such report, authored by Johns Hopkins University researchers Joseph Kennedy, a gynecologist, and Roger Donahue, a medical geneticist, had been published in the journal *Science* in June 1969. Kennedy and Donahue, following up on studies with mice which had been performed by their Hopkins colleague John Biggers, had cultured 426 oocytes in several media, including Krebs-Ringer medium, a saline solution to which pyruvate had been added; and a standby of cell biologists known as Ham's F10 medium, to which bovine serum albumin, a protein derived from cow blood, had been added. Almost half the oocytes suspended in the modified Krebs-Ringer and Ham's F10 media geared up and resumed meiosis; a significant portion even reached metaphase II, normally attained right before ovulation. Apparently, the requirements of cleaving embryos were not terribly exotic. They needed ample nourishment, but it could be supplied through commercially available media supplemented with a few easily obtainable substances.

Early in 1970, Steptoe corralled a group of forty-nine infertile women who had agreed to assist his research, hoping one day to conceive a child with the aid of his unorthodox methods. He submitted them to the gonadotropin regime and was able to retrieve eggs from twenty-nine of them. Steptoe and Purdy, having set up a small lab on the Oldham premises, then took over, incubating the women's eggs for up to four hours, washing them twice in one of the culture media, and placing them in droplets of medium containing follicular fluid and sperm gathered from the women's respective husbands, which had been centrifuged to get rid of cellular debris. Within a few hours, the cumulus cells surrounding the eggs drifted away. After another ten to thirteen hours, Edwards and Purdy placed the fertilized eggs in one of several test media whose formulas seemed suited to meeting the metabolic requirements of cleaving eggs. Altogether, thirty-eight underwent cleavage, eleven of these to the eight-cell stage. Eight others contained eight to sixteen cells when examined, and three had more than sixteen.

This achievement brought Steptoe and Edwards one giant step closer to their goal. It now seemed eminently feasible to transfer a lab-made embryo and expect it to survive, although whether such an embryo would progress normally after implantation remained unknown. Despite the uncertainties, Edwards and Ruth Fowler, in a December 1970 article written for *Scientific American,* billed IVF as a means of alleviating both infertility and fetal abnormalities. In vitro fertilization, they asserted, opened the way for couples to make eugenic decisions regarding embryos. They imagined that several embryos would be grown for each couple and then selected for replacement according to their characteristics. They pointed out that "choosing male or female blastocysts is one possibility that has already been achieved with rabbits. . . ."[17]

To sex a blastocyst by methods developed in the Physiological Laboratory in Cambridge, a researcher had to hold it in place under the microscope by positioning a pipette against the zona pellucida and applying mild suction. Then, using another pipette, he would pluck out a small segment of the zona pellucida and trophoblast. After staining the trophoblast cells, he examined them under the microscope to see whether they contained sex chromatin. Edwards and Fowler wrote that this procedure could help eliminate sex-linked genetic diseases like hemophilia. "Placing female embryos in the mother would avoid the birth of affected males, although half of the female children would be carriers of the gene."[18]

Although identifying genetically faulty embryos posed major challenges which had yet to be addressed, Edwards and Fowler implied that one day only those embryos found upon inspection to be free of defect would make the transition from lab dish to uterus. They envisioned other bioengineering feats, such as the splitting of blastocysts (yet a third type of "cloning," along with nuclear transplantation and vegetative propagation), and the colonization of one embryo with trophoblasts taken from another, which would presumably produce a fetus combining select traits of both. While aware that such experiments would present "challenges to a number of established social and ethical concepts," Edwards and Fowler contended that the main criterion for assessing the work should be "the rewards that [it] promises."[19]

A round this time, Edwards and Steptoe intended to shift their clinical setup from Oldham to a hospital near Cambridge. This would have greatly simplified Edwards's life, sparing him all those miles on the highway, but it was not to be. In order to pull off the move, they required funding from Britain's Medical Research Council, but their application had been flatly rejected. The council informed them that their proposal to couple IVF with embryo transfer was "premature in view of the lack of preliminary studies on primates and the present deficiency of detailed knowledge of the possible hazards involved."[20] Members of Parliament, unwilling to accept the continual assertions by Steptoe and Edwards that their only intent was to help infertile women realize their dreams of having a baby, declaimed against the work, insisting that it receive no government support. Facing a financial squeeze, the pair sought other backers. The local Oldham health authority came through, providing lab space and treatment rooms at Dr. Kershaw's Cottage Hospital in the suburb of Royston. In addition, several American foundations began underwriting the research, and Steptoe funneled the fees he collected from performing abortions into the effort. But Edwards would still be forced to shuttle back and forth to the Midlands.

Throughout 1971, opposition to IVF and embryo transfer grew. Medical ethicists, theologians, lawyers, and even some physicians raised objections in professional forums as well as in the popular media. For instance, Luigi Mastroianni, head of obstetrics and gynecology at the University of Pennsylvania School of Medicine, who was himself involved in embryo research, maintained that fertilization in vitro "must be looked upon solely as a biological experiment."[21] As far as he was concerned, no one was yet in a position to carry out IVF and embryo transfer in humans. He told a *Washington Post* reporter,

> It is my feeling that we must be very sure we are able to produce normal young by this method in monkeys before we have the temerity to move ahead in the human. . . . Then we can describe the risk to the patient and obtain truly informed consent before going ahead. We must be very careful to use patients well and not be presumptuous with human lives. We must not be just biologic technicians.[22]

James Watson, the biologist who with English physicist Francis Crick had in 1953 divined the molecular structure of DNA, stunned the scientific community with his Cassandran pronouncements regarding advances in embryology. At an October panel discussion sponsored by the Kennedy Foundation in Washington, D.C., Watson stared straight into Edwards's face and declared the inevitable losses of embryos in his experiments tantamount to infanticide. Watson also felt there was an element of dishonesty in IVF and embryo transfer. As he quite rightly pointed out, these procedures were not therapeutic. Through their auspices, infertile women might have a child, but they would remain infertile. Thus, at a fundamental level, physicians pursuing this line of work could be said to be engaged purely in experimentation.

Watson wasn't the only one on the panel to attack Edwards, who sat on the podium with a mounting sense of indignation. Medical ethicist Leon Kass, himself an M.D. and a member of the National Research Council's Committee on the Life Sciences and Social Policy, also roundly questioned the acceptability of IVF and embryo transfer. The following month, Kass reiterated his argument in the *New England Journal of Medicine*. Kass reasoned that these procedures could not be considered therapeutic because they did not cure women of infertility. Furthermore, while impaired fecundity— that is, the incapacity of a man to inseminate a woman, or of a woman to conceive and carry a pregnancy to term—might be a symptom of disease, past or present, the inability of a couple to produce a biological child should by rights be considered a social, not a medical, problem. What IVF and embryo transfer "treat" is the "perfectly normal and unobjectionable desire" of a woman to bear a child. But such a desire hardly justified "the use of the new and untested technologies for initiating human life."[23] From Kass's point of view, IVF researchers had crossed the line. They were using women primarily, and their spouses secondarily, as experimental subjects, along with the "potential human subjects" who were being concocted in the lab, who could not possibly give informed consent. Since the Nuremberg Trials, practitioners have had the obligation to fully apprise patients concerning the nature and possible risks of any medical treatment. Ignoring or skirting this responsibility constitutes a major ethical breach.

Kass contended that "serious questions can be raised about the safety of the manipulations" which eggs and embryos underwent. No one could say with any certainty that these would not result in deformities or malformations in a fetus conceived as a consequence of them. Kass thought that Steptoe and Edwards, Landrum Shettles, and others in the field seemed insufficiently concerned about this prospect. In fact, Kass feared that a race was on among scientists to achieve the first embryo transfer, and that, as a result, "the swift will throw caution to the more sober, and will trust to luck that their victory in the race does not issue in a deformed or retarded child."[24]

Edwards and others pointed to results in lower mammals as evidence that embryos crafted in vitro would weather transfer into the uterus and one day emerge as bouncing babies, but such evidence struck Kass as inadequate. No one had fully assessed the risks of the technologies—in fact, given the state of embryological knowledge, there was probably no way of "finding out in advance whether or not the viable progeny of the procedures of in vitro fertilization, culture and transfer of human embryos" would emerge healthy and intact, without physical or mental impairments or genetic abnormalities.[25] Consequently, Kass wanted to see those in the field enact a self-imposed moratorium until the safety issues were settled through extensive studies with primates and other mammals. By no means should decisions regarding IVF and embryo transfer be left in the hands of scientists and physicians alone.

The matter, touching on the entire human family, warranted deliberation by as wide a spectrum of people as possible.

At the Washington, D.C., meeting, Princeton University theologian Paul Ramsey seconded the remarks of Watson and Kass, arguing that IVF and embryo transfer were misbegotten and unethical. He asserted that "it is not a proper goal of medicine to enable women to have children and marriages to be fertile by any means," especially if those means might entail risks to the "child not yet conceived." To hold the would-be parents' right to have children above the right of the future child to be born healthy was an untenable position.[26]

In vitro fertilization, Ramsey said, should properly be considered a manufacturing process, with its focus an end product, subject to quality control. Ramsey wondered, "If medicine turns to doctoring *desires* instead of medical conditions . . . is there any reason for doctors to be reluctant to accede to parents' desire to have a girl rather than a boy, blond hair rather than brown, a genius rather than a lout, a Horowitz in the family rather than a tone-deaf child . . . ?"[27] This would be "the triumph of manufacturing over parenthood." Because researchers could not exclude the possibility that IVF would harm the "child-to-be," Ramsey further demanded that IVF be prohibited, for then and anon, on moral grounds—a clarion call echoed in May the following year by the American Medical Association.

Edwards had no truck with such absolutist thinking. Ramsey, he considered, "abused everything I stood for."[28] Howard Jones, the Hopkins gynecologist who had provided Edwards with aid and comfort for those six critical weeks in 1966, got to his feet to deride Ramsey's proposal for a moratorium. Engaging in a bit of histrionics, he compared the public trials of Edwards and Steptoe to "the tribulations of Galileo."[29]

Edwards, a self-described "blunt Yorkshireman," exhibited no difficulty defending himself. Asked to reply to his critics, he accused Ramsey of "taking up an ethical stance that is about one hundred years out of date and one that is totally inapplicable to meet the difficult choices raised by modern scientific and technological advance." According to Edwards, he had gone on to say that "dogma that has entered biology either from Communist or from Christian sources has done nothing but harm . . . ," when the audience leaped to its feet in thunderous applause.[30]

With that clapping perhaps still echoing in his ears, Edwards made his way to Oldham a week or so later. He and Steptoe agreed that the time had come to essay a transplant. They proceeded immediately, drafting another volunteer. By January, they knew that the experiment had failed. The woman wasn't pregnant. Mulling over what went wrong, they concluded that the gonadotropin regime they had their patients on was at fault. With each cycle, the endometrium, or uterine lining, undergoes cellular changes necessary for implantation. The gonadotropins were apparently speeding up the process, so that the endometrium had gone past its prime and was about to be shed at around the same time that Steptoe was transferring the embryos. This represented a dilemma: Without the gonadotropins, only one egg would mature each month; with them, the chances for implantation appeared unlikely. Steptoe came up with an alternative: Retrieve eggs from a gonadotropin-primed Mrs. A, fertilize them with the sperm of Mr. B, then transfer the resulting embryos to a steroid-free Mrs. B. Today, this technique is known as egg donation and offers the best hope for older women who cannot conceive with their own eggs. At the time, Edwards found the idea clever but too controversial, and nixed it.

Another possibility presented itself. A researcher at University College in London, David Whittingham, had been exposing mouse embryos to -196°C (-384.8°F) temperatures, then thawing and implanting them in foster females to get live young: Maybe he could do the same with human embryos. Edwards went to the bother of shipping an embryo down to London by fast train, to no avail. To eliminate lag time, he purchased an automated cryo-preservation machine and tried freezing the embryos himself at Kershaw's, but had no better luck than Whittingham. A few years later, Whittingham, having moved to the Physiological Laboratory at Cambridge, performed the neat trick of freezing several dozen mouse embryos there and shipping them to the Jackson Laboratory in Bar Harbor, Maine, where they sat in storage for as long as eight months before being thawed and transferred to receptive females, who bore a total of eleven young.

Over the next four years, Edwards and Steptoe tried a combination of drug protocols to get around the endometrium problem, with no success. Meanwhile, the race was indeed on. Dozens of other researchers around the world, especially in Australia and the United States, were pursuing the same goal as the two Britons. In the land down under, physician Carl Wood of Monash University and Melbourne's Queen Victoria Medical Centre recruited a crew of researchers who were interested in IVF, including clinicians and biologists working with animals. For a time, they engaged in a collaborative effort that encompassed far-flung sheep stations and the busy urban obstetrics and gynecology department of the Royal Women's Hospital in Melbourne. In Virginia, Howard Jones, retired from Johns Hopkins, and his wife, Georgeanna Jones, also a gynecologist, founded the Jones Institute with the express purpose of achieving an IVF/embryo transfer birth.

Although most of the researchers tended to follow the broad path laid out by Steptoe and Edwards, they to one degree or another blazed their own side trails. For example, several scientists and physicians in Santiago, Chile, and New York returned to Heape's method of flushing the uterus with saline solution to recover eggs. This had the advantage of being faster and less invasive even than laparoscopy. Their early attempts with humans encouraged those directly involved in animal breeding to contemplate flushing fertilized eggs from the uterus of prime females and gestating them in lesser females. Here was a rare instance in which the technology transfer went against the customary grain, from humans to animals rather than vice versa. By 1981, embryo transfer in cattle was a $20 million a year business, with total births from the procedure nearing one hundred thousand in the United States, and twice that worldwide. Shipped in the uteri of rabbits, embryos crossed the oceans in the holds of airliners, to be implanted in local dams, who would convey to the growing fetuses immunity to local pathogens.

Partway through 1975, excitement gripped the Steptoe and Edwards group. A volunteer's pregnancy test had come back positive. But soon Steptoe realized something was wrong. The embryo had lodged in the fallopian tube and had to be aborted. Everyone was demoralized.

After a second pregnancy failed, Edwards wondered whether something might be wrong with their culture media. Testing them, he discovered that liquid paraffin, which was used in almost all embryo work and seemed harmless in other species, was toxic to human embryos. In a remixed medium, he grew an embryo for nine days in vitro, until it was about one-sixteenth of an inch across.

By 1977, Steptoe and he had finally decided to jettison the fertility drugs. They would let the woman's natural cycle unfold, settle for one egg, and trust that they could produce one viable embryo which would take hold. Steptoe, suffering from degenerative arthritis, planned to retire in June the following year, and Edwards felt the pressure acutely. Would they accomplish their goal before then?

Enter Lesley Brown.

Lesley Brown came from a broken home. At sixteen, at a sailor's café on the Bristol docks, she met John Brown, a sometime truck driver who was estranged from his first wife, with whom he had had two daughters. Lesley lived with John for six years before they married. At age twenty-two, she visited a fertility specialist, depressed because she had not conceived after years of trying. The diagnosis was tubal blockage, and the doctor told her the chances were a million to one against her ever having a baby. Perversely, Lesley became from that time on obsessed with children. She read about them, tuned in shows about them on TV, considered adoption. On a referral, on a gamble, she made the long trip to Manchester to see Patrick Steptoe at his office. Manchester's St. John's Street, the equivalent of London's tony Harley Street, where the private practitioners cater to those who can afford to pay for medical care instead of relying on the National Health Service, cowed Lesley and John. So did the distinctly non-working-class price Steptoe quoted for the two "small operations" that might bring her a child. In the end, they were only able to afford to go ahead because John hit the football pools.

Steptoe was elliptical. He didn't inform Lesley and John that IVF had never yet worked. Lesley later recounted, "I just imagined that hundreds of children had already been born through being conceived outside their mothers' wombs."[31] The intricacies of reproduction were not within Lesley's immediate bank of knowledge, and John was momentarily befuddled by Steptoe's use of the word "semen."

To ensure that no ectopic pregnancy resulted, Steptoe performed a tubal ligation on Lesley in August 1977. A few months later, sworn to secrecy by Steptoe, she and two other women entered Dr. Kershaw's Cottage Hospital for egg retrieval. Steptoe found no follicle in the first ovary and had trouble reaching the second, but persevered, and a nurse carried the fluid that had drained from the follicle into the adjacent embryo lab, where Edwards and Purdy confirmed, yes, they had gotten an egg. John Brown did his manly duty, and within hours, egg had become embryo.

A few days later around midnight—Edwards and Steptoe had this theory that bodily rhythms were most propitious for implantation at the witching hour—Lesley was wheeled into the operating room. Supine on the table, knees up and spread, legs in stirrups, she was conscious during the procedure. Steptoe threaded a duckbill speculum into her vagina, spread the blade, and exposed her cervix. He swabbed the mucus from it with culture medium. This was to ensure that the catheter penetrated cleanly. (Sometimes cervical mucus is so viscous that it stretches and doesn't break, and the embryo never makes it into the uterus.) Meanwhile, the embryo was being loaded into a catheter. Steptoe then threaded this long, thin, flexible plastic tube through Lesley's cervix, pushing it up into the cavity of her uterus until the tip lay only a short distance from its arching roof. There he paused, and slowly depressing the catheter's plunger, deposited the embryo. He let the catheter remain in that position for a few moments, then cautiously withdrew it. The catheter was hurriedly checked under a microscope to be sure the embryo had been expelled. Lesley said, "That was a marvelous experience," and then was wheeled back to bed.[32]

Back at Cambridge afterward, Edwards got bulletins on Lesley's condition. At two weeks, her urine and blood tests looked good. After twenty days, Edwards thought he spied a blood-stirring blip in progesterone levels, a faint sign that the embryo had taken. On December 6, the results were unequivocal:

> Dear Mrs. Brown,
>   Just a short note to let you know that the early results on your blood and urine samples are very encouraging and indicate that you

might be in early pregnancy. So please take things quietly—no skiing, climbing, or anything too strenuous including Xmas shopping![33]

Lesley Brown, hardly a candidate for impromptu schussing or rappelling, read Edwards's letter over and over, elated.

Edwards, while thrilled, was almost more concerned with cementing the team's reputation. During the first fortnight of 1978, he and Purdy screened sixteen additional women for IVF and embryo transfer, in a rush to establish at least three more pregnancies before Steptoe's official departure.

Steptoe, meanwhile, kept close tabs on Lesley Brown. He tried to get her to stop smoking, and although she cut back, she still cadged the occasional cigarette. At week sixteen, without preparing her ahead of time, Steptoe performed amniocentesis, which revealed no chromosomal abnormalities and measured normal levels of alpha-fetoprotein, a sign that the fetus's neural development was proceeding without a hitch. Steptoe had fully intended to urge an abortion upon Lesley if any anomalies had been evident. Lesley felt "tricked" by his failure to let her know that she would be having amniocentesis, but was at the same time satisfied that Steptoe had acted in her best interest.

Secrecy, though, seemed to be the theme of the hour. In San Francisco for a meeting of laparoscopists prior to the amniocentesis, Steptoe kept mum about the pregnancy. Ostensibly to prevent leaks by staffers to the tabloid press, which had been alerted to the pregnancy by an article in the *New York Post* and had descended on the hospital offering cash for insider information, he had Lesley Brown checked in and out under the assumed name "Rita Ferguson." He also falsified her official medical docket and kept the legitimate notes on the case in a private diary.

At six months, Steptoe put his arm around her and said, "You're going to do it for us, aren't you, Lesley?"[34] By that time, three other women had conceived after embryo transfer at Kershaw's, although two of the pregnancies would end unhappily, one by miscarriage at five months. Steptoe aborted the other fetus because of Down syndrome. Reporters from the daily press continued to besiege the hospital and tracked down the Browns at home. On Steptoe's counsel, the couple offered the London *Daily Mail* an exclusive on their story for a sum of around twenty thousand pounds, much of which went into a trust fund for their yet-to-be-born child.

As Lesley's due date approached, Steptoe decided not to risk a natural delivery. The day before her caesarean section, she was in her room in Oldham's maternity ward, watched over by Jean Purdy, when Steptoe came in. Quoting Telly Savalas's television cop, Kojak, he said to Lesley, "Who loves ya, baby?" At eleven-thirty the next night, July 25, 1978, Steptoe hoisted a five-pound twelve-ounce baby girl out of the incision in Lesley's abdomen. A miracle, Lesley and John pronounced her. They named the baby Louise and asked Steptoe and his wife to choose her middle name. For Steptoe that was easy: He could think of one word that about said it all.

Edwards and Steptoe held a press conference announcing the birth of Louise Joy Brown. Caught in the lens of a photographer for the *Daily Telegraph*, Edwards, wearing a white shirt with a long-pointed collar and a wide tie whose paisley pattern is strangely evocative of sperm, looks a bit of the didact. Mouth open, he holds one finger to heaven, as if lecturing the crowd. He wears his dark hair, starting to gray, longish, over the ears. To his right, Steptoe stands, an elder Michael Caine, complete with wavy hair and sixties-era black plastic spectacles. Flush with their success, they appear nonetheless a bit beleaguered, on the defensive. Perhaps a truer read on their emotions can be had from a photo taken in the operating room moments after Louise came sideways into the world. Swathed in surgical scrubs, caps, gloves, and masks, Steptoe and Purdy gaze down at the blanketed infant, who is being held somewhat gingerly by Edwards. Edwards stares out at the photographer, his eyes glittering with triumph.

The letter to the editor of the *Lancet* was old news when it was published on August 12. "Sir," wrote Steptoe and Edwards, "we wish to report that one of our patients, a 30-year-old nulliparous married woman, was safely delivered by caesarean section on July 25, 1978, of a normal healthy infant girl weighing 2700 g." Already, the whole world knew.

Hard on the heels of their victory, Steptoe received a communiqué from the Barren Foundation. The Chicago-based foundation, a major supporter of fertility research, was rescinding an award which Steptoe had been slated to accept in October. Given the strength of their collective high, Steptoe's reaction to the news was understandable. He told *Time* magazine that the foundation had engaged in "the most utterly disgraceful exhibition of bad manners I've come across in the scientific world." More bad manners were forthcoming in the form of hectoring from colleagues who disputed Steptoe and Edwards's claims. Scientists' acceptance of any experimental result always depends on a review of the data, and the Oldham team was charged with being too slow in writing up a proper account of its work. Oral presentations, at the Royal College of Obstetricians and Gynaecologists in January 1979 and before one thousand colleagues at the American Fertility Society annual meeting in San Francisco, did not quite suffice.

Infertile couples cared not at all for niceties of scientific protocol. To them, Louise Joy Brown constituted sufficient evidence that IVF and embryo transfer worked. Letters flooded into Oldham from around the world from women asking for information and pleading for help. In Edwards's view,

> many of the stories are sad, even pathetic, others are occasionally ridiculous and Patrick naturally finds it painful to have to disillusion patients who have not fully understood the implications of our

work. It was embarrassing, for instance, for him to have to disillusion the occasional lady several years past her menopause that there was no chance of her becoming pregnant.[35]

As it turns out, the "occasional lady" showed greater prescience, greater understanding about the implications of his work, than Edwards did. Through hormones and egg donation, physicians would, not too far in the future, enable postmenopausal women to deliver children.

Years before, while a medical student at St. George's Hospital, Steptoe had heard a patient with tubal damage who longed for a pregnancy ask an innocent enough question of her gynecologist. She had said, "Doctor, can't the fallopian tubes which you say are blocked be bypassed?" The gynecologist had replied, "Oh no," and looked at Steptoe, who had shaken his head vigorously in agreement.[36]

13

Cowboys and Controversy

The week before Louise Brown was born, a strange scene unfolded in U.S. district court in New York. Doris Del Zio, a thirty-four-year-old housewife from Fort Lauderdale, Florida, sat sobbing in the stand before a jury of her peers, agonized over the loss of her embryos. According to Doris and her husband, John Del Zio, a fifty-nine-year-old dentist, the embryos had been destroyed by Raymond Vande Wiele, chief of obstetrics and gynecology at Columbia-Presbyterian Medical Center. For the damage inflicted by Vande Wiele, the Del Zios were asking the court to award them $1.5 million.

The true protagonist in this unprecedented suit was not, in fact, Vande Wiele, but Landrum Shettles, the same medical researcher who had so enthusiastically sought to sort sperm for sex selection purposes and who had been among the earliest to attempt in vitro fertilization of human eggs. Doris Del Zio had been his patient. Doris had one child from a previous marriage, but damaged fallopian tubes were preventing her from bearing a child with John. She came to Shettles for help, and in 1972 he suggested that she let him attempt IVF and embryo transfer. The Del Zios agreed to the procedure, and after eggs were retrieved from Doris by a Cornell Medical School gynecologist who operated at Shettles's behest, John hand-carried them across town to Columbia-Presbyterian, where he produced a semen sample.

Shettles was incubating the resulting embryos in his lab and no doubt would have proceeded with his stated plan had he not been thwarted by his department head Vande Wiele, who, like the University of Pennsylvania's Luigi Mastroianni and other cautious physicians, considered it far too early in the game to implant lab-fertilized embryos in a patient. Without consulting the Del Zios, Vande Wiele removed the embryos from the incubator.

Although Vande Wiele's defense lawyers questioned whether Shettles even had the expertise to produce the embryos at issue, the jury was convinced by the Del Zios' claims that the department chief's action had caused them physical and psychological harm, and it ruled in their favor, although it assessed Vande Wiele not $1.5 million but $50,000 in damages.

Shettles, who left Columbia-Presbyterian shortly after the suit was filed and set up a fertility research foundation in New York City, was an odd case. A respected Johns Hopkins–educated physician, he also displayed a peculiar attachment to hugger-mugger. During interviews with journalists, he had a habit of revealing dark secrets concerning his clinical exploits. For example, he told David Rorvik, on assignment for the *New York Times Magazine* in 1974, that he had started implanting lab-made embryos in women as early as 1963. He further disclosed to author Vance Packard that he had "fully mastered" the procedure by 1968, but had kept his accomplishment to himself to avoid controversy.[1] In retrospect, even Shettles's naming of his 1960 monograph, which presented photomicrographs of human eggs, sperm, and embryos fertilized in vitro, looks a bit off. Evocative of earlier eras when physicians wielded Latin to keep the common people at bay, the title *Ovum Humanum—The Human Egg*—appears to be yet another manifestation of Shettles's fondness for cloaking his work in a bit of mystery.

Shettles never substantiated his claims of embryo transfers. Neither did Douglas Bevis, a University of Leeds gynecologist who at the July 1974 annual scientific meeting of the British Medical Association issued a press release announcing that three infants had been born through IVF and embryo transfer. Peppered with questions, Bevis disclosed little except that the three births had come after thirty unsuccessful transfers and that the parents of the babies, who appeared to be normal, lived in the United Kingdom and Western Europe. In refusing to reveal the names of either the physicians or patients involved, he provoked a swirl of innuendo and denouncements. Steptoe reportedly dismissed him out of hand: "I am astounded that Professor Bevis would have made this statement. As far as I know, no one in this country or anywhere else has yet succeeded in this technique."[2] True or not, Bevis's enigmatic statement added to the disquiet surrounding IVF, and shortly afterward, the Medical Research Council moved to suspend human embryo research in Britain for a time.

Shettles and Bevis failed to play by the rules of research, which require a full public accounting of experimental methods and results.[3] So, in the annals, Steptoe and Edwards go down as having achieved the first IVF and embryo transfer in humans—but not by much. A team in Melbourne, Australia, which had been pulled together by physician Carl Wood of Monash University, had in fact come very close in 1973 to besting their British rivals.

Like Steptoe, Wood had approached infertility from the surgical side, having been intrigued by accounts of failed turn-of-the-century attempts to overcome tubal blockage by grafting ovaries into the uterus. Wood himself attempted in 1969 to replace faulty fallopian tubes with plastic prostheses, but like the early transplants he had read about, these operations invariably led to infections and had to be scrapped. However, in the process of working with artificial tubes, Wood had performed some crude in vitro fertilization attempts. An encounter with Neil Moore, a Sydney University reproductive biologist who wanted to perfect IVF and embryo transfer for the sheep industry, set Wood to organizing an effort to do the same in humans. He drafted physicians at both the Queen Victoria Medical Centre, which was affiliated with Monash, and the University of Melbourne's Royal Women's Hospital. Although Wood did not have great expectations, he hoped the team would at least gain insights into abnormal embryogenesis and perhaps learn something that would lead to improved contraceptives. One of the members, Alex Lopata, soon made a research foray to Germany, England, and the United States, gathering all the information he could on IVF.

By 1973, the Melbourne group had swelled to almost a dozen physicians and reproductive biologists who were eager to carry out embryo transfers on volunteers. It looked as if they had experienced an extraordinary bit of beginner's luck when two patients appeared to conceive right off the bat. These so-called chemical pregnancies showed up as hormonal blips in urine tests but terminated within weeks. The researchers then failed to produce even such illusory pregnancies for the next seven years. At one point, despondent, the team tried to induce human sperm and eggs to fertilize in a sheep oviduct, to no avail. Wood later saw that this had probably been for the best: "In some ways we were relieved at the failure of this experiment as it may have been difficult to convince the community that a sheep was an appropriate place for human fertilization and early embryo development."[4] The unflappable Wood was given to such understatement.

Dissatisfied with the team's progress, Wood began looking for someone who might bring something extra to the embryological work and found his man in Alan Trounson. Trounson's primary interest as an undergraduate had been agriculture: He had ambitions to be a farmer. But sensing his opportunities might be limited, he had pursued a doctorate at Sydney University in embryology, then gone on to join several senior investigators who were freezing and thawing cattle embryos at Cambridge University. In 1977, Wood lured Trounson back to Australia with the offer of a job.

July 1978, of course, brought word that Steptoe and Edwards had succeeded where the Aussies had failed, and Wood and Trounson set to revamping their quality control, bringing greater systematization to the harvesting of sperm and eggs and the production of embryos. Meanwhile, Lopata, at Royal Women's Hospital, was becoming more proficient at growing embryos in culture and more reluctant to share his findings with others in Melbourne. In June 1980, Candice Reed was born at Royal Women's Hospital—Australia's first, and, by general reckoning, the world's third, IVF baby—care of Lopata and several others.

Despite the Melbourne effort's resounding success, a schism soon opened between those working at Royal Women's and at Queen Victoria. Lopata and others broke off and formed their own program at Royal Women's, while Wood stayed at Queen Victoria with Trounson at the helm of the embryology lab. Trounson further modified the handling of sperm, eggs, and embryos, while Wood developed a protocol for drug stimulation, moves which led to a series of pregnancies at the hospital in 1980.

The Australian press gave the births enormous coverage, much of it tinged with nationalistic pride, and the citizenry awarded IVF a solid vote of confidence in a poll taken the following year. Of 1,000 men and women over fourteen who were queried, 77 percent approved of the procedure, while only 11 percent disapproved. (Two polls taken in the United States in 1978 found Americans somewhat less enthusiastic. A Gallup poll of men and women revealed that 93 percent had heard or read about England's test-tube baby and 42 percent could correctly explain how IVF and embryo transfer were done.[5] Fifty-three percent said that they would resort to IVF if they were married and infertile, while 36 percent said they would not. Sixty percent agreed that IVF should be available to anyone who needed it. A similar Harris poll of 1,501 women which was conducted for *Parents Magazine* asked whether the respondents would choose IVF or adoption if faced with infertility. Fifty-seven percent said they would adopt, while just 21 percent would choose IVF. Asked whether they believed IVF should be banned until scientists had proof that it would not contribute to birth defects, 63 percent said yes.)[6]

As of 1982, Wood and his team had a backlog of some two thousand couples who were willing to wait up to two years to undertake IVF; by August the following year, seventy-five IVF babies had been born. The Monash/Queen Victoria program's successes had caught everyone's attention. Physicians from dozens of countries sought the expertise of Wood and his colleagues, eager to establish IVF programs of their own. Wood attempted to inject a note of sobriety, remarking, "It is to be hoped that publicity about programs around the world will quote success rates in terms that do not give infertile couples a distorted and overly optimistic picture."[7] However, the stampede was on, among patients and physicians alike.

Despite the popular appeal of IVF, a strong Catholic faction in the Australian government lobbied against it. Politicians who nominally supported the procedure left Wood and Trounson to duke it out by themselves, and the pair faced down hostile parliamentarians, church representatives, and feminists, who put forward multiple objections. Wood's position was that IVF was unexceptionable. To argue against its "unnaturalness," he said, was to argue against the entire enterprise of medicine, which was, after all, "always about overcoming nature's defects." Neither did Wood have qualms about embryo research, as only abnormal embryos, which lacked the ability to implant or develop, underwent study at Monash. As for religious critics, Wood took heart in a second 1981 poll which showed that around half the Catholics who had been asked believed that IVF was acceptable if employed by couples who would not otherwise be able to have children. Anglicans concurred.

Trounson, for his part, grappled with the ethicality of embryo research in 1983 during a working sojourn in Italy. Colleagues there arranged for him to talk about the issue with three or four professors at a Vatican university, all priests. Although not himself Catholic, Trounson found that the priests helped him come to terms with his concerns. However, they did not persuade him to adopt the Catholic point of view. Quite in opposition to the church's stance, Trounson concluded that the technology he was helping to develop would not be a malign force in the world, but rather a benefit to the infertile.

With full confidence, then, Wood and Trounson proceeded to expand IVF's scope. Straight IVF addressed the needs of women with tubal blockage. But what about women who had no eggs, whether because of premature menopause or some other reason? Would it not be possible to obtain eggs from a healthy woman, fertilize them, and transfer them to the infertile woman? Wood and Trounson petitioned for, and received, permission from the Queen Victoria ethics committee to attempt such a scheme. The government waged an effort to block the endeavor by imposing an across-the-board ban on the donation of eggs and sperm, but backed down after Trounson threatened to resign.

Wood and Trounson established a protocol for egg donation which was widely adopted. Today, egg donors go through a standard hormonal routine to prod their ovaries into ripening multiple eggs, then receive a shot of HCG agonist, a drug that for all intents and purposes behaves like HCG at the cellular level, to bring them to the brink of ovulation. Thirty-four to thirty-six hours later, they are anesthetized, their eggs are aspirated, or sucked out by a needle poked through the wall of the swollen follicles. From these eggs, embryos are concocted. The woman who receives the embryos is given progesterone to facilitate implantation and help maintain a pregnancy if one ensues. At Melbourne, a woman whose identity was cloaked went through egg retrieval in early 1983, and in November, a second woman gave birth to a baby boy whose genetic mother was this anonymous donor.

Wood and Trounson also implemented a policy of freezing embryos as a way of getting more mileage out of each egg harvesting (and sparing women extra laparoscopies). Superovulation might yield a dozen or more eggs, and a corresponding abundance of embryos, most of which went to waste if they had to be used immediately. Since 1974, when Edwards's Physiological Laboratory cohort David Whittingham had shipped his frozen and thawed mice embryos off to Maine and gotten a transAtlantic litter, reproductive physiologists had scored similar successes with goat, cow, and rat embryos, and Wood and Trounson saw no reason that the same might not be done for humans. Accordingly, they instituted a policy of freezing any embryos that could not be transferred immediately and thawing them for subsequent IVF cycles if the first one, with "fresh" embryos, did not yield a pregnancy. Still, attrition was high: Out of 230 two- to eight-cell embryos Trounson assigned to the liquid nitrogen tanks prior to 1984, 40 were thawed. Of these, only 23 remained intact enough for implantation, a die-off rate of 57 percent.

Nonetheless, one of Trounson's embryonic survivors went on to implant and develop. In March 1984, Zoe Leyland was delivered prematurely by caesarean section—and bruited as the world's first "freeze-thaw" baby. The Melbourne team pointed to her robustness as evidence that embryos were resilient enough to withstand freezing and thawing without undergoing genetic damage. The British magazine *New Scientist* ran a story about the birth above an irreverent cartoon in which a squint-eyed, bespectacled man holding a paper titled "Embryo Research Programme" tells a buck-toothed female lab assistant, "It'll be OK with the Church so long as we freeze 'em in holy water."[8] The magazine also reported that two more pregnancies from frozen embryos were under way among Wood's patients. Shortly afterward, the state of Victoria, alert to the growing need of addressing legal issues raised by IVF, set up a committee headed by a law professor, Louis Waller, who announced that his first charge was to determine "whether freezing or thawing of embryos should proceed at all."[9] Meanwhile, the National Health and Medical Research Council suggested that no embryos should be frozen for longer than a decade.

A few months later the importance of the issue was heightened by the Rios case. A brace of frozen embryos sitting in cryopreservation tanks at Queen Victoria became the focus of an international brouhaha after the wealthy American couple who had consigned them to the custodianship of the center died in a plane crash in Chile.

Mario Rios and Elsa Rios had come to Melbourne for IVF in 1981. The Wood team had crafted three embryos from their sperm and eggs, and transferred one, which failed to take. The Rioses departed, thinking that they might return to try again. The air accident intervened. This left Queen Victoria with the orphaned embryos, and the dilemma of deciding their fate. Should the embryos go to another couple or serve as objects of research? Should they be thawed and disposed of, or kept on ice indefinitely? There was wild speculation that if the embryos were given to another couple and the woman carried them to term, the children would be rightful heirs to the

$8 million Rios estate; this, despite the fact that Australian law provided that the children would be considered the legitimate offspring of the recipient.

Catholic leaders and right-to-lifers squared off against the hospital, demanding that it appoint guardians for the embryos. They declared that under no circumstances should the embryos be thawed: Like abortion, this would be the equivalent of murder. Flying in the face of this opposition, a committee mustered by the state of Victoria suggested that the researchers go ahead and take the embryos out of the deep freeze. However, parliamentarians settled the matter by requiring the hospital to donate the embryos to another couple or keep them on ice indefinitely.

Wood and Trounson next came under siege from one of their own when Robyn Rowland, who had served as head of the committee coordinating research at Monash, Queen Victoria, and the affiliated Epsworth Hospital, resigned in a highly public way. Rowland charged that Wood and his colleagues were rushing pell-mell into the future without taking the time to assess the impact of the new reproductive technologies. Her reservations had been mounting and had not been diminished by Wood's recent claims that the twenty-five children produced through IVF at Monash and Queen Victoria displayed greater intelligence, motivation, and sociability than "normal" children; nor by his hints that IVF might one day be employed to effect further "improvements," to breed better babies by design. These eugenic speculations had given her pause, but the last straw had been the team's newly expressed intention to perform embryo flushing of the sort that had been used with livestock for several years.

The chief difference between Melbourne's modern version of embryo flushing and that of Walter Heape was the addition of hormones to synchronize and direct the cycles of the two females involved, the donor, from whom embryos were obtained, and the recipient, into whose womb one or more embryos would go. The Australians proposed to inseminate the donor with sperm from an infertile woman's husband, wait a few days, flood her uterus with a saline solution, then drain the fluid out with a catheter, thus capturing an embryo. This procedure would appeal particularly to women who knew they were carriers of diseases like hemophilia which were caused by faulty genes located on the X chromosomes. These so-called X-linked diseases manifest themselves only in sons; daughters, who possess two X chromosomes, generally receive a good copy of the problem gene from their fathers and so do not fall prey; however, they, too, may pass the bad gene on to their sons. Obtaining an egg from a woman without this genetic legacy would not only eliminate the threat to male offspring but also break the chain of malign inheritance.

Notwithstanding this benefit, Rowland did not see how embryo flushing could be construed as acceptable. She concluded that "the technique was morally reprehensible and could not be justified by the desire to have a child."

Just as reproduction in cattle had been transformed into a quasi-industrial process, with the goal being to turn out whole herds of animals having the same weight, speed of maturity, and overall characteristics, so, too, Rowland feared, were women being reduced to "living laboratories," mere vehicles for the creation of a "super-race." In the months that followed, Trounson was often left to fend off Rowland's criticisms, but he soon gave up trying, he later said, because "you can't debate someone like that who's all over the field."[10] Rowland would become a much-heeded voice in the mounting feminist reaction to the new birth technologies.

Embryo flushing was not the sort of thing that gave the Seed brothers pause. Around 1980, the two Chicago entrepreneurs, Richard and Randolph, began contemplating the $32 million market for embryo flushing in cattle, and got the idea that there might be profit in the "adoptive pregnancy" business for humans. Richard had founded a successful cattle breeding company, and figured that transferring the technology into the medical arena wouldn't entail much more than raising a little venture capital, hooking up with suitable physicians, doing some PR, drafting women willing to be inseminated by a stranger's sperm and then have their wombs washed out, and finding space for the thirty-odd clinics nationwide which he planned to open. It wasn't long before the Seeds had incorporated as Fertility and Genetics Research, enlisted an investment banker, Larry Suscy, as chairman, and made an arrangement with a UCLA reproductive endocrinologist, John Buster, to perform pilot embryo flushings. Buster, working out of the Harbor-UCLA Medical Center, gathered a team that included Maria Bustillo and John Marshall, chairman of the obstetrics and gynecology department there.

The chief selling point of artificially inseminating a third party and then retrieving the embryo was that it did not require surgery. Although this impressed physicians, the Seed brothers seemed to have severely overestimated the altruistic urges of women. In fact, Harbor-UCLA and a Fertility and Genetics Research clinic in Chicago had a difficult time recruiting women willing to act as stand-ins for an infertile wife, and, not incidentally, to chance pregnancy or abortion should an embryo implant itself rather than float out during the lavage with saline solution. These, Buster and his colleagues found, were genuine risks, because more often than not, the first round of lavage did not yield up an embryo. Moreover, even when embryos could be picked out of the saline outflow, they were rarely in excellent shape.

Still, assiduous application of the technique finally yielded two pregnancies, which were reported by the Harbor-UCLA team in a letter to the *Lancet* in July 1983. Buoyed by the news, Richard Seed proceeded to make a severe faux pas by announcing his intention to patent the procedure, which he said he liked to think of "in terms of industry, where large numbers of process patents have been granted."[11] Seed's pecuniary aims were viewed as distasteful at best by many in the medical field, where long-standing protocol demands that advances in surgery and treatment be shared freely among all for the good of humanity.

Besides, the prestige and popularity of IVF militated against a wholesale embrace of a common stock-breeding technique. By 1983, the acknowledged leader in the American infertility business was the Jones Institute, where savvy combined with status to create a mystique unlike that anywhere else in the country. And at the Jones Institute, IVF was the method of choice.

The institute was the brainchild of Robert Edwards's supporters at Johns Hopkins University, Howard and Georgeanna Jones, who had kept in touch with him, and in 1979, after moving to Eastern Virginia Medical School in Norfolk, started an in vitro program of their own with the blessing of the hospital's committee on human experimentation and research. A measure of the Joneses' stature in the gynecological community was that Steptoe took the time and effort to pay the pair a visit late that first year, as did Carl Wood and Alex Lopata. In discussing how they might best set up their program, Steptoe advised the Joneses to eschew superovulatory drugs and stick with natural cycles, and also to inseminate eggs as soon as possible after retrieval, preferably within minutes. As it turned out, the Joneses ultimately rejected both suggestions.

Before the Jones group could begin to transfer embryos, the hospital had to obtain a document known as a certificate of need from the Virginia health department, which required physicians to seek clearance before using any new procedure on patients. The hospital fully expected the department to rubber-stamp the request; however, local residents, among them members of the Tidewater Chapter of the Virginia Society for Human Life, an antiabortion group, got wind of the application, and forced hearings. After a five-month delay, the certificate was granted, and the institute opened for business in February 1980. Jones considered most of the public objections "trivial."[12]

No half measures for Norfolk: The researchers plunged in, attempting forty-one egg harvests over the next twelve months. The Joneses insisted on a tight screening policy, accepting only couples who were married, physically healthy, and childless into their VIP—for Vital Initiation of Pregnancy— program. The institute also required that the marriage of couples applying for VIP treatment be "clinically" stable, and that the wife fall between ages twenty-five and thirty-five. The institute's paternalism was complemented by its sharp eye for the bottom line: Couples also had to demonstrate an ability to pay the $4,000 fee for the procedure, which was not covered by insurance.

Unlike most other IVF clinics, the Jones Institute had as one of its stated goals the treatment of male infertility—an aim that had been singled out by some ethicists as one of the unacceptable potential applications of the technology. But the Joneses felt strongly that subfertile men should have recourse to IVF, which would afford them the opportunity to have genetically related offspring. In a lab dish, poor-quality sperm stood a better chance of penetrating an egg than in the female reproductive tract. This meant that fertile women would have to undergo the same somewhat risky egg retrieval operation as subfertile or infertile women; however, the Joneses decided to offer the service and let the affected women decide whether they found it acceptable or not.

By Christmas, with no pregnancies to report, the institute shifted away from the natural cycle approach that Steptoe and Edwards had preferred and began superovulating their patients. In short order, the next twelve patients who came in for IVF had their ovaries bombarded with gonadotropins, their eggs gathered, and embryos replaced. None conceived. However, the thirteenth patient to be superovulated, Judith Carr, underwent a successful embryo transfer in early 1981 and in December gave birth to Elizabeth, the first IVF baby in the United States.

A few months later, responding to continuing criticism, Howard Jones defended the IVF enterprise from the bully pulpit of the editor's page in the journal *Fertility and Sterility*. Jones reprimanded those having religious scruples about the new technologies for attempting to impose their beliefs upon others. He charged them with inviting "tyranny" by failing to respect that "keystone of contemporary societal organization," the separation of church and state.[13]

Against the argument that life begins at conception and the ancillary contention that embryos therefore have the moral status of individuals and deserve the full protection of the law, Jones argued that all life is a continuum and that there is no moment of conception per se, rather a gradual process leading to individuation. Sperm and eggs are alive, but cannot be considered individuals and are not treated as such. Even a blastocyst cannot be said to be an individual, since the cells making up the trophoblast are destined to transform into the extraembryonic membranes, placenta, and umbilical cord, which, while having the same genetic composition as the fetus, are discarded after birth. Jones even rejected the idea that embryos can be thought of as *potential* persons: "A chassis with four wheels attached at the beginning of the assembly line is potentially an automobile, but no one would buy it for such until it was developed into an object that could be driven away from the end of the line."[14]

To those like Kass and Ramsey who had represented IVF as unethical experimentation on the unborn, Jones pointed to the healthy children who had resulted from the procedure and made the defensive suggestion that couples over thirty-five who undertook to reproduce without submitting to amniocentesis were more ethically suspect than physicians carrying out IVF, because such couples knowingly exposed an unborn child to an elevated risk of Down syndrome. To those who objected that IVF violated the sanctity of marriage, Jones responded:

Sexual intercourse might well be an act of love and sanctify marriage, but it certainly does not guarantee a happy family environment. This is too self-evident to require documentation. The process of in vitro fertilization demands sacrifice on the part of both husband and wife far beyond anything required for normal procreation. Sacrifice is possible only by stable couples who would create an ideal family environment for the rearing of children. To hold otherwise is to be unfamiliar with couples who are infertile and who are willing to go the last mile to overcome this physical handicap.[15]

Jones was here making assertions for which he had little hard evidence: Even today, there have been no substantive studies to demonstrate whether couples undergoing IVF are somehow more "stable" than others. Nor is there any evidence that upon becoming parents, they "create an ideal family environment."

Jones held that most questions about the ethicality of IVF and embryo transfer were simply alarmist, the work of nervous "ethicists of all stripes who oppose the program for whatever reasons." He contended that raising concerns about embryo modifications, surrogate motherhood, and cloning was out of line because such possible manipulations "have little relation to in vitro fertilization."[16]

But this was a fatuous claim on his part, given that the technique of in vitro fertilization is the necessary prerequisite to such future endeavors. It was something like contending that Nagasaki and Hiroshima bore little relation to Ernest Rutherford's—Ruth Fowler's grandfather's—bombarding of the nuclei of helium and nitrogen with uranium rays to produce the first atomic reactions ever initiated in the laboratory. What Jones seemed to want to say was that such potential applications of the technology bore little relation to IVF as it was being applied by infertility specialists at that time, but even this was a feeble defense, since many of his colleagues had expressed a patent interest in the eugenic uses of IVF, and surrogate motherhood arrangements were already being pursued with vigor by some practitioners as a key part of their programs.

Jones signaled more clearly his attitude regarding embryos in 1987, when he refused to allow a couple who had undergone four embryo transfer attempts at the institute to reclaim a single frozen embryo so that they might try again at another clinic. The New Jersey couple, Steven York and Risa Adler-York, had begun trying to conceive in 1984, a year after their marriage. By the spring of 1986, they had resorted to the Jones Institute, which, according to documents they later filed with the court, told them that Risa stood a 20 percent chance of becoming pregnant with the transfer of one embryo, a 28 percent chance with two embryos, and a 38 percent chance with three embryos.[17] After being accepted into the VIP program and signing various consent forms, the Yorks, who had in the months since first consulting the institute moved to California, made four separate cross-country trips to Norfolk.

Before their last visit, from May 17 to June 5, 1987, they signed another consent form agreeing that if more than five embryos resulted from the upcoming egg retrieval, the extras would be frozen. The consent form stipulated that the Yorks had responsibility for the embryos and could withdraw their consent at any time. Moreover, in the event of divorce, they must include the embryos in any property settlement.[18] If they decided they no longer wanted the embryos, they could donate them anonymously to another couple, release them for research, or have them thawed. All these stipulations came into play, because Risa's ovaries yielded up six eggs. Five embryos were transferred—fruitlessly, as it turned out—and there was one embryo left over, destined for cold storage.

A year later, the Yorks decided to try again, this time with Richard Marrs, whose clinic at the Hospital of the Good Samaritan in Los Angeles had gained a good reputation. Accordingly, they wrote the Jones Institute in May, asking whether Steven might fly out and bring the embryo back to L.A. in an insulated container. Although the consent form also absolved the institute of any liability in the event of damage to the embryo, a Norfolk physician replied to the Yorks denying their request. On June 18, figuring to prevail upon Howard Jones one professional to another, Marrs wrote asking for the embryo, and he, too, was rebuffed.

The Yorks responded by filing suit in U.S. district court in Virginia, charging Jones, the institute, and the hospital with which it was associated with breach of contract and other complaints. Lawyers for the institute argued for a tight interpretation of the consent form, contending that only the three options specified by it were permitted the Yorks; however, Justice J. Calvitt Clark ruled that the institute had clearly accepted that the embryo was the "property" of the Yorks, and required the release of the embryo.

Notwithstanding the adverse publicity, the Jones Institute continued to attract thousands of patients, growing into the largest program in the country for a time, and building up a ten-year backlog. Jones assumed a leadership role in shaping American Fertility Society policy toward IVF. As head of a committee to set nationwide standards, he pressed for a multidisciplinary approach in which programs were advised to have at minimum a reproductive endocrinologist, a reproductive biologist, a gynecologist proficient in laparoscopy, and an andrologist, or specialist in male infertility. In 1986, the Jones Institute won what was then the largest award ever made by the government to a single biomedical institution, a $28 million Agency for International Development grant to devise cheap and reliable birth control for the Third World.

By 1983, estimates put the number of physicians, clinics, and large medical centers in the United States offering IVF services at one hundred to two hundred. No one knew for sure, because American IVF efforts constituted backdoor biomedicine. IVF researchers proceeded without government financial support, and therefore without the oversight of the state. This situation had emerged not by plan but by default, as a consequence of a failure of will among legislators.

So charged was the public debate over abortion in the early 1970s that most federal agencies responsible for funding scientific and medical research became reluctant to deal directly with requests by those desiring to carry out experiments involving human embryos. Too, the vocal opposition to IVF specifically by the likes of James Watson gave bureaucrats pause. This unofficial reticence took concrete form in 1974, when Congress forbade the Department of Health, Education, and Welfare (HEW) to grant funds for any research on living fetuses, except that which might assure the survival of an individual fetus. The next year, regulations which had been in the works since late 1973 were promulgated by HEW; these suspended government funding of IVF research until such time as an Ethical Advisory Board (later renamed the Ethics Advisory Board) had made a thorough investigation of the field and advised the department's secretary whether it considered such research ethically acceptable. Four years would pass before a board was appointed.

Legislators had not banned IVF outright, but cutting off federal funding pending the board's review served as a de facto moratorium.[19] A good many early IVF researchers had received support from private philanthropies such as the Ford and Rockefeller foundations, the Population Council, the Planned Parenthood Federation, and the Carnegie Corporation. But they had also relied on grants from the National Institutes of Health, the U.S. Public Health Service, the National Research Council, and other entities, as well as on money channeled to them indirectly from the federal government via their own medical or educational institutions. Congress's action forced researchers to rely solely on private sources, or to turn their IVF efforts into cash-generating enterprises, which many physicians attempted to do.

By fence-sitting, the government lost the ability to guide and regulate the course of research: It is by now old hat to point out that the federal granting process exerts a tremendous influence over how science proceeds, determining which fields flourish and which flounder, which lines of inquiry will be pursued and which neglected. Scientific freedom is merely a potent myth, trotted out whenever scientists feel their privilege is under attack.

As it turned out, the enormous demand for IVF services provided a ready flow of income. A well-to-do segment of the American population was desperate enough not only to allow physicians to use them as experimental subjects but also to foot the bill for it. These private-sector dollars largely eliminated the need for future federal involvement and gave researchers far more latitude than they would otherwise have had. Those working out of institutions receiving HEW grants in other research areas involving human subjects did have to pass proposals through their institutional review boards, but these boards tended to acquiesce fairly readily to plans for IVF. Private clinicians could do whatever they pleased, answerable only to their patients and accountants.

In this light, the actions of the Ethics Advisory Board (EAB), when it finally got up and running in late 1977, appeared somewhat superfluous. Headed by the prominent San Francisco lawyer James Gaither, the EAB had thirteen members in all, of whom only two were women. Its mandate broadened somewhat by HEW, the board busied itself commissioning reports from physicians, ethicists, geneticists, social scientists, lawyers, and theologians. It staged a series of eleven public hearings across the United States and fielded two thousand pieces of correspondence. Eventually, it even considered an application which had sat in limbo since its receipt in early 1977, a bid by Vanderbilt University researcher Pierre Soupart to study lab-manufactured embryos in an attempt to quantify possible genetic defects.

The EAB's report, issued in May 1979 and published in the June 19 *Federal Register*, drew some thirteen thousand comments. The report provided historical highlights of IVF and embryo transfer in animals and humans, detailed the state of the technology, and surveyed the ethical arguments pro and con that had been raised to date. It addressed the goals of IVF research, its possible benefits and adverse consequences, the issues surrounding the use of egg and sperm donors, and the status of the embryo, leaning heavily on a paper which had been prepared at the behest of the committee by LeRoy Walters, director of the Center for Bioethics at Georgetown University. (Walters's survey of the literature regarding IVF was subsequently published in the journal *Hastings Center Report* and remains the most comprehensive overview of the ethical battle that raged from the early 1970s to the time of the EAB report.)

The EAB had solicited the opinions of two legal analysts, Dennis Flannery and Barbara Katz, who advised them about existing statutes pertaining to IVF (one at the federal level, none at the state), artificial insemination, and fetal research. Flannery and Katz also addressed constitutional issues raised by the notion of regulating IVF research.

According to the two scholars, a compelling case might be made that citizens possessed a constitutional right to employ IVF and other reproductive technologies. In *Skinner* v. *Oklahoma*, a 1942 U.S. Supreme Court decision "striking down Oklahoma's compulsory sterilization law, [the Court] held that individuals have a right to be free from unwarranted governmental interference with procreative capabilities. . . . This might be termed the 'right to procreation.'" A second Supreme Court decision, in the 1965 case *Griswold* v. *Connecticut*, established the right of married couples to privacy (the same protection upon which *Roe* v. *Wade* is based). The Court said, "The entire fabric of the Constitution and the purposes that clearly underlie its specific guarantees demonstrate that the rights to marital privacy and to marry and raise a family are of similar order and magnitude as the fundamental rights specifically protected." Thus, the courts might recognize that couples had a fundamental right to "choose whether, and in what manner, to achieve pro-creation," and could reject legislation or force a retraction of federal regulations that restricted access to new technologies.[20]

Scholars who have devoted further attention to the legal issues raised by IVF, including most notably Boston University professor of health law George Annas and University of Texas law professor John Robertson, have embroidered and expanded upon this basic line of argument. However, other legal analysts have seen hints that the high court might be quite reluctant to expand the umbrella of protection to cover assisted reproduction.[21]

The legal analysts and EAB did recognize that several legitimate grounds for banning or restricting clinical uses of IVF might be found, that is, grounds which allowed the state to infringe upon citizens' fundamental right to reproduce. The state could declare that it had a compelling interest in the protection of embryos, and move to shut down research involving them. It could probably prevent single women from obtaining access to new reproductive technologies by asserting its interest in promoting marriage and legitimacy within its populace. Furthermore, it could intervene if it could demonstrate that institutions or individuals were employing the new technologies to influence the genetic makeup of offspring. It could also outlaw surrogate motherhood because of the "insuperable legal problems" it stood to create. Finally, the government could shut the technology down if it created risks for women and offspring outweighing those inherent in natural conception and pregnancy.[22]

After gathering and sifting through masses of information and opinion, the EAB arrived at a handful of conclusions. On the broadest level, it deemed IVF ethically acceptable and worthy of federal dollars, and considered an outright prohibition on the technology "neither justified nor wise."[23] While the human embryo was "entitled to profound respect," it did not qualify for treatment as a person, either legally or morally; thus, IVF research for the sake of furthering embryo transfer should be supported, as long as investigators sought permission from institutional review boards, fully informed those providing gametes about the use to which their embryos would be put, and suspended any and all examinations of embryos once they had reached fourteen days old. As for embryo transfer, board members thought that it should be attempted only with married couples.

Although the EAB hoped that federal involvement would help guide IVF research, it also recognized that the genie was well out of the bottle, and so recommended that "every effort be made to collect whatever information may be elicited from [medical] practitioners in this country and abroad." Ideally, the National Institute of Child Health and Human Development would solicit and analyze data from everyone in the field, a massive task but one that "would increase the opportunity for investigators, clinicians, and prospective patients to be fully informed."[24]

For all this, the EAB's labors essentially came to naught. In 1980, the newly renamed Department of Health and Human Services allowed the board's charter to lapse. Supposedly filling the gap was a Presidential Commission for the Study of Ethical Problems in Medicine and Biomedical and Behavioral Research, appointed by Jimmy Carter. After two senators asked the commission to look over the EAB report, its chairman duly advised Senator Edward Kennedy that the board's conclusions appeared well considered.[25] The commission went on to produce a report on genetic screening and counseling, but was disbanded by Ronald Reagan before probing IVF further. Meanwhile, support for IVF research continued to be withheld. Given the preexisting law requiring an EAB to approve any grant proposals involving embryos, and given that the EAB was defunct, the net result of some five years of bureaucratic exertions was a continuance of the de facto moratorium on IVF research under the federal aegis.

Other countries addressed the issues raised by IVF head-on. Between 1980 and 1987, seven Australian and two British government committees considered whether to permit embryo research and IVF, as did a single officially sanctioned committee each in Canada, West Germany, Spain, the Netherlands, and France. Four of these bodies rejected embryo research outright, no matter its purpose. Of the ten which accepted its permissibility, six decided that the creation of embryos strictly for scientific study should not be allowed. Only certain embryos left over from IVF might be experimented with, and those for a maximum of fourteen days. Two committees set the limit lower: The French National Ethics Committee said that research should halt at seven days; the German committee, after the first cleavage.[26]

Perhaps fittingly, given the leading role of British IVF researchers, the report which attracted the most international attention—and drew the heaviest fire—was that produced in 1984 by the United Kingdom's Warnock Commission, chaired by Dame Mary Warnock. Warnock's brief was to advise on policy, and it did not shy from that charge. It proposed that bench scientists be required to go before a licensing board before engaging in embryo research. This licensing agency would also regulate infertility clinics. The government followed the recommendation by setting up the Voluntary Licensing Authority (VLA) in 1985. One of its early moves was to limit to three the number of embryos clinics were allowed to transfer into women to lessen the chances of multiple pregnancies, which were by 1987 being "selectively reduced." Essentially, the physician injects the unwanted embryos with potassium chloride, and their remains are resorbed or ejected along with the full-term fetus. There are indications that the growth of remaining embryos may be retarded as a result.

The VLA also frowned on egg donation by sisters, because a child whose aunt was his or her genetic mother might be confused as to his or her identity or place in the family. Anonymous egg donation seemed to protect the best interests of the child. By 1986, Christopher Chen, at Flinders Medical Centre in Adelaide, South Australia, had had success freezing eggs and produced one embryo from a frozen egg which had resulted in a pregnancy. However, the VLA strongly objected to further experiments of this sort, since another Australian, Alan Trounson, had gleaned evidence that cryopreserving eggs could induce chromosomal abnormalities in them.

Some activities Warnock recommended disallowing altogether, including free-market sale or purchase of human sperm, eggs, or embryos, and the procuring of women willing to gestate embryos not their own. In fact, commercial surrogate arrangements should be criminally punishable, declared the commission. (Today, commercial surrogacy is banned in Britain and fourteen other countries, although not in the United States.) Of the sixteen members, thirteen echoed the view expressed by the EAB that embryos did not qualify for personhood, yet should be accorded respect. Three members dissented and maintained that the government should eradicate all research involving embryos, whether for the sake of treating infertility or pursuing basic knowledge.

Howls of protest rose from all quarters of the British IVF community at Warnock's recommendations. Basic scientists howled loudest. Many derided the report as muddled and moralistic, and objected that it unfairly limited them while granting leeway to clinicians pursuing their IVF work. While allowing that the commission had done a good job reviewing the history and status of infertility research, critics dismissed the report's analysis of future possibilities, such as cloning, gestating fetuses in artificial wombs or in other species, nuclear transplantation, embryo biopsies for sex selection or genetic screening, and germ-line gene therapy, that is, alterations of an embryo's chromosomes. Exhibiting either ignorance or disingenuousness, some Warnock debunkers proclaimed that the commission had engaged in emotionalism and cant when addressing these potential applications—all of which had actually been positively discussed at various times by scientists in a position to make judgments about their feasibility. Peter Braude, a senior research assistant in the gynecology department at Cambridge University, told the *New Scientist* that "the first part of the [Warnock] report is practical and sensible because it was based on at least 10 years of experience. When you come to the regulations on research it draws on science fiction and so is tinged with hysteria."[27]

In truth, when it came to regulations on research, the problem was that Warnock had recommended them at all. Ask biomedical researchers anywhere whether they would prefer to submit to oversight or be subject to their own conscience, to be forced to justify their actions or be able to follow a line of inquiry wherever it leads, and what do you think their answer will be? Going as a supplicant before a licensing board—this was not a prospect that appealed in the least. The Warnock report, felt many scientists, set a bad precedent. International attitudes might be swayed by it, and there could be a clamp-down all over.

Indeed, the latter half of the 1980s did see a general tightening of controls, although this trend could not be pinned entirely on Warnock. The Australian state of Victoria in 1984 adopted legislation closely regulating not only IVF and embryo experimentation, but also embryo freezing and artificial insemination by donor. It outlawed surrogate motherhood for pay, as well as commercial sperm, egg, and embryo enterprises. Germany ultimately banned all embryo research. The conservative mood was further reflected in a set of rules devised by the Council of Europe in 1986. This influential body, drawing its membership from parliaments across Europe, signaled its displeasure with the more extreme forms of manipulation, voting to adopt bans on cross-species experiments in fertilization, embryo transplant, or gestation; on the creation of chimeras, bearing the genes of one or more different species; on sex selection, whether of embryos or fetuses, for nontherapeutic reasons; on attempts at vegetative cloning or artificial twinning of human embryos by division of the blastocyst; and on attempts to bring a fetus to term in an artificial womb.

Around the time that the council made its views known, the premier French IVF researcher announced that he would no longer employ the technology for any reason other than to assist infertile couples. Jacques Testart, research director at the National Institute of Health and Medical Research in Paris, head of an IVF lab and progenitor, along with gynecologist René Frydman, of the first test-tube baby in France, declared in an article in *Nature* that IVF was getting out of hand. Testart had trained as a soil scientist, then carried out superovulation and embryo transfer research in cattle. From cows, he shifted to humans, joining a medical center outside Paris and eventually collaborating with Frydman on the work that would lead to the birth in 1982 of the country's first test-tube baby, Amandine.

Despite his intense involvement with IVF, Testart came to question whether scientists had the right to continue research in this area. In *L'oeuf transparent (The Transparent Egg)*, a book describing the events surrounding Amandine's birth, Testart cast himself and his colleagues in a transgressive drama: Toying with godhood, they contravene taboos and engage in a heady "violation of love and intimacy."[28] The transparency Testart refers to is that of the dividing embryo under the microscope, a metaphor for scientists' sense that they can see through all of nature's mysteries. But this is a delusion, he said, because to understand an embryo is not to comprehend the process by which cells become human, gaining a soul.

Testart renounced the push to routinize IVF and further manipulate the embryo. Within ten years, he predicted, the technology would have gained widespread popularity among fertile couples, who would prefer it to natural procreation because of the opportunity it afforded for genetic screening and sex selection of embryos. He believed that "people will ask for a baby completely as they want it—girl or boy, completely normal. With genetic progress, the way is open for eugenics."[29]

In Testart's view, IVF at its simplest is benign, easing the plight of the infertile. However, the thrust of ongoing research has unsavory undertones:

> The way for science is not to invent new needs for people, but just to resolve the true need of people. Doctors say, we need this technique, you have to help us. But science has its own responsibility. We must think about the consequences before we do the research, not after.[30]

Testart's colleagues on the international scene largely pooh-poohed his concerns, which they dubbed not so much inaccurate (indeed, his predictions that fertile couples would flock to IVF clinics has not yet come to pass) as misbegotten. Anne McLaren, a British embryologist schooled at the University of Edinburgh who had long played an active role in discussions of reproductive technologies and had sat on the Warnock Commission, told the *New Scientist* that "[Testart] is absolutely right that IVF *will* be used for diagnosis of genetic diseases. But in my view that is an important application."[31] McLaren, of course, joined the long line of thinkers who believed that one of the highest accomplishments of science would be to usher in a day of genetic perfection.

Upon retiring from the presidency of the American Association for the Advancement of Science in 1971, Bentley Glass had left his colleagues with the ringing proclamation that all children have a right "to be born with a sound physical and mental constitution, based on a sound genotype," and added that "no parents will in that future time have a right to burden society with a malformed or a mentally incompetent child."[32] That same year, a biologist quoted in a CBS program on reproductive technology remarked that "if we cull down the lazy type that is not interested to contribute to society, I think we have done a great deal. We do that in race horses and farm animals. We select the best dairy cow, we select the fastest horse, and we select sheep for their wool. I think we can do a little bit of selection on the human level."[33] This was the old familiar refrain given new potency, for now it should be possible to carry out that sorting in a petri dish. In fact, scientists could and did contend that they had done away with the superfluous aspects of reproduction. As Clifford Grobstein wrote, "In physiological terms, parental intercourse is merely a way to bring egg and sperm into the proximity necessary to

undergo fusion. It is the cellular fusion that has important biological consequences, not the procedure to assure it."[34]

S cience had effected what amounted to a total objectification of procreation, and this struck many feminists as ominous. From the mid-1980s
onward, they formed or joined a number of organizations designed to
examine, monitor, and speak out on the new reproductive technologies.
These groups included the Feminist International Network Against Reproductive Technology (FINNRET), the Coalition for Responsible Genetics,
the Reproductive Rights National Network, and the women's branch of the
Green Party.

For one thing, these groups acted on behalf of consumers, identifying IVF
as a "failed technology." They pointed to the low success rates: As of 1985,
only 14.1 percent of cycles resulted in a pregnancy. After another year, the rate
had improved to 16.9 percent, but that still meant that on a per-cycle basis 83
percent of those undergoing the procedure failed to become pregnant. The
industry liked to represent chemical pregnancies as successes, but to a couple
wanting a child, a positive urine test meant little. When one looked at the
number of live births resulting from each egg retrieval, the statistics looked
quite unpromising. Only 8.9 percent of egg retrievals ended in a live birth.[35]

Although the across-the-board rate of live births per IVF cycle has now
risen to 21.2 percent, there are indications that the average take-home baby
rate may never edge substantially over 50 percent unless embryos are further
screened before transfer. Studies of women in general have revealed that with
normal intercourse between fertile couples very few conceptions actually
translate into births. (Even in IVF, where sperm and egg are placed in direct
proximity, fertilization fails to occur in 15–20 percent of cases.) If conception
does occur naturally, fully two-thirds of embryos may be spontaneously
aborted. The high rates of spontaneous abortion in normal pregnancies have
been linked to chromosomal abnormalities in the embryos, which suggests
that as long as IVF practitioners work with embryos that have been adjudged
sound simply because they have cleaved and appear well formed, they may be
constrained by a built-in biological limit. By this logic, only the use of
younger, more genetically sound eggs and of preimplantation genetic screening, which would identify flawed embryos before transfer, stand to substantially improve IVF odds.

By the mid-1980s there were also hints that women might be incurring
health risks by undergoing ovarian stimulation with gonadotropins. The
ovaries of some women—and no one could predict which ones—responded
badly to the superovulatory drugs. These women could develop cysts, which
were usually noncancerous, but the question was raised whether physicians

were increasing women's chances of sprouting malignant tumors by injecting them with fertility drugs. Nor was IVF entirely without danger in itself: To date, two women had died as a result of egg donation, one in Israel, one in Brazil, and three others had died in IVF clinics in Spain and Australia.[36]

Expanding upon statements she had made upon resigning from the Monash program, a disaffected Robyn Rowland charged that researchers were parlaying the desperation of infertile couples into dollars which they then used to support investigations with a larger eugenic design. She declared that physicians failed "to differentiate between research to aid infertility and research to change and control conception and the genetic balance," and charged medicine with exploiting women's bodies for experimentation, capitalizing on "their 'need' (social and otherwise) to have babies."[37] Rowland hit upon an incontestable truth, one that Testart also alluded to: IVF research by its very definition was ongoing, and led inexorably to previously untried manipulations of gametes and embryos.

Concerned that IVF and other reproductive technologies were detrimental to women's best interests, feminists from around the world gathered at the Second International Interdisciplinary Congress on Women, held in Groningen, Holland, in April 1984. Out of the conference emerged a book of essays, *Man-Made Women,* dealing with the "urgent and life-threatening dimensions that these technologies pose for women." Those represented in the volume included Rowland, journalist Gena Corea, British professor of biology and women's studies Renate Duelli Klein, biologist Helen Holmes, and founding member of FINNRET Janice Raymond, all of whom had already spoken and written critically about new reproductive technologies. Like all feminists, these thinkers did not hew to a unitary philosophy; however, they were in unanimous agreement that these technologies represented an unacceptable extension of male control over women's bodies.

The party line of physicians was that women were fully informed about the experimental nature of IVF and submitted to it willingly for the sake of a personal or larger societal goal. Many of those attending the Groningen conference argued that while women might think they were exercising choice when deciding whether to proceed with chancy and invasive infertility treatments, they were not in fact operating in a realm in which free choice was possible because they had been socially conditioned to believe that without children they were worthless. As long as they felt that childlessness was a stigma, they would be willing to try just about anything to alleviate that condition.

Some attendees contended that while the Pill and other gynecological innovations had expanded women's choices and their ability to exert control over childbearing, the new technologies threatened in the long run to close down women's opportunities as a group, mainly due to the widespread use of sex selection. Here, the culprit was not IVF but the screening methods that had arisen alongside it during the 1960s and 1970s: amniocentesis and sonography. Already in India, conference participant Madhu Kishwar, a professor at Delhi University, reported, amniocentesis was being used by couples in the relatively prosperous states of Punjab, Haryana, and western Maharashtra to identify female fetuses so that they could be aborted. Amnio clinics advertised "through leaflets, newspapers, and magazines: 'Come for this test so that you don't have an unwanted daughter born to you.' "[38] One clinic in Amritsar, in northwestern Punjab, bid for clients with a billboard reading, "Know the sex of your unborn child with the aid of the latest imported equipment and sophisticated scientific techniques."[39] Although amniocentesis for the purposes of sex selection had been nominally forbidden by the Gandhi government in 1983 after heavy lobbying by women's advocates, loopholes in the law, as well as haphazard enforcement of it, permitted the practice to continue. Many gynecologists in India justified the procedure on the grounds that it was a legitimate family planning tool demanded by women, and argued that if it was declared indefensible, then so, too, must conventional abortions be.

Thus, the feminists assembled in Groningen found themselves in the unusual position of arguing against reproductive rights. Declared Rowland, " 'Choice' and 'freedom' as a continuing ideological base in the area of reproductive technology may eventually entrap women further and limit their choice to say 'no' to increased male control of the reproductive process."[40]

Again the following April, at a conference in West Germany, some two thousand women condemned the new reproductive technologies in a resolution declaring them a violation of women's dignity and a tool of eugenicist and racist ideologies.[41] That July, at the Women's Emergency Conference in Sweden, those fighting the new technologies registered frustration. In the introduction to their 1989 volume *Test-Tube Women,* molecular biologist Rita Arditti, Renate Duelli Klein, and biologist Shelley Minden explained that in 1984 they had been "skeptical," but by 1988, they were "angry and outraged about the continued experimentation on women's bodies, about the infliction of violence and pain, about the perpetuation of lies, about the increasing control of reproduction."[42]

In the span of those four years, accusations that the new technologies exploited women as "rent-a-wombs" and that childbearing was being sullied by commerce had gained new credibility. The whole world had watched, for example, the unfolding of the so-called Baby M case, in which Mary Beth Whitehead, having agreed to a surrogate gestation arrangement with a New Jersey couple, the Sterns, decided not to give up her infant and was sued by them. That the state supreme court ultimately allowed Whitehead to retain her child did not negate the trauma she had undergone in her fight to keep the baby. More than just feminists now evinced concern that poor women would become factories for the production of the children of the wealthy. Others declared that surrogacy was baby selling by any other name.

Furthermore, many feminists had been alarmed to observe that the use of amniocentesis for sex-selection purposes had continued apace, while in China and India a new cottage industry had burgeoned. Entrepreneurs carting sonograph machines around in vans to the remotest villages had turned the elimination of female fetuses into a profitable enterprise. In the First World, there were signs that amniocentesis and sonography were increasingly being employed as "quality control" devices. Initially intended for use only in high-risk pregnancies, especially those of older women, amniocentesis had in the two decades since its introduction become routine for virtually all pregnancies. The average age of those having their amniotic fluid sampled dropped from forty in the late 1960s to thirty-five in 1985.[43]

Critics decried the medicalization of birth. Both the author Barbara Katz Rothman and the biologist Ruth Hubbard, whose reputation for lucid analysis of the sociology and politics of science extended far beyond feminist circles, argued that the technologies for screening fetuses fundamentally altered the relationship of a woman to her pregnancy, injecting into it an unavoidable provisionality. A woman agreeing to amniocentesis also tacitly agreed to the possibility of aborting the fetus were it found to be flawed—or at least of having to decide whether to abort. In forcing women to "confront further decisions" about whether to carry a fetus to term, Hubbard noted, medicine had once again reconfigured the way in which women related to their pregnancies.[44] Gynecology as a discipline had tended to pathologize pregnancy, treating it more as a disease state than a normal manifestation of female physiology. But now, by engaging in the wholesale screening of fetuses, which had the look and feel of an industrial process, it appeared to be turning babies into another commodity, like cars or luxury items. As Testart did, Hubbard envisioned a time when people would opt for IVF because " 'in-body fertilization' will not only have come to seem old-fashioned and quaint, but downright foolhardy, unhealthful and unsafe."[45] (In 1986, Robert Edwards prognosticated similarly, telling the *New Scientist* that "we will not [by the year 2000] have reproductive cycles as we know them now. We are still trapped in ancient reproductive cycles; we can't remain in that condition much longer.")[46]

The rise of the fetal rights movement had left feminists convinced that the state might one day intervene and force all women to undergo screening of embryos or fetuses. This emerging legal philosophy argued that fetuses who were harmed in utero by the actions of their mothers, for example, as a result of heavy drinking or drug use during pregnancy, or by the negligence of physicians, could sue for damages. Women who appeared to be conducting their pregnancies in a fashion which might inflict damage could be compelled by the state to desist. Legal theorist John Robertson has written that if a woman

> decides to go to term, one may reasonably argue that she has a duty
> to accept minimally invasive or minimally risky medical treatments

that will prevent severe harm in offspring. . . . Neither her procreative liberty nor right of bodily integrity give [sic] her the right to cause or avoid preventing reasonably avoidable harms to offspring that she chooses to bring into the world.[47]

Extending this basic line of thought to its logical conclusion, some critics of assisted reproduction arrived at the position elaborated by Bentley Glass, that bearing a child who wasn't "perfect" might one day become tantamount to a crime. Those in the disability rights movement construed such talk as a kind of retroactive death sentence on themselves, a declaration that they did not belong in the world and that the world of the future would be better off without people like them. Hubbard argued that the effect of genetic and fetal screening societally would be to shift responsibility for children's disabilities squarely onto the shoulders of the mother. Rather than blaming fate, society would blame women; to avoid censure, women and their families would begin "implementing the society's prejudices, so to speak, by choice."[48] Science writer Tabitha Powledge echoed this thought in observing that "where it initially appeared that government would impose test-tube babies and genetic engineering on society, the great irony is that we now look to government to protect us from the Brave New World."[49]

Regardless of whether federal regulation was necessary, it was not forthcoming. Instead, the American IVF industry moved to police itself. Under American Fertility Society auspices, the Society for Assisted Reproductive Technology (SART) instituted a voluntary reporting system in 1986, and gathered information on chemical and full-term pregnancies achieved through IVF and embryo transfer. Physicians Ricardo Asch and Jose Balmaceda, and others at the University of Texas Health Science Center in San Antonio, had in 1985 introduced GIFT and ZIFT, and SART registered instances of, and computed success rates for, these procedures, as well as for egg donation and surrogate gestation. Those sending data to SART followed the same steps in calculating their success rates so that clinicians and consumers alike could make ready comparisons between programs. This seemed an especially apposite tactic, since rumblings had started in the popular press about the overselling of the technologies. To stave off charges that incompetent clinicians were gulling the public, SART opted to throw the field open to inspection. It also formed a laboratory accreditation committee, which drew up rigorous standards for sperm, egg, and embryo handling, both to deflect criticism on that score and to bring up to par any programs that might be iffy in this regard.

In the United States, defenders of IVF adopted a strategy of playing up the technology's importance to individuals: They rehearsed the agonies and sorrows of infertility, and charged that those who would curtail IVF showed a heartless lack of sympathy for couples unable to have children. Left unsaid in these plaints was the fact that the only people who had access to the technology were white, highly educated, and affluent men and women, a percentage of whom (12 to 15 percent in a 1983 study of the Yale infertility program) already had biological children. Also unsaid was the fact that the average price tag for an IVF baby (as of 1989) was $50,000. Since this was coming out of patients' pockets for the most part, it was not germane to suggest that those health care dollars might have better served pressing medical needs of children already born. However, the highly restricted nature of the IVF patient population did tend to undercut notions that any limitation on its availability would do irrevocable harm to the infertile. Already, the vast majority of the nation's infertile couples had been shut out, as had, at most programs, single women.

European researchers, too, had realized that they must capture the high ground in the debates over embryo research, and accordingly formed the European Society of Human Reproduction and Embryology, which promulgated guidelines for the future. On the whole, members had resigned themselves to some kind of government oversight, while hoping that embryos would continue to be construed for legal purposes as chattel.

In the end, no watchdog body or legislation might have stopped Cecil Jacobson, the physician who in 1992 was found guilty of fifty-two counts of fraud and perjury, fined $116,000, and sentenced to five years in prison without parole. Jacobson, who had helped introduce amniocentesis to the United States and carried out IVF studies in the early 1970s while at the reproductive genetics unit of the George Washington University medical school, operated a private infertility clinic in Alexandria, Virginia. According to a receptionist who testified at his trial in U.S. district court, Jacobson would routinely schedule artificial insemination appointments for late afternoon, and on those days would retire to the rest room for long periods. The jury was convinced by the prosecution's argument that during these visits, Jacobson was masturbating in order to produce sperm, with which he inseminated as many as seventy-five women from 1976 to 1988, after having told them that they were receiving the sperm of anonymous donors. Indeed, Jacobson's records did not indicate that he had ever paid any sperm donors or obtained frozen sperm from a bank. According to DNA tests, at least nine children had been endowed with the genetic material of the portly fifty-five-year-old.

The prosecuting attorney, Randy Bellows, argued in his opening remarks that Jacobson's "deceitful, cunning, and above all else cruel betrayal of trust" was motivated largely by greed.⁵⁰ Jacobson's income had fallen markedly, from $475,000 in 1982 to $300,000 in 1986, and the prosecution cast his actions as an attempt to up his income. For his part, Jacobson claimed that he was only trying to guard his patients against infection with HIV or other viruses by providing them with sperm he knew was clean.

Marian Damewood, a Johns Hopkins University gynecologist who took the stand, had combed through thousands of patient files from Jacobson's practice and determined that the physician also deceived hundreds of women by announcing to them that they had miscarried. These cases had been the subject of civil actions taken against Jacobson in 1989 by the Federal Trade Commission, and during the criminal trial it was found that he perjured himself in these earlier proceedings. Jacobson apparently injected women with hormones which produced all the signals of an early pregnancy, then withdrew the drugs and conveyed the bad news to them. One former patient, Vicki Eckhardt, testified that Jacobson had gone so far as to show her her fetus on a sonogram only to inform her a few weeks later that the fetus had died in utero. Eckhardt said, "He sat my husband down and he said, 'Let's join hands and join in prayer.' "⁵¹

Only by accident was Jacobson's malfeasance discovered, and even then, he maintained that his only intention had been to help infertile couples. Stripped of his license by the Virginia medical board, Jacobson retreated to Utah while awaiting trial in district court. He was said to be engaged in genetic research. It seemed all too fitting. A cowboy in cattle country, dreaming of "fathering" genetically perfect humans.

14

Tomorrow's Children

O n a short tether, the stallion is led around the paddock past an aromatic array of mares in estrus. After a few minutes, the circuit has done its work, and the stallion, wild-eyed, its penis unfurled, is bucking and screaming as it is coaxed toward its intended target—not one of those exciting mares, but a yard-long leather sheath lined with a giant condom. This receptacle is being proffered by a female veterinary student, alert and a little jumpy alongside a leather gymnastic horse, which the stallion should willingly mount, since it has been smeared with estrous secretions. Before the woman hoisted the outsized contraption and positioned herself for the task, the onlookers, all male, teased her. But it was clearly an old tired joke, and she had taken it in stride. Now, with the stallion throwing himself upon the fake mare, the blue-jeaned student braces herself, and guides the sheath onto its penis until the shaft is enclosed and her shoulder is kissing the horse's heaving flanks. After the stallion has spent itself, the vet student retreats with her trove: a condom full of prime semen, which wins her general acclaim from her fellow students and professors at the Colorado State University animal reproduction lab there in the dusty lee of the Front Range.

This is merely step one for the Colorado researchers. The year is 1983, and George Seidel, a physiologist and biophysicist, has set out, along with reproductive biologist Edward Squires, to produce cloned horses. Blastocyst division is the sort of cloning they intend to perform, and so, after the semen is collected, it is used to inseminate a mare. Embryo flushing yields an embryo, which is carried into the neighboring embryology lab, where a steady-handed postdoctoral fellow, Tetsuo Takeda, sits at the controls of a micromanipulator.

Peering into the microscope, Takeda inserts a microsurgical blade through the zona pellucida and cleaves the cluster of one hundred or so cells. He then evicts one clump of cells and guides them into a waiting zona pellucida which has been emptied of its contents. This is actually a cow zona pellucida, but it will function just like that of a horse. Despite the rents, the zonae will remain intact when they are transferred surgically into the uteri of two separate brood mares. The cells within will resume dividing, and eleven months later, the mares will foal, yielding two coal black twins with white blazes running from forelock to nostrils, colts with identical genetic material who will be named Question and Answer. They will join a mini-menagerie of artificially produced identical twins around the world—mice, sheep, pigs, cows—many of them the handiwork of the Animal Research Station in Cambridge, England, the site of so many other advances by which humans have wrested control of reproduction.

In the wake of the horse cloning, several ethicists around the country said that while the same technique could be easily adapted for use with humans, they believed it highly unlikely that anyone in the United States would ever attempt the procedure. They said there was simply too much opposition within ethical review boards and other institutional oversight bodies to permit it. They were wrong. In October 1993, Robert Stillman, a physician at the George Washington University Medical Center fertility clinic, quietly announced to his convened fellows at the annual meeting of the American Fertility Society in Montreal that a team headed by his colleague Jerry Hall had cloned human embryos, using a technique similar to that employed with animals.

Hall worked with surplus embryos from the IVF clinic, seventeen of which were "genetically abnormal," having undergone penetration by more than one sperm. He divided the embryos and then, rather than inserting the cell clusters into the emptied zonae of human embryos, he coated them with a gel to form an artificial sac. He allowed the resulting forty-eight embryos to cleave for six days before terminating the experiment.

Hall hoped to perfect the cloning technique to give IVF practitioners a way of manufacturing more embryos—identical ones—from a couple's zygotes, since the more embryos available for transfer, the merrier. He further asserted that he had intended to spur ethical debate: "It was clear that it was just a matter of time until someone was going to do it, and we decided it would be better for us to do it in an open manner and get the ethical discussion moving."[1]

Journalist Gina Kolata, writing about the development in the *New York Times,* reported that "while some doctors who do in vitro fertilization said they would never clone human embryos, others said they would offer it to their patients as soon as the technology was ready"—which was to say, as soon as their embryologists and lab technicians could get up to speed on the technique.[2] As usual, partisans of IVF found nothing worrisome in cloning, seeing it as a tool that would improve infertile couples' chances of bearing a genetically related child, and as an aid to genetic diagnosis. Given a clutch of cloned embryos, one could be sacrificed for preimplantation screening, and if everything checked out, the remaining clones could be transferred. However, this prospect struck some bioethicists, including Arthur Caplan, an outspoken critic of reproductive technologies who is dismissed by many in the IVF field as a gadfly, as insupportable. Caplan told *Science* magazine that "creating embryos solely for the purpose of genetic diagnosis is morally suspect."[3]

Cloning of this sort introduces a number of bizarre possibilities, including that of a couple opting to raise a family composed of clones born several years apart or of a woman giving birth to her own clone, which perhaps accounts for the American Society for Reproductive Medicine's rather fussy insistence that the procedure be referred to as blastomere separation. The word "cloning" brings to mind too many uncomfortable sci-fi associations and is a reminder of the scandal involving author David Rorvik. In 1978, Rorvik attempted to pass off *In His Name,* the story of the vegetative cloning of a human being, as nonfiction; however, upon publication, the book was roundly denounced by scientists as untrue.

That said, ASRM finds cloning rather unexceptionable. Its sixteen-member Ethics Committee noted in a revision of the society's ethical guidelines that the procedure was "arguable," yet downplayed the disturbing side of cloning by discussing it merely as another "micro technique," like intracytoplasmic sperm injection, which should not be practiced except in "laboratories that have experience with animal preembryos and gametes," by those skilled with micromanipulators.[4]

Certainly, public sentiment registered more than enough disquiet for everyone: A week after the George Washington results made the papers, a joint Time/CNN poll of five hundred adults in the United States revealed that 75 percent opposed cloning and 46 percent favored making it illegal.[5] Fifty-eight percent agreed that it was morally wrong. However, unless cloning is banned by law internationally, the world's first delivery of artificial twins will doubtless occur within the next few years. If it does, the probity of performing cloning will become the subject of raucous disputations.

A lready, surrogacy and egg donation, once thought to be beyond the ethical pale and highly unlikely to ever become more than rare practices, have edged toward the mainstream.

Few believed that many couples would find surrogate arrangements appealing after the Baby M drama and a less well known but equally disturbing 1990 case, *Johnson* v. *Calvert,* in which Californian Anna Johnson gestated an embryo formed from the sperm of Mark Calvert and an egg retrieved from Crispina Calvert, his wife, who had undergone a hysterectomy. Midway through the pregnancy, relations between Johnson and the couple soured, and she decided to keep the baby. Ultimately, the state supreme court upheld an Orange County Superior Court decision enforcing the surrogacy contract and requiring Johnson to give up the infant.[6]

Yet despite legal imbroglios like these, the United States boasts a dozen or more businesses which offer to enlist surrogate mothers. These are generally private concerns staffed by lawyers and mental health professionals who act as middlemen between would-be parents and women who are paid to be inseminated with a man's sperm. Increasingly, gay male couples and single men are considering surrogate arrangements.[7] In addition, some seventy-five fertility clinics in the United States offer gestational surrogacy, in which a woman brings to term an embryo crafted from a couple's own gametes.

Like artificial insemination by donor, such third-party arrangements challenge social norms and do not sit well with some infertility specialists, psychologists, ethicists, and legal commentators. The New York State Task Force on Life and the Law, in making public policy recommendations on surrogacy, saw that gestating children for a fee represents a "radical departure from the way in which society understands and values pregnancy. It substitutes commercial values for the web of social, affective and moral meanings associated with human reproduction."[8] In the interest of couples' desire to have genetically related offspring, women serve essentially as incubators, and may do so primarily for mercenary reasons. Any possible bond they may develop with the infant they have borne must be torn asunder.

Conversely, should surrogates drink, smoke, or otherwise expose fetuses to harm during a pregnancy, couples may be put in the uncomfortable position of taking action against the surrogate. If as a result of the surrogate's actions, the child is born impaired, or if it has a physical or mental defect due simply to the vagaries of development, who bears responsibility for its care? Roughly three-quarters of the states lack any legislation regarding surrogacy, and as long as the law remains in flux, both would-be parents and surrogates may find themselves tangled up in custody fights, their profound attachments and yearnings subordinated to the banal requirements of contract law. Furthermore, children thus delivered may, if they discover their origins, want to gain information about the surrogate, an eventuality that poses the same sorts of psychological disruptions and legal impediments as when AID children seek their roots.

Egg donation, another third-party arrangement, carries its own set of complications, although the psychological risks to offspring due to familial secrecy may not be one of them. Houston psychologist Patricia Mahlstedt, speaking at a 1994 workshop, pointed out that egg donation emerged in a more liberal cultural era than sperm donation. As a result, it has been surrounded by less shame and a more concerted effort to address the best interests of everyone involved. According to Mahlstedt, egg donor couples generally discuss the means of conception openly with family, friends, and the child and use nonanonymous donors whose medical histories are therefore known. Moreover, most donors and recipients go through psychological screening, a step recommended by the ASRM guidelines (although, as Mahlstedt noted, no such screening is recommended for donor insemination). Mahlstedt opined that "since dealing with known difficulties in honest, thoughtful ways usually promotes growth in families, children conceived through egg donation are more likely to be born into families in which their parents are at peace about the means of conception."[9]

That said, egg donation is not without ethical problems. Critics suggest that the rights of egg donors are being insufficiently safeguarded. Within families, for instance, pressure may be put on siblings or other relatives to donate eggs. Clinical social worker Jean Benward, who counsels infertile couples at her private practice in the Bay Area, describes a couple, Sam, forty, and Karla, thirty-five, who were seeking an egg donor due to Karla's infertility after cancer treatment. Karla first arranged with her sister to provide eggs, but when the sister changed her mind, Karla turned to her brother's eighteen-year-old daughter, Mary. Karla told Mary that the procedure entailed no risk, and that the hormones she would be taking were all right because they were "natural." Mary had had an abortion at seventeen, which she reported to Benward had been anxiety producing, and although Mary's mother and boyfriend objected to her plans to aid her aunt, the teenager insisted that it was "something she really wanted to do." However, she intended to hide what she was doing from her father and brother, at least for a time if not for good.[10] While the whole experience might ultimately prove positive, there were enough signs that Mary was complying with her aunt's wishes for unhealthy reasons to give Benward pause.

In this instance, the person soliciting the donation of eggs intentionally failed to mention, or was ignorant of, studies which have revealed possible links between fertility drugs and ovarian cancer. Although further investigation is necessary to determine whether stimulating the ovaries does indeed render them more prone to malignancies, even a hinted connection imposes an ethical obligation on the part of clinicians to inform patients, and introduces an uncertainty that egg donors must decide whether they feel comfortable about. While the ASRM Ethics Committee found nothing highly objectionable about oocyte donation, it did express "concern that the acceptance by a third party of a risk associated with ovarian stimulation . . . might be in excess of the expected benefit to the recipient couple."[11]

Most IVF programs have been scrupulously careful to avoid giving critics any grounds for accusing them of being involved in the purchase of oocytes. Yet there is another ethical issue: Donor oocytes are increasingly being seen as the solution to the infertility of older would-be mothers, those over forty years old, whose chances of conceiving through IVF with their own eggs are virtually zero. As Jean Benward has observed, there is a trend afoot in which increasingly younger donors are being roped in for the sake of increasingly older donors. The astonishing births to women who might be (or are) grandmothers, master-minded primarily by Mark Sauer and Richard Paulson at the University of Southern California IVF program, and by Severino Antinori in Rome, have only been accomplished thanks to the vigorous eggs of young women.

Benward has pointed out that between 1990 and 1993 the number of IVF births involving donor eggs rose from 122 to 800 annually. Whereas donors averaged around thirty at the beginning of that period, the average age had by the end hovered in the early twenties. At the University of Miami, coeds who have seen ads offering payment for donated eggs joke with their boyfriends about making extra cash, while at private clinics, callow girls wearing University of Hawaii sweatshirts spread their legs for transvaginal ultrasound scans, in which a probe is placed into their vaginas so that the progress of their hormonally prodded ovaries can be gauged. "Lots of big juicy follicles," remarked a physician while performing one such scan. A few days later, those follicles would be emptied through needle aspiration, now also done almost exclusively with the aid of transvaginal ultrasound, which eliminates the need for laparoscopy. Inserting the needle through the vaginal wall, the physician watches what is going on inside by way of a monitor and vacuums the eggs out follicle by collapsing follicle.

In vitro fertilization programs are nowadays known in part for the relative robustness of their donor pool. One physician has attributed the success of his donor egg program to the ready availability of "altruistic types"—what another physician called "ultra Earth Mothers"—living in the immediate vicinity. Others admit that the volunteers they get are women for whom the $2,000 payment (for time and inconvenience) is "a substantial amount of money." Said Richard Paulson of USC, "You don't get somebody who makes two thousand dollars a week in their profession, because for them it's not cost effective, no matter how altruistic they may feel about it."[12]

For those in the fertility business the "hot" meeting of the year, with more cachet and currency than the annual ASRM meeting, is the UCLA School of Medicine's annual program on IVF and embryo transfer, which is organized by David Meldrum—who teaches in the UCLA obstetrics and gynecology department in addition to directing his IVF clinic in Redondo Beach—and planned by a tireless team of administrators from the Office of Continuing Medical Education. Held at the four-star Four Seasons Biltmore in Santa Barbara, the four-day conference is all eucalyptus, cool breezes off the Pacific, and elegant three-course lunches alfresco on the impeccably preened grounds. Inside the packed conference room, an international audience sucks on hard candies from cut-glass bowls (a different flavor is provided each day) and soaks up the latest research from a bevy of stars.

On the roster is the soft-spoken Aussie Alan Trounson, looking more outback rancher than lab denizen. He's flown in from Singapore, where he is on sabbatical from Monash University. Zev Rosenwaks, trim, fastidious, with a bit of an autocratic European air, jets in for the Sunday afternoon and Monday morning sessions, then hurries back to New York Hospital–Cornell Medical College, where he is director. Rosenwaks has just weeks before unceremoniously ousted Jacques Cohen, the wry Frenchman who formerly headed the embryology lab of the Cornell program; at least publicly, Cohen exhibits sangfroid about it all. He has landed at the St. Barnabas Medical Center in West Orange, New Jersey. St. Barnabas, unlike Cornell, where some three thousand IVF cycles are initiated yearly, has not thus far made a name for itself in IVF. Cohen's new position elicits good-spirited jibes from some colleagues, who ask, "St. Barnabas? What's that?" albeit out of earshot of Cohen. In the wake of Cohen's departure, another Cornell physician, Jamie Grifo, has left to become head of the New York University IVF program, and the entire genetics team, led by Spaniard Santiago Munné, has also moved elsewhere. To help fill the gap at Cornell, Rosenwaks has imported a big gun from the Jones Institute, the embryologist Lucinda Veeck, who is also a presenter at Meldrum's conference. Perfectly coiffed, dulcet voiced, she commands an extraordinary amount of respect for someone who is not a reproductive endocrinologist. Also on hand is Gianpiero Palermo, Cornell's wizardly ICSI man, who at lunchtime shows comely modesty about his accomplishments but reveals a zeal for *la dolce vita* in his table talk with Bill Yee, director of the Long Beach Memorial Medical Center, who has just merged his private practice with that of Meldrum and his partner Arthur Wisot. Yee seems intent on reminding Palermo of some divertissements they enjoyed while attending a conference in Korea, home base for the man sitting at Yee's elbow, Kwang-Yul Cha, whose IVF program at Younsei University Medical College rivals Cornell's in size.

For all the chatter during lunches, coffee breaks, and buffet breakfasts, for all the trade in the latest gossip, there is not actually much breaking news at Santa Barbara. The conference is billed as a comprehensive update, but has the feel of a mopping-up exercise. More than anything, it accentuates the split in the IVF community which is becoming more pronounced with time, a split between those who do IVF strictly to alleviate infertility (these are the clinicians who worry foremost about success rates and hormonal protocols, who hunger to know what to do about refractory ovaries and want to take home the latest on cryopreservation regimes) and those who are paddling hard to catch the next big wave. And the next big wave, barely in evidence during the conference's formal presentations, is definitely preimplantation genetic diagnosis, that is, the screening of embryos—or what researchers nowadays prefer to call preembryos—while they are in the petri dish, to make sure they have the proper chromosomal makeup and to spot specific genes known to cause disease.

Preimplantation diagnosis is where molecular biology, genetic medicine, and IVF converge. If science were an ocean, the wave thrown up by this confluence would be a monster, a towering, four- or five-story killer like the Maverick's which sometimes rise up a couple hundred miles north of Santa Barbara off Half Moon Bay. This particular scientific surf began with groundswells back in the mid- to late 1980s, when researchers began developing techniques for removing material from cleaving embryos, subjecting them to certain molecular tests, and thereby diagnosing abnormalities.

Originally, a few people working with animal embryos, including Alan Handyside, then at Yale University, looked at DNA taken from polar bodies ejected during meiosis. They developed a means of assessing an embryo's chromosomal contents from the polar body by using polymerase chain reaction, or PCR (the same test that played such a large role in the O. J. Simpson trial). In PCR, researchers take a segment of DNA and load it, along with certain primers, chemical reagents, and buffers, into a machine that repeatedly heats and cools in rapid succession. Presto: The machine cooks up millions of copies of the DNA segment, which then exists in sufficient quantities to be analyzed. With PCR, it was possible to identify embryos that carried single genes responsible for hereditary disorders such as beta-thalassemia, Duchenne muscular dystrophy, sickle-cell anemia, and cystic fibrosis.

More recently, another molecular technique, fluorescent in situ hybridization, or FISH, has come to be favored because of PCR's susceptibility for contamination (and thus for false results). With FISH, single-stranded DNA probes tagged with a fluorescent substance are added to DNA taken from embryos. Like puzzle pieces, these probes will only mesh with those lengths of DNA possessing a complementary "shape," which in the case of DNA means a complementary sequence of the four bases which form the rungs of the ladderlike DNA helix. When the probes attach, or anneal, it is a signal that a certain gene or DNA sequence is present. Under a fluorescence microscope, the annealed probes glow like deep-sea creatures.

Researchers at several programs around the country have now begun to yoke preimplantation genetic diagnosis with IVF. Under Rosenwaks's direction at Cornell, Jamie Grifo, who trained under Handyside, and geneticist Santiago Munné pioneered the use of blastomere biopsy in a clinical setting. Instead of looking at the first polar body, they would remove a single blastomere from a six- to ten-cell embryo. (As animal studies have shown, the removal of one cell appears not to harm the embryo at all.) Fluorescent in situ hybridization then revealed whether a chromosome or known gene was present or absent. The technique is especially good for identifying aneuploidy, in which too many chromosomes are present.

Through blastomere analysis of this sort, Grifo screened the embryos of one couple who feared having a hemophiliac son and eliminated those bearing a Y chromosome; only X-bearing embryos went to transfer, and the couple gave birth to a girl. However, the Cornell team made a misdiagnosis in another case. When standard prenatal screening revealed that the fetus was affected by the genetic ailment the couple had sought to avoid, they chose an abortion, which Grifo remorsefully says was devastating for everyone involved.

By 1994, Cornell tallied a total of six children who had been born after transfer of embryos screened for various flaws. Worldwide, some thirty children had been born after preimplantation screening. Grifo and most others involved in the research acknowledge its controversial nature.

For one thing, it brings to the fore, once again, the debates over the status of the embryo and about what research on embryos, if any, is permissible. That most zealous defender of embryonic sanctity, the Catholic Church, has tacitly rejected the morality of preimplantation genetics. In 1987, the Vatican promulgated its *Instruction on Respect for Human Life in Its Origin and on the Dignity of Procreation, Replies to Certain Questions of the Day*, otherwise known as *Donum Vitae*. The document, composed by the Sacred Congregation for the Doctrine of the Faith, voiced again the church's blanket condemnation of abortion, AID, surrogate motherhood, and the creation of embryos for research purposes alone. Although it allowed that prenatal diagnosis was morally licit if it "respects the life and integrity of the embryo and the human fetus and is directed toward its safeguarding or healing as an individual," it argued that such screening, when carried out "with the thought of possibly inducing an abortion depending upon the results," was "a gravely illicit act."[13] Since it also found IVF in general to be morally illicit, preimplantation screening was, by extension, unacceptable:

> The connection between *in vitro* fertilization and the voluntary destruction of human embryos occurs too often. This is significant: through these procedures, with apparently contrary purposes, life and death are subjected to the decision of man, who thus sets himself up as the giver of life and death by decree. This dynamic of violence and domination may remain unnoticed by those very individuals who, in wishing to utilize this procedure, become subject to it themselves. . . . [T]he abortion mentality which has made this procedure possible thus leads, whether one wants it or not, to man's domination over the life and death of his fellow human beings and can lead to a system of radical eugenics.[14]

Indeed, it is hard to counter the contention that preimplantation screenings, like amniocentesis and sonography for purposes of aborting defective fetuses, partakes of eugenics. In effect, it is the fulfillment of race hygienist Alfred Ploetz's aspiration to push eugenic selection to the level of the germ plasm. From the point of view of the ASRM Ethics Committee, this is a plus, not a minus, in that this form of screening "will preclude the need for a couple to undergo later prenatal diagnosis and terminate a pregnancy in case of a positive result."[15] As has often been noted, the psychological impact of deciding to abort a damaged conceptus or one carrying deadly disorders can be profound, especially if quickening has occurred, as it generally does around the time women undergo amniocentesis. In the weeks between the sampling of the amniotic fluid and the arrival of results from the test, the fetus will have begun to move, which for many women marks the beginning of recognition of the fetus as an individual.

Havelock Ellis once said, "The superficially sympathetic man flings a coin to a beggar; the more deeply sympathetic man builds an alms house for him so that he need no longer beg; but perhaps the most radically sympathetic of all is the man who arranges that the beggar shall not be born."[16] Following this logic, the proponents of preimplantation screening would arrange that "children who would experience severe health problems that would diminish their quality of life and longevity" and whose illnesses would impose psychological, financial, and other hardships upon their families and those close to their families shall not be born.[17]

While many parents or would-be parents see in prenatal screening a humane alternative to raising a severely deformed or diseased child, sociologists have found that this outlook is not necessarily universal. Women and couples within certain cultural communities may see amniocentesis or CVS as an invasion of their bodies and lives, an imposition requiring them to make life-and-death decisions that they would rather not be forced to make.

Against these deeply felt personal responses, modern science and medicine positions itself as the champion of human values, arguing that to fail to act to eliminate the vagaries of nature is to be inhumane. Bentley Glass, writing in the *Quarterly Review of Biology* in March 1993, stated this position clearly:

> Idealism among scientists is not dead, and so long as the genetic load is a danger to the well-being of the population of any country, or of the whole world, and so long as there is a prospect that the human genome and its environment can be improved beyond their present state, eugenics in its broadest sense will continue to attract the ideals of geneticists, physicians, environmentalists, and in fact every body else. For the ideal is simply that of human betterment, for every class, color, and creed. The history of eugenics to the present day merely reflects human failure, prejudice, and the imminent danger of hasty prejudgment in the absence of sufficient knowledge.[18]

In fact, historically, prejudice and hasty prejudgment have animated more eugenic dreams than idealism. Most eugenicists who have succeeded in translating their beliefs into action have not been interested in creating compassionate, fair-minded, color-blind, and nonsexist societies. Instead they have been driven by the conviction that some select group or groups—of a certain class, race, or nationality—are better than others. In this sense, then, Glass has unintentionally identified the root causes for the failure of eugenics, while trying to excuse that failure.

At base, the new eugenics is as misbegotten as the old. The old eugenics argued that a person's characteristics were in a one-to-one correspondence with the quality of his or her "blood" or "germ plasm." The new eugenics establishes the same equation, replacing characteristics with phenotype and "blood" with genotype. Preimplantation screening holds out the possibility of testing not just for disease-causing abnormalities but also for other characteristics. Basically, it could be used to screen the presence or absence of any gene whose function and sequence had been identified. Would-be parents could thus reject embryos for purely cosmetic reasons, or demand that they be given embryos of only one gender.

Lest one think that this is far-fetched, physicians from IVF programs across the country can attest to the frequency with which they have been asked to perform sex selection, usually because patients want to guarantee that they have a boy. Around the world, people almost universally express a preference for boys as first children. Several clinicians told me that, as with AID, it is not unusual for couples involved in donor egg IVF to attempt to ensure that their offspring will have a certain skin color. For instance, black couples have sometimes requested white egg donors because they want a light-skinned child, and there was the highly publicized case in Italy when a white man and black woman opted for a white donor in hopes, they said, of minimizing the racial bias the child would encounter. One physician told me, "We've had a lot of mixed-race couples—Asian and white, black and white, Asian and black—all different combinations, and we've chosen a different gamete depending on what their particular need is." The same physician went on to add that he has frequently listened to couples express an interest in obtaining, as it were, other traits in their offspring. Infertility counselors report that many IVF patients they see have a fantasy of a perfect baby, and often do not see beyond the pregnancy. Unlike people who choose to adopt, they often seem to be fixated on having, in the sense of possessing, a child, rather than on being a parent.

It must be admitted that some things improve. The smog in Los Angeles, for instance, seems less awful than it used to be. But it still settles over the sprawling USC Medical Center, shrouding it from a distance, and, once you are inside one of the buildings there, dissolving anything further away than the freeways slicing past on either side. In W. French Anderson's USC lab, cynicism has been banished. Anderson, a sinewy, fourth-degree black belt in Tae Kwon Do with an infectious can-do attitude, is a man for whom the concept of a goal might as well have been invented. Maintaining a lab at the National Institutes of Health as well as at USC, he is arguably the country's preeminent gene therapy researcher.

Almost since the day that English biologist Lionel Penrose confirmed that some afflictions are indeed passed from parents to children, physicians have sought to vanquish these genetic scourges. Until recently, they could do little besides amassing information. Today, thanks to molecular biology, geneticists know of at least four thousand specific mutations associated with disease and can test for the presence of fifty or more. Through techniques which allow them to copy and insert genes into living cells where they are expressed, that is to say, turned on so that they produce a given protein, scientists are gradually gaining the ability to reverse some of these ailments; through germ-line gene therapy, in which they would insert or delete genes from sperm and eggs, or from embryonic cells, they hope one day to eliminate them altogether.

Before receiving approval for the first test of gene therapy on a patient, W. French Anderson argued his case in person and on paper before seven different regulatory bodies, who reviewed the proposal fifteen times. Finally, in January 1989, the NIH and Food and Drug Administration gave Anderson and his team the go-ahead to perform a procedure in which a single gene was inserted into a type of cell called a tumor-infiltrating lymphocyte, taken from a patient's cancer tumor. The gene was smuggled into the DNA of these lymphocytes by a retrovirus. Retroviruses, of which HIV is one, possess the capacity to infiltrate cells and insert themselves into the DNA contained within. Most gene therapy experiments rely on this retroviral talent: Any snippet of DNA piggybacked on the retrovirus will get imported into cells and incorporated. The bit of DNA Anderson and his team inserted coded for the protein interleukin-2, which is normally produced by the body and boosts immune system function. Having attached a marker to the altered tumor-infiltrating lymphocytes, the researchers were able to ascertain that the cells survived when restored to the patient's body.

This experiment did not cure anyone, nor was it meant to. It was just the first step on the long road toward the genetic treatment of disease. In 1991, Anderson followed up this test with the first real attempt to actually supply a patient with cells containing a gene he or she lacked, and thereby to reverse the patient's disease. The disease was adenosine deaminase deficiency (ADA), in which a genetic glitch blocks production of an enzyme essential to immune function. With NIH colleagues Michael Blaese and Kenneth Culver, Anderson treated two girls suffering from ADA by retrovirally inserting the missing gene into their white blood cells and then returning the cells to their bodies. As expected, the altered cells multiplied, the girls' ADA levels rose, and their immune systems began to work with a semblance of normalcy. In time, though, the new cell lines died out, and the girls required additional treatments.

While a number of genetic ailments might one day be controlled or eliminated by gene therapy, scientists will probably still be laboring well into the next century to refine the technology. Take the delivery system used by Anderson early on: Retroviruses, it turns out, may promote certain forms of cancer and thus prove undesirable in the long run. Disabled adenoviruses, a family to which the mutable viruses that cause the common cold belong, may be more benign conveyances. Anderson and his teams are throwing enormous brainpower into devising delivery systems that will be efficient and harmless.

In the meantime, researchers elsewhere are already at work on techniques for performing germ-line alterations. University of Pennsylvania researchers announced in late 1994 that they had successfully inserted genes into the primordial sperm cells of sterile mice, placed them in the animals' testes, and thereby produced offspring which also bore the gene. Alarmed ethicists, reading the writing on the wall, called for a national review of germ-line gene therapy before it had advanced any further. Writing on the op-ed page of the *Times,* John Maddox, editor of the journal *Nature,* branded those who did so "doomsayers [who] ignore the technical difficulties of applying to humans what has been done in mice." He went on to suggest that even if germ-line therapy became possible, it would in all likelihood prove benign, or even ineffectual, considering that the gene pool is unimaginably diverse. Indeed, the estimated 50,000 to 80,000 genes distributed among the forty-six chromosomes, known collectively as the human genome, come in many varieties. Each version, or allele, may or may not translate into phenotypic difference in a given generation, but somewhere down the line, combined with another allele, might convey benefits or detriments to the possessor. Maddox wrote that

> it is an illusion to think that a genetically "pure" society could ever be created. Many disease-linked genes arise spontaneously in apparently normal families, including Huntington's disease and Fragile X syndrome (which is one of the most common genetic causes of inherited mental defect).[19]

Maddox also took issue with the doomsayers for exaggerating the risks of one, not yet fully formed, technology while ignoring another which already had come on line: "They overlook the potential for genetic improvement through existing fertility techniques. And they forget that eugenics programs like those of the Nazis would require governmental compulsion that is unlikely."[20] Yet later in the same editorial, Maddox allows that the Chinese government was planning a nationwide program to prevent those citizens with heritable diseases from procreating. In addition, Maddox did not mention, but surely should have known about, the Chinese birth control program, which has attempted to limit each couple's family to one child. In villages, couples who attempt to evade this edict have been rousted from their homes by local officials who have proceeded to level the houses. Moreover, in the United States, welfare agencies have in certain locales required women receiving government money to have the time-release contraceptive Norplant inserted into their arms. So one wonders which governments Maddox could be thinking of that lack the ability or will to compel their citizens on matters having to do with reproduction.

Those within the fertility community who have championed the "reproductive rights" argument have tended to adopt a laissez-faire attitude toward most of the projected forms of genetic selection. Constitutional scholar John Robertson, who laid out all sides of the argument in his 1994 work, *Children of Choice*, but generally can be characterized as an advocate for the reproductive technologies, wrote:

> Although it may not count as part of a core procreative liberty, non-therapeutic enhancement may nevertheless be protected. A case could be made for prenatal enhancement as part of parental discretion in rearing offspring. If special tutors and camps, training programs, even the administration of growth hormone to add a few inches to height are within parental rearing discretion, why should genetic interventions to enhance normal offspring traits be any less legitimate? As long as they are safe, effective, and likely to benefit offspring, they would no more impermissibly objectify or commodify offspring than postnatal enhancement efforts do.[21]

W. French Anderson can see a certain logic in such an argument, and understands how parents might feel the necessity of making such interventions. Suppose there were a gene that could be shown to improve memory? Children possessing it might do better in school. Wouldn't parents be extremely desirous of giving their child that extra edge? But Anderson, for all his promotion of gene therapy, has repeatedly cautioned against germ-line enhancements. Finally, they do not belong in the same class as "postnatal enhancement efforts," as ballet classes and after-school math tutors. It is worth quoting Anderson at length on this topic:

> Medicine is a very inexact science. We understand roughly how a simple gene works and that there are many thousands of house-keeping genes, that is, genes that do the job of running a cell. We

predict that there are genes which make regulatory messages that are involved in the overall control and regulation of the many housekeeping genes. Yet we have only limited understanding of how a body organ develops into the size and shape it does. We know many things about how the central nervous system works—for example, we are beginning to comprehend how molecules are involved in electric circuits, in memory storage, in transmission of signals. But we are a long way from understanding thought and consciousness. And we are even further from understanding the spiritual side of our existence.

Even though we do not understand how a thinking, loving, interacting organism can be derived from its molecules, we are approaching the time when we can change some of those molecules. Might there be genes that influence the brain's organization or structure or metabolism or circuitry in some way so as to allow abstract thinking, contemplation of good and evil, fear of death, awe of a "God"? What if in our innocent attempts to improve our genetic make-up we alter one or more of those genes? Could we test for the alteration? Certainly not at present. If we caused a problem that would affect the individual or his or her offspring, could we repair the damage? Certainly not at present. Every parent who has several children knows that some babies accept and give more affection than others, in the same environment. Do genes control this? What if these genes were accidentally altered? How would we even know if such a gene were altered?[22]

Still, there are those who hold, as biologist Clifford Grobstein, that humanity is on the verge of a revolutionary transition, from chance to purpose, from genetic roulette to genetic determinism. As the long line of eugenicists did before them, they argue that humans can no longer, as Grobstein has written, "shift responsibility, whether to Divinity, Chance, or Unkind Fate."[23] We have to become the creators of ourselves.

Such arguments are invariably contrapuntal, pitting helplessness against control, unintentional actions against forethoughtful ones. An assumption is made that we can now grasp human evolution, which proceeded previously in a miserable fashion, and turn it to our own ends. Even providing that we could, should we want to?

Such arguments further assume that genes equal fate. Hereditarians have contended that IQ correlates strongly with worldly success, and the most extreme proponents of this view, including Charles Murray, the late Richard Herrnstein, and the British behavioral geneticist Hans Eysenck, have argued that everything from intelligence to religious and political beliefs are owed primarily to inheritance, and that economic and class differences in society are owed to inborn differences. Eysenck, in an introduction to a 1978 edition of Galton's *Hereditary Genius,* proclaimed that those who write about society "without reference back to what is known about behavioural genetics are writing in sand; they fail to mention the most important factors responsible for the particular shape human organizations take, and are indeed bound to take."[24] Yet nearly all contemporary brain research is showing that it is not genes so much as the developmental stimuli children receive during gestation and early childhood which shape their future selves. The brain is extremely plastic, meaning that it is continually forging neuronal connections as a child grows, connections that depend to a large degree on what the child experiences. What he or she sees, hears, and feels during those first crucial years will heavily influence his or her emotional and mental makeup. Recent studies have even strongly indicated that aggression in young men is caused by neuronal factors that are owed directly to childhood abuse and neglect.

The hereditarians, for the moment, hold center stage in the scientific world. The fanfare surrounding the ambitious $3 billion Human Genome Project, which aims to lay bare the entire genetic makeup of the species, attests to the powerful appeal of behavioral genetics, as does the resurgence of eugenic attitudes regarding poverty, violence, and other social ills. Even the marketers have gotten into the game. For example, analysts Don Peppers and Martha Rogers have salivated over the opportunities presented by the Human Genome Project:

> As sweeping and pervasive as the media technology revolution may be, the pace of discovery and innovation in biotechnology is even more startling. The University of Minnesota studies of identical twins—in which identical twins separated at birth frequently smoke the same brands of cigarettes and buy the same brands of clothing—strongly indicate that understanding the human genome could offer insights into the marketplace. . . . In an era of targeted marketing, might not the human genome be the best database of all?[25]

It seems to make little difference, in this case, that the statistical methodology and assumptions of twin studies have repeatedly been shown to be faulty and their conclusions scientifically bogus by Stephen J. Gould, Richard Lewontin, and Ruth Hubbard, among others. Nor do enthusiasts pause to reflect upon the fact that already employers and insurance companies have shown a disposition to use genetic information as a way of discriminating against "high-risk" employees and clients.

With regard to offspring, the overemphasis on genes and on the primacy of a genetic tie pushes into the background an essential factor in the parent-child equation: the quality of the parenting. In the end, what difference does a genetic link make if children are the repeated targets of a mother's or father's rage, indifference, or self-serving manipulation? And what good is a supposedly superior genotype if a child must daily battle to survive poverty, nonexistent or inadequate medical care, abysmal schooling, and threats of physical harm? Only those aspects of character we deem bravery and persistence, combined with a large dose of luck, will carry such children through, and these are unlikely to be conveyed by genes.

What children need is not "good genes"—whatever those are—but love, physical affection, food, clothing, and shelter, rules, discipline, moral instruction, acceptance, mental stimulation, and a sense that in their lives justice, fair play, humor, friendship, and a valued place in their families and communities are distinct possibilities. Genetic tinkering will not provide these, only a concerted commitment by parents and families, schools, churches, and other social organizations. Greater government expenditures on prenatal care, day care, health care, and education for children, and greater corporate and government support for childbearing and child raising are also vital.

Children do not need perfection. Perfection is not about children. It's about our own needs.

Of any human endeavor it can be asked, Who does it profit? And how so? Emotionally? Materially? Intellectually? The motives of the individual actors in any community enterprise, too, can be assessed. Why do they do it? For the love of the game? For the money? For the glory? And, of course, these motives may be mixed in any one person, and also susceptible to outside influence, as in clinical science, where the physician-cum-researcher operates, literally and figuratively, within the framework of a hospital, private practice, or academic department—and, in some cases, all three at once—and is subject to the demands and aims of these institutions.

Those who have collectively devised the technologies for manipulating sperm, eggs, and embryos serve under the rubric of science, by which I mean not pure knowledge, but rather the set of economic, sociological, and intellectual relationships that constitutes modern research. It is fair to say that the driving forces behind modern science are money and politics: findings that will bring in dollars (or marks or yen) and that will cater to a perceived need of a government, no matter how distant from the realities of daily life of its citizens that perceived need may be (the Star Wars scheme of Ronald Reagan comes immediately to mind). For a scientist, new findings represent not only a triumphant further insight into the enigmas of the natural world, but also direct fiscal and social gains: more grant money, greater status among one's peers, perhaps public acclaim and a lucrative international prize. Like journalists, whose paramount aim is always to get the scoop or catch the next big trend right as it breaks, scientists adore the cutting edge: It's what they live for; it's what redeems all the hours in the lab watching experiments fail.

Like quantum physicists, reproductive endocrinologists and geneticists and embryologists have taken it upon themselves to fiddle with some pretty fundamental stuff. The stuff of life. As such, they have entered an area which concerns us all, scientists and nonscientists alike. Is it any wonder that their efforts have drawn such fire? Often, those in the field of assisted reproduction have expressed perplexity or impatience or anger at the response of the common run—that undifferentiated "media" or "public" which asks them to justify what they are doing. This is not naiveté on their part, but rather an ingrained elitism: Like the fundamentalist who knows himself to be about God's work, the clinical scientist perceives himself as a righteous upholder of not only Hippocrates but also Prometheus. Why not steal fire from heaven if it can be had?

I would hardly argue that we are better off without the fire. But that transmogrified fire, that fire of Trinity, with Siva's mushroom cloud over the white sands—certainly, we did not need that.

This is not a question of obliterating technology. It is a matter of recognizing the dark impulses which have guided our species vis-à-vis reproduction, of recognizing the unsavory fantasies adults have regarding children. Upon the unborn, and then upon the born, we impose images of perfection—whatever those may be for us, whether physical, moral, intellectual, or social. We want our children to be what we cannot: above the mundane world, immortal, ourselves incarnate. Just like nuclear power, they should provide us with clean energy endlessly for pennies.

On a practical level, we want them safe, worry-free, always triumphant. We desire that no one should suffer, those closest to us least of all. And yet we do, all, suffer. The culture of physicians, like that of Jains (whose tenderness extends even down to the merest myrmidon or microbe), holds any suffering anathema. Yet on an individual basis, physicians can be callous, unfeeling, indifferent to their patients as often as they are attentive, respectful, and kind. Thus, assisted reproduction may in principle alleviate the suffering of the childless but on another basis altogether perpetuate harm.

Can this not be said of almost any enterprise? Yes, but the essential and inescapable difference is that the object of action here is life: the creation of life and the fiddling with fate. Here the powers of science fail. Here its determinative reach exceeds its grasp.

We are out on the border between the known and the unknown, trying to decide what to do. Our only reference point is history, and history tells us that we have not, as a group, acquitted ourselves well with regard to children and the weak, much less with regard to more capable, stronger members of our group. So the question becomes, given our propensities for overreaching and overexpectation, should we undertake to rearrange the basic stuff of our being, with the intent of improving ourselves? Should we do it in the name of eliminating suffering? The choices are already being made. Couples, with their doctors, are making them. It is time that tighter controls were put in place. Whether governmentally contrived or community based, oversight committees should be instituted, seating people from diverse walks of life with the goal of ensuring that the new reproductive technologies are not abused by those with misguided eugenic notions. Indeed, in November 1995, the ASRM itself averred in a press release that "now is the time to consider establishing an independent licensing authority," akin to the one in Britain that regulates embryo research.

In the modern era, we have commodified kids, to a far greater degree than previously. We want them to be delivered in the model and color of our choice. In earlier times and places—say, Renaissance Europe—people deigned to consider human only those children who survived the rigors of wet-nursing and parental neglect. Now we want offspring whose genotypes have been declared free of defect.

Every age perpetuates its own style of cruelty. And every age attempts to compensate itself for what it lacks. So we, in an age billed as the epitome of history, in which all nature has yielded to our mastery, compensate ourselves by demanding absolute control over that most concrete expression of futurity, our progeny. Compensate, because in fact, all the advertisements aside, we lack fundamentally in this age of runaway population, transnational pollution, potential global annihilation, any sense of control. At the most critical level, we do not know if we, as a species, will survive.

The advocates of the Human Genome Project tell us that their way, the way of ultimate genetic knowledge, will set us free. Those who would tinker with our genes tell us that society must simply "catch up" with science and medicine, must look to future rewards rather than present-day reservations. Even if we presume that this is a decent method for making societal decisions—and it clearly is not—there is no reason to think that modern eugenics will be any more exalted an enterprise than the old version. Humans have long since possessed the tools for crafting a better world. Where love, compassion, altruism, and justice have failed, genetic manipulation will not succeed.

NOTES

INTRODUCTION

1. Steve Fishman, "Inconceivable Conception," *Vogue*, December 1994, p. 306.
2. Ashley Montagu, *Man's Most Dangerous Myth* (New York: Columbia University Press, 1945), p. 62.
3. Philippe Ariès and Georges Duby, eds., *A History of Private Life*, vol. 1, *From Pagan Rome to Byzantium*, ed. Paul Veyne and trans. Arthur Goldhammer (Cambridge, Mass.: The Belknap Press of Harvard University Press, 1987), p. 20.
4. Hermann J. Muller, "Human Evolution by Voluntary Choice of Germ Plasm: This procedure should be more acceptable and effective than differential control over family size," *Science* 134 (1961): 643.
5. Quoted in Troy Duster, *Backdoor to Eugenics* (New York: Routledge, 1990), p. 46.

CHAPTER I

1. See, for instance, Patricia Mahlstedt, "The Psychological Component of Infertility," *Fertility and Sterility* 43 (1985): 335; or "Psychological Issues of Infertility and Assisted Reproductive Technology," *Urologic Clinics of North America* 21 (1994): 557.
2. Mahlstedt, "The Psychological Component of Infertility," op. cit., p. 336.
3. Ibid., p. 341.
4. Judith Lorber and Lakshmi Bandlamudi, "The Dynamics of Marital Bargaining in Male Infertility," *Gender and Society* 7 (1993): 33.
5. Ibid., p. 42.
6. Anjani Chandra and William D. Mosher, "The Demography of Infertility and the Use of Medical Care for Infertility," *Infertility and Reproductive Medicine Clinics of North America* 5 (1994): 293.
7. In 1992, the Fertility Clinic Success Rate and Certification Act, known as the Wyden Law, was passed by Congress. It will regulate embryo labs and require reporting of success rates through the Centers for Disease Control, and was in the process of being implemented as this book went to press.
8. Marilyn Strathern, *Reproducing the Future: Anthropology, Kinship, and the New Reproductive Technologies* (New York: Routledge, 1992), p. 31.
9. Ibid., p. 31.

CHAPTER 2

1. William L. Langer, "Infanticide: A Historical Survey," *History of Childhood Quarterly* 1 (1974): 353.
2. Laila Williamson, "Infanticide: An Anthropological Analysis," in Marvin Kohl, ed., *Infanticide and the Value of Life* (Buffalo, N.Y.: Prometheus Books, 1978), p. 61.
3. Alexander Morris Carr-Saunders, *The Population Problem: A Study in Human Evolution* (London: Oxford University Press, 1922), p. 262.
4. John Wymer, *The Paleolithic Age* (New York: St. Martin's Press, 1982), p. 171.
5. Fossilization may also occur through a process in which atoms of minerals replace

those of an organism one for one. Commonly, marine fossils have been formed under intense pressure in deposits where oxygen is almost absent. In this case, rock layers bear a highly detailed impression of the organism, often glazed with carbon.

6. See Henri V. Vallois, "The Social Life of Early Man: The Evidence of Skeletons," in Sherwood Washburn, ed., *Social Life of Early Man* (Chicago: Aldine Publishing Co., 1961), p. 227.

7. Thomas McKeown, *The Modern Rise of Population* (New York: Academic Press, 1976), p. 157.

8. Charles Darwin, *The Origin of Species and The Descent of Man* (New York: The Modern Library, 1936), p. 611.

9. Vallois, op. cit., p. 225. Admittedly, contemporary archaeologist Jeffrey Schwartz has argued that in fact it is easy to miscategorize adult skeletons. Among Arctic groups, for instance, female skulls develop masculine features whereas male pelvises are quite feminine. In sub-Saharan Africa, male skulls look female and female pelvises look masculine. Schwartz says that it is nearly impossible to sex younger individuals by skull, hip, or pelvis, since boys and girls are virtually identical until pubescence. Personal communication and Jeffrey Schwartz, *What the Bones Tell Us* (New York: Henry Holt & Co., 1993), p. 59.

10. Vallois, op. cit., p. 225.

11. See Mildred Dickemann, "Infanticide in Humans: Ethnography, Demography, Sociobiology, and History," in Glenn Hausfater and Sarah Blaffer Hrdy, eds., *Infanticide: Comparative and Evolutionary Perspectives* (New York: Aldine Publishing Co., 1984), p. 428.

12. Birdsell in Richard Lee and Irven DeVore, eds., *Man the Hunter* (Chicago: Aldine Publishing Co., 1968), p. 239. He also said (p. 236) that "systematic infanticide may be assumed to have characterized human populations throughout the Pleistocene. Its probability of being preferential female infanticide is strengthened by data from recent hunters."

13. Some computer models have indicated that a population practicing selective female infanticide would doom itself to extinction in the space of a century or so; however, in 1974, one researcher used a Monte Carlo simulation to show that as long as several local groups affiliated for the purpose of mate selection, forming a larger entity numbering at least 175—what Birdsell called the effective breeding unit—they could "ride out any stochastic fluctuations in sex ratios, mortality and fertility" and survival would be assured. See Clive Gamble, *The Palaeolithic Settlement of Europe* (Cambridge England: Cambridge University Press, 1986), p. 50.

14. Joseph B. Birdsell, "Some Population Problems Involving Pleistocene Man," *Cold Spring Harbor Symposia of Quantitative Biology* 22 (1957): 68.

15. For an account of this line of argument, see especially Nancy Howell, "Toward a Uniformitarian Theory of Human Paleodemography," in R. H. Ward and K. M. Weiss, eds., *The Demographic Evolution of Human Populations* (New York: Academic Press, 1976), pp. 30–32. Note that Howell does not begin with the premise, as do so many interpreters, that these figurines were fertility figures, but rather comes up with a plausible argument for why they might have been. This is a subtle but important distinction.

16. See Albert J. Ammerman and L. L. Cavalli-Sforza, *The Neolithic Transition and the Genetics of Populations in Europe* (Princeton, N.J.: Princeton University Press, 1984), pp. 64–66, for a version of this argument. Upswings in population can also be spurred by or tied to falling death rates, but this does not seem to have been the case in the Neolithic run-up. Ammerman and Cavalli-Sforza wrote (p. 64), "By and large . . . no major change in the expectation of life at age 15 seems to have accompanied the transition to agriculture. Until the last three centuries, it would appear that no major change occurred in mortality rates of human populations. One might expect a decrease in mortality to be associated with the transition, if agriculture ensured a more reliable food supply. But, as mentioned earlier, the nutrition of hunters and gatherers is in most cases qualitatively and quantitatively quite satisfactory."

17. Atkinson has more formally declared this view, which for him is an item of faith. For example, in a 1986 book review, Atkinson expressed his philosophy in commenting on the author's attempt to "build hypotheses about contemporary social structure" on the Orkney Islands during Neolithic times. Atkinson wrote, "It must be admitted . . . that the inferences which can *validly* be drawn (that is, are *compelled* by the evidence, without ambiguity or the exercise of pure preference or prejudice) are incommensurate with the labour so admirably expended on the collection and marshalling of the [Neolithic] evidence itself; but this is an inherent and inevitable limitation of all studies of pre-literate remains which cannot speak for themselves. . . . The fact is that in prehistory, in Orkney as elsewhere, social structure is a will-o'-the-wisp, endlessly pursued but for ever out of reach." See R. J. C. Atkinson, *Antiquity* 60 (1986): 68.

18. On this issue, the archaeologist and author Evan Hadingham is illuminating. See his *Secrets of the Ice Age* (New York: Walker & Co., 1979), p. 16. Also see Frank Poirier, *Understanding Human Evolution* (Englewood Cliffs, N.J.: Prentice-Hall, Inc., 1993), p. 53.

19. See, for example, Hausfater and Hrdy, op. cit., pp. xxviii, 4, 449.

20. See, for example, Kohl, op. cit., p. 63; Hausfater and Hrdy, op. cit., pp. xviii, 446. In *Death Without Weeping* (Berkeley: University of California Press, 1992), anthropologist Nancy Scheper-Hughes mentions this sort of rationale as operating in impoverished communities of the Nordeste in present-day Brazil.

21. Herbert Aptekar, *Anjea: Infanticide, Abortion and Contraception in Savage Society* (New York: William Godwin, Inc., 1931), p. 88.

22. See, for example, Hausfater and Hrdy, op. cit., p. 493.

23. Darwin, op. cit., p. 609.

24. Quoted in Lee and Devore, op. cit., p. 81.

25. See Marvin Harris, *Cannibals and Kings* (New York: Random House, 1977), p. 16.

26. Hausfater and Hrdy, op. cit., p. 427.

27. Aptekar, op. cit., p. 170.

28. In Kohl, op. cit., p. 64.

29. See Hausfater and Hrdy, op. cit., p. 493; Aptekar, op. cit., p. 67.

30. Quoted in Hadingham, op. cit., p. 134.

31. Harris, op. cit., p. 4.

32. Quoted in David Rindos, *The Origins of Agriculture: An Evolutionary Perspective* (New York: Academic Press, 1984), p. 2.

CHAPTER 3

1. Plutarch, "Life of Lycurgus," in *The Lives of the Noble Grecians and Romans,* trans. John Dryden, Great Books of the Western World Series, ed. Robert Maynard Hutchins, vol. 14 (Chicago: Encyclopaedia Britannica, Inc., 1952), p. 39.

2. Ibid., p. 40.

3. See W. K. Lacey, *The Family in Classical Greece* (Ithaca, N.Y.: Cornell University Press, 1968), p. 196.

4. In Mary R. Lefkowitz and Maureen B. Fant, *Women's Life in Greece and Rome: A Source Book in Translation* (Baltimore: Johns Hopkins University Press, 1992), p. 84.

5. Plutarch, op. cit., p. 40.

6. Ibid., p. 39.

7. Ibid., p. 40.

8. Sophocles, *Oedipus Rex,* in *Three Theban Plays,* trans. Henry H. Stevens (Yarmouthport, Mass.: The Register Press, 1958), p. 36.

9. Although, as Claude Lévi-Strauss warned, one must be careful about assuming that the myths of a people reflect their social reality, it hardly seems far-fetched to assume that mythic incidents of infanticide had resonance in the minds of a people who so widely practiced infanticide. Greeks being no strangers to the concept of guilt, I take the liberty of inferring that they may have experienced a degree of this emotion in

regard to the systematic killing of babies, even in the face of the numerous social justifications offered for the habit.

10. See John M. Riddle, *Contraception and Abortion from the Ancient World to the Renaissance* (London: Harvard University Press, 1992), pp. 7–9. The translation of the Hippocratic oath given by *Stedman's Medical Dictionary* flatly declares, "Nor will I give a woman a pessary to procure abortion."

11. Riddle, op. cit., p. 28.

12. J. T. Hooker, in A. Powell, ed., *Classical Sparta* (Norman: University of Oklahoma Press, 1988), p. 136, has argued that Plato may have modeled his ideal state on the so-called Lycurgan reforms of Sparta.

13. *Great Dialogues of Plato*, ed. Eric H. Warmington and Philip G. Rouse and trans. W. H. D. Rouse (New York: The New American Library, 1956), p. 257.

14. Ibid., p. 259.

15. Ibid., p. 258.

16. Ibid., p. 259.

17. Ibid.

18. Ibid.

19. In an unconscious implementation of this Platonic scheme, the Catholic Church in Italy employed unwed women who had just given birth to nurse infants in its foundling homes during the eighteenth and nineteenth centuries. Officials of the homes took care to prevent these mothers from identifying their infants.

20. Historian Sarah Pomeroy, in *Women in Hellenistic Egypt* (New York: Schocken Books, 1984), p. 126, pointed out that the word *paidiskoē* refers to either a young girl or a female slave, while *pais* or *paidarion* refers to either a child or a male slave of any age.

21. Ibid., p. xviii.

22. Aristotle, in *Politics*, Book III, Chapter 19, wrote, "For the noble are citizens in a truer sense than the ignoble, and good birth is always valued in a man's own home and country. Another reason is, that those who are sprung from better ancestors are likely to be better men, for nobility is excellence of the race." In Book III, Chapter 14, he argued that "barbarians, being more servile in character than Hellenes, and Asiatics than Europeans, do not rebel against a despotic government." Because these ethnic characteristics were thought to derive from the blood, there was a danger posed by intermixing. *Introduction to Aristotle*, ed. Richard McKeon (New York: Random House, 1947), pp. 602, 607.

23. Pomeroy, op. cit., p. xvii.

24. Aristotle, op. cit., p. 617.

25. Ibid., p. 623.

26. Ibid., p. 617. Plutarch, several centuries later, echoed Aristotle in his *Moralia*, where he wrote, "Carry the discussion back to primitive mankind, to those whose women were the first to bear, and whose men were the first to see a child born; they had neither any law which bade them rear their children, nor any expectation of gratitude or of receiving the wages of maintenance 'lent to their children when they were young.'" Plutarch also quoted Plato, who in his Laws asserted that rearing children is a thankless task. Plutarch went on, "He that plants a vineyard in the vernal equinox gathers the grapes in the autumnal; he that sows wheat when the Pleiades set reaps it when they rise; cattle and horses and birds bring forth young at once ready for use; but as for man, his rearing is full of trouble, his growth is slow, his attainment of excellence is far distant and most fathers die before it comes." And he suggested that mainly poor men choose not to rear their children: "Since they consider poverty the worst of evils, they cannot endure to let their children share it with them, as though it were a kind of disease, serious and grievous." *Plutarch's Moralia*, trans. Frank Cole Babbitt (Cambridge, Mass.: Harvard University Press, 1936), pp. 349, 351, 357.

27. Aristotle, op. cit., p. 617.

28. Ibid., p. 625.

29. Ibid., p. 617. He further said (p. 625) that "in all the states where it is the local custom

to mate young men and young women, the people are deformed and small of body. And again young women labour more, and more of them die in childbirth." In fact, Aristotle's observations are borne out in part by modern findings. Women under eighteen and over forty do display significantly higher rates of chromosomal abnormalities, especially Down syndrome, than those between eighteen and forty, and more commonly deliver premature and low-birth-weight babies.

30. Ibid., p. 625.

31. In Gen. 22:2, God tests Abraham by telling him to take his son Isaac to Moriah and "sacrifice him there as a burnt offering on one of the mountains I will tell you about." In Exod. 22:29–30, the Hebrews are advised: "You must give me the firstborn of your sons. Do the same with your cattle and your sheep. Let them stay with their mothers for seven days, but give them to me on the eighth day." Martin S. Bergman, in *In the Shadow of Moloch* (New York: Columbia University Press, 1992), p. 85, cites M. Wolf (1945), who suggested that "the original god of the Israelites was a sun and fire god called El. During the spring the firstborn of the flock, as well as the firstborn son, was sacrificed to this deity."

32. See, for example, Lev. 18:21; 20:2–3; and 22:21–31; as well as 2 Kings 17:16–17.

33. Exod. 1:10.

34. In Lefkowitz and Fant, op. cit., p. 187.

35. Quoted in Pomeroy, op. cit., p. 44.

36. Ibid., p. 45.

37. Pseudo-Plato, Tertullian, Diodurus Siculus, and Plutarch all made this charge against the Carthaginians.

38. Quoted in Jeffrey H. Schwartz, *What the Bones Tell Us* (New York: Henry Holt & Co., 1993), p. 28.

39. Ibid., p. 39.

40. Cited in Lefkowitz and Fant, op. cit., p. 94. From ed. S. Riccobono et al., *Fontes Iuris Romani Antejustiniani* (Florence: S.A.G. Barbera, 1940–43).

41. William Lecky, *History of European Morals* (New York: D. Appleton & Co., 1900), p. 27. Edward Gibbon also made this observation in Chapter 44 of *The History of the Decline and Fall of the Roman Empire*.

42. Suetonius, *The Lives of the Caesars*, trans. J. C. Rolfe (New York: G. P. Putnam's Sons, 1914), p. 409. See also Tacitus, *The Annals of Imperial Rome*, trans. Michael Grant (Middlesex, England: Penguin Books, Ltd., 1973), p. 117.

43. Cited in Philippe Ariès and Georges Duby, eds., *A History of Private Life*, vol. 1, *From Pagan Rome to Byzantium*, ed. Paul Veyne and trans. Arthur Goldhammer (Cambridge, Mass.: The Belknap Press of Harvard University Press, 1987), p. 11.

44. *Terence*, ed. E. Capps, T. E. Page, and W. H. D. Rouse and trans. John Sargeaunt (New York: G. P. Putnam's Sons, 1912), vol. 2, pp. 163ff.

45. David Kertzer, in *Sacrificed for Honor: Italian Infant Abandonment and the Politics of Reproductive Control* (New York: Beacon Press, 1993), p. 6, cited these figures from John Boswell's *The Kindness of Strangers*.

46. Cited in Ariès and Duby, op. cit., p. 11.

47. Quoted in ibid., p. 9.

48. Seneca, *Moral Essays*, trans. John W. Basore (Cambridge, Mass.: Harvard University Press, 1970), in the essay "On Anger," p. 145.

49. Suetonius, *The Lives of the Caesars*, trans. Robert Graves (Baltimore: Penguin Books, 1957), p. 96.

50. Pomeroy, op. cit., p. 126.

51. Quoted by Keith Bradley, "Wet-Nursing at Rome: A Study in Social Relations," in Beryl Rawson, *The Family in Ancient Rome* (Ithaca, N.Y.: Cornell University Press, 1986), p. 218.

52. Peter Brown, "Late Antiquity," in Ariès and Duby, op. cit., p. 241.

53. Pliny, *Natural History*, trans. H. A. Rackham (Cambridge, Mass.: Harvard University Press, 1938), p. 539.

54. Brown in Ariès and Duby, op. cit., p. 308. This view persisted to a degree among the upper classes in the Middle Ages, although gradually people came to believe, as the church argued, that conception came under the governance of God and was therefore far more unpredictable than previously imagined.
55. Pliny, op. cit., p. 539.
56. Ibid., p. 535.
57. Ibid., p. 553.

CHAPTER 4

1. See Michel Rouche in Philippe Ariès and Georges Duby, eds., *A History of Private Life*, vol. 2, *Revelations of the Medieval World*, ed. Georges Duby and trans. Arthur Goldhammer (Cambridge, Mass.: The Belknap Press of Harvard University Press, 1988), p. 455.
2. See, for example, I Cor. 2:6–14; 7:1–14, 25–34.
3. Saint Jerome, "Letter to Eustochium," in Emilie Amt, ed., *Women's Lives in Medieval Europe* (New York: Routledge, 1993), p. 23.
4. Saint Jerome, "Against Jovinian," in ibid., pp. 24, 25.
5. Such works proliferated in the later Middle Ages along with the growth of monastic life. For example, the thirteenth-century treatise *Holy Maidenhood* decried intercourse as "a nauseous thing to think thereon"; rather than provide marital contentment, it partakes of "the filth of the flesh or worldly vanity, which turns all to sorrow and care in the end." Pregnancy brings only bodily miseries, withering beauty, and the strong possibility of death, and caring for children is tedious and produces little joy and much maternal anxiety over the possibility that the child will be harmed or will die. See ibid., pp. 90–94.
6. Canon law drafted later in the Middle Ages, in the twelfth century, while asserting that "childbirth is the sole purpose of marriage for women," allowed that some leeway existed under the sacramental umbrella of the institution: "Immoderate conjugal union is not an evil of marriage, but a venial sin, because of the good of marriage...." However, contraceptives and abortifacients still had the power to negate that good. "Those who obtain drugs of sterility are fornicators, not spouses," declared the church. See Gratian's *Decretals*, excerpted in ibid., pp. 79–83.
7. See Saint Augustine, "On Marriage and Concupiscence," in ibid., pp. 26–28. His reasoning is somewhat circular, but clearly, the premise underlying his argument is not that procreation is the goal of human existence, rather that service to God is the goal, and chastity a key element in achieving that goal.
8. Rouche in Ariès and Duby, op. cit., pp. 468, 472. Also see Eveleyne Patlagean in the same volume (p. 598), who said that in tenth- and eleventh-century Byzantium, people had an "obsessive fear of 'mixing of the blood,' which is to say, of incest." Here, clearly, there was a push not only to differentiate between insiders and outsiders but to enforce boundaries within the family having to do with acceptable and unacceptable intercourse.
9. See Helen Lemay, "Women and the Literature of Obstetrics and Gynecology," in Joel T. Rosenthal, *Medieval Women and the Sources of Medieval History* (Athens: University of Georgia Press, 1990), pp. 189–209.
10. Plato dilated upon this in the dialogue *Timaeus:* "Now the opinion that the cause of male and female is heat and cold, and that the difference depends upon whether the secretion comes from the right side or from the left, has a modicum of reason in it, because the right side of the body is hotter than the left; hotter semen is semen which has been concocted; the fact that it has been concocted means that is has been set and compacted, and the more compacted semen is, the more fertile it is." Quoted in Amt, op. cit., p. 228.
11. Trotula in ibid., pp. 98ff.
12. Ibid.

13. Howard Adelmann, in *Marcello Malpighi and the Evolution of Embryology* (Ithaca, N.Y.: Cornell University Press, 1966), vol. 2, p. 739, discussed a general misunderstanding that has arisen regarding Aristotle's views of female semen. Adelmann wrote, "He is usually interpreted as denying that the female produces semen, and it is true that he usually does insist that the female does not contribute semen to the generation of the offspring. The secretion in females corresponding to the semen of males is the menstrual blood or catamenia, he says. Yet (and here lies the source of the prevalent confusion) he also explains that the catamenia are semen too, but semen which is impure and not thoroughly concocted, and is thus in need of further elaboration. He says, further, that there is semen *in* the catamenia."

14. See Pliny, *Natural History*, trans. H. A. Rackham (Cambridge, Mass.: Harvard University Press, 1938), p. 26.

15. Technically, these terms refer strictly to theories propounded by physicians and early scientific researchers from the seventeenth to nineteenth century, but are occasionally applied retroactively to earlier postulates which fall within the same parameters as the later, experimentally based formulations.

16. Tertullian made this case in *Veil of the Virgins*, according to Ariès and Duby, op. cit., p. 12.

17. Quoted in Arthur Meyer, *The Rise of Embryology* (Stanford, Calif.: Stanford University Press, 1939), p. 36.

18. Ibid., p. 33.

19. LeRoy Ladurie, *Montaillou: The Promised Land of Error*, trans. Barbara Bray (New York: George Braziller, Inc., 1978), p. 291.

20. See Lemay in Rosenthal, op. cit., p. 195.

21. Cited by Lemay in ibid., p. 201. Both stories were related by Anthonius Guaninerius in *Tractatus de matricibus*.

22. See, for instance, Sue Sheridan Walker, "Widow and Ward: The Feudal Law of Child Custody in Medieval England," in Susan Mosher Stuard, *Women in Medieval Society* (Philadelphia: University of Pennsylvania Press, 1976), pp. 159–72, for the importance of lineage. The emphasis given lineage varied from place to place and depended upon one's economic status. Ladurie, op. cit., p. 48, explained that in rural southern France, "the *domus* cannot be understood without its genealogical links, which connected it with other related, living *domus* through consanguinity *(parentela)*. These bonds also linked the *domus* with the past, under the auspices of the lineage *(genus)* of the family, which was the *domus* looked at against the background of the past four generations at the most.

   "Some authors have seen lineage as one of the most important values of ancient societies. This is certainly true in the case of the nobility. But, as regards Montaillou, the sense of lineal continuity was a local and rural one, not of primary importance.... The sense of *genus* was quite vivid, but no more than that: the peasants spoke of someone belonging to a race of priests, a race of liars, a race of heretics, a race of curmudgeons, or a race of lepers. (*Genus* is here translated by the word 'race,' though it would be more correct and scientific to talk of '*lines* of priests,' etc.)" Among the Norse, the societal emphasis on the importance of lineages is apparent in literature, specifically the sagas of Sigurd the Volsung. See Jenny Jochens, "Old Norse Sources on Women," in Rosenthal, op. cit., p. 165.

23. Battles over whom one could marry and whether divorce was permissible dominated much of church debate and law, much of it with the aim of determining which heirs might be considered legitimate. See Jo-Ann McNamara and Suzanne F. Wemple, "Marriage and Divorce in the Frankish Kingdom," in Stuard, op. cit., pp. 95–124.

24. Paul Veyne in Ariès and Duby, op. cit., p. 120.

25. This notion was discussed by Aristotle, Empedocles, Democritus, and other early Greek thinkers, and was reiterated by Pliny, who also thought that twins invariably resulted from not one but two separate inseminations.

26. Lemay in Rosenthal, op. cit., p. 200.

27. See Emily Coleman in Stuard, op. cit., p. 59.
28. Peter Hoffer and N. E. H. Hull, in *Murdering Mothers* (New York: New York University Press, 1981), p. 4ff., contended that in England infanticide was not often prosecuted, either in ecclesiastical courts or in assize courts until the late 1500s. They cited figures given by historian Barbara Hanawalt on criminal homicides in Norfolk, Northamptonshire, and Yorkshire from 1300 to 1348. Of 2,933 cases, only one was for infanticide. Extreme violations of the sense of public decorum did draw attention. For instance, Hoffer and Hull reported that in 1249, "the hundred of Chippenham [an administrative division of the county of Wiltshire] was fined [by the court] for letting Basilia of Wroxhall flee. She threw her infant son into a ditch to die, and a dog carried the corpse through the town." Appallingly, discarded infants often fell to dogs or rutting pigs, both of which wandered free in many medieval villages and towns.
29. Maimonides declared murder a capital offense "whether one kills an adult or a one-day-old child," according to Immanuel Jakobovits, in Marvin Kohl, ed., *Infanticide and the Value of Life* (Buffalo, N.Y.: Prometheus Books, 1978), p. 23. But here, the definition of who was considered human was key: A fetus did not qualify until birth, as specified. Babies born prematurely had to survive thirty days; if they died before then "whether by violence or from natural causes," the birth was retroactively classed as a miscarriage, the killer could not be charged with capital punishment, and mourning was not required. The same exemption applied if an infant was deformed, in which case infanticide was considered noncapital homicide. The offender could not be tried, Maimonides declared, but infanticide was still an act "for which the offender is accountable before the heavenly tribunal."
30. Rouche in Ariès and Duby, op. cit., p. 460.
31. Quoted in Maria Piers, *Infanticide* (New York: W. W. Norton & Co., 1978), p. 52.
32. Quoted in Amt, op. cit., p. 105.
33. David Herlihy and Christiane Klapisch-Zuber, *Tuscans and Their Families* (New Haven: Yale University Press, 1985), pp. 135ff. These authors further contended (pp. 147ff.) that "fathers were well aware of the risk of placing babies outside the home. They often railed against the wetnurses, and warned that through their carelessness they might cause the deaths of children placed in their charge. Baby boys, nursed at home more frequently than their sisters, thus gained an inestimable advantage in the struggle [to survive]. And they probably continued to enjoy better care throughout childhood. According to Paolo de Certaldo, who wrote at the end of the fourteenth century, fathers were expected to supervise the diet of a boy more attentively than the nourishment of a girl."
34. Barbara Tuchman, *A Distant Mirror* (New York: Alfred A. Knopf, 1978), p. 49.
35. Ibid., p. 52.
36. Ibid., p. 92.
37. Herlihy and Klapisch-Zuber, op. cit., p. 69.
38. Henry IV quoted in Alexander Morris Carr-Saunders, *The Population Problem: A Study in Human Evolution* (London: Oxford University Press, 1922), p. 21.
39. Botero, quoted in Lamont C. Cole, "Sketches of General and Comparative Demography," *Cold Spring Harbor Symposia of Quantitative Biology* 22 (1957): 3.
40. John Graunt, "Natural and Political Observations Mentioned in a Following Index, and Made upon the Bills of Mortality," reprinted in Kenneth Weiss and Paul Ballonoff, *Demographic Genetics* (Stroudsberg, Penn.: Dowden, Hutchinson & Ross, Inc., 1975), p. 13.

CHAPTER 5

1. Alternate spellings of his given name include Gabriello and Gabrielus, and, of his surname, Fallopia, Falloppio, and Fallopius.
2. Galen knew about the fallopian tubes, but had no idea what the function of the ovaries was, hence did not understand the role the fallopian tubes played.

3. According to the *Oxford English Dictionary*, the word was introduced by Whewell in the introduction to his *Philos. Induct. Sci.*, in which he wrote, "We need very much a name to describe a cultivator of science in general. I should incline to call him a Scientist." The *OED* also reports that "Aquinus also uses *scientificus* for 'expert in science, learned,' a sense which still survived in sixteenth c. Latin. The lateness of the first appearance of the word in English [ca. 1589] is remarkable."

4. This story is told by Howard B. Adelmann of Jacopo Berengario da Carpi in *Marcello Malpighi and the Evolution of Embryology* (Ithaca, N.Y.: Cornell University Press, 1966), vol. 2, p. 753.

5. Quoted in William Harvey, *On the Generation of Animals*, trans. Robert Willis, in the Great Books of the Western World Series, ed. Robert Maynard Hutchins (Chicago: Encyclopaedia Britannica, Inc., 1952), vol. 28, p. 348.

6. Ibid., p. 263. In point of fact, even Harvey never entirely succeeded in escaping from the trap of Scholasticism. He, too, refers back continually to Aristotle, Fabricius, and others.

7. Ibid., p. 419.

8. Ibid., p. 336. It must be added, however, that Harvey himself was not immune to lapses of this sort, as, for example, when he wrote (p. 347) without fully exploring the matter that "women occasionally become insane through ungratified [sexual] desire, and to such a height does the malady reach in some, that they are believed to be poisoned, or moon-struck, or possessed by a devil. And this would certainly occur more frequently than it does, without the influence of good nurture, respect for character, and the modesty that is innate in the sex, which all tend to tranquilize the inordinate passions of the mind."

9. Ibid., p. 431.

10. Ibid., p. 433.

11. Ibid., p. 455.

12. For example, Adelmann, op. cit., vol. 2, p. 773, quoted Sir Kenelm Digby, a friend of Harvey's and occasional biologist (with highly distinctive orthography), who insisted that "generation is made of the bloud, which being dispersed into all the partes of the body to irrigate every one of them; and to convey fitting spirits into them from their source or shoppe where they are forged; so much of it as is superaboundant to the nourishing of those partes is sent backe againe to the hart to recover the warmeth and spirits it hath lost by so long a iourney." As blood journeyed through the body, the strengths and weaknesses of the organs were impressed upon it. The purest blood, Digby wrote, "is reserved in convenient receptacles or vessels till there be use of it: and is the matter or seede, of which a new animal is to be made; in whom, will appeare the effect of all the specificall vertues drawne by the bloud in its iterated courses, by its circular motion, through all the severall partes of the parents body." Thus the "vertues" and vices of the parent will be transmitted to offspring. "The yong animal come into the world savouring of that origine; unlesse the mothers seede, do supply or temper, what the fathers was defective or superaboundant in; or contrariwise the fathers do correct the errors of the mothers." Although Digby, writing in 1644, had no knowledge at all of genes, he did anticipate in this last sentence an aspect of recessive genetic diseases, whereby a child receiving one faulty copy of a gene from a parent and one good copy will not develop a disease, one parent's "seede" having thus supplied or tempered the other's.

13. Harvey, op. cit., p. 493.

14. Ibid., p. 492.

15. Ibid., p. 385.

16. Ibid., p. 412.

17. Ibid., p. 393.

18. Ibid., p. 391.

19. Ibid., p. 395.

20. Ibid., p. 384. He also viewed the egg as "a mean between the animate and the inanimate world; for neither is it wholly endowed with life, nor is it entirely without vitality."

21. Ibid., p. 395.
22. Ibid., p. 473.
23. Ibid., p. 479.
24. Ibid., p. 451.
25. Harvey missed entirely the role of the ovaries in egg production. He wrote in *Genera-tion of Animals* (p. 476) that "the female testicles as they are called [i.e., the ovaries], whether they be examined before or after intercourse, neither swell nor vary from their usual condition; they show no trace of being of the slightest use either in the business of intercourse or in that of generation." However, this did not prevent him from advancing his theory regarding the egg. See also p. 470.
26. Ibid., pp. 412ff.
27. Some scholars make a further distinction between preformation and preexistence, arguing that the former, a seventeenth-century concept, places the preformed embryo in the parent only, whereas the latter, an eighteenth-century notion, holds that preformed embryos have existed since God created the universe. However, some histori-ans feel this is hairsplitting, and I have chosen to follow their lead. See Shirley A. Roe, *Matter, Life, and Generation* (Cambridge, England: Cambridge University Press, 1981), p. 174fn.
28. Arthur Meyer, in *The Rise of Embryology* (Stanford, Calif.: Stanford University Press, 1939), pp. 28–29, described a number of scientists, including the English physician H. Charlton Bastian in 1872 and the French scientist F.-A. Pouchet in 1864, who insisted that experiments they conducted proved that spontaneous generation occurred. But most people must have by then discarded the idea. Meyer (p. 43) also quoted George Cheyne, who in the fourth edition of *Philosophical Principles* (1734) wrote: "No Body now-a-days, that understands any thing of Nature or *Philosophy*, can so much as imagine, that any Animal, how abject soever, can be produced by an *equivocal* Gener-ation, or without the conjunction of Male and Female Parents, in the same, or in two different Individuals. And very few, who have considered the Matter, but own, that every Animal proceeds from a pre-existent Animalcul; and that the Parents conduce nothing but a convenient Habitation, and suitable Nourishment to it, till it be fit to be trusted with the Light, and capable of receiving the Benefit of the Air." Specula-tion about spontaneous generation was finally put to rest by the work of Louis Pas-teur, who showed that "organic corpuscles" or "germs" in the air cause putrefaction.
29. Adelmann, op. cit., vol. 2, p. 831. See also pp. 828ff.
30. Ibid., p. 824.
31. Adelmann, op. cit., vol. 2, p. 865. Malpighi actually uses the term "cicatrix" rather than "blastoderm," but I have chosen to avoid confusion by using the preferred modern syn-onym, "blastoderm." As Adelmann pointed out (p. 938fn.), "Cicatricula (a little scar), the tread or treadle of English popular speech, because it was and still is often believed by the uninformed to be the cock's semen or tread—for the same reason the chalazae are also sometimes so called—the *Narbe* (scar) or *Hahnentritt* (treadle) of the German, still lingers in embryological literature as a not-uncommon term for the blastodisc of the ovarian and oviductal egg and for the blastoderm."
32. Ibid., p. 867.
33. Roe, op. cit., p. 5.
34. See ibid., pp. 6ff., and Meyer, op. cit., p. 60.
35. Leeuwenhoek's letter is reprinted in Meyer, op. cit., pp. 139–41.
36. Quoted in ibid., p. 139.
37. Both Grew's letter and Leeuwenhoek's response are quoted in ibid., pp. 143, 144.
38. It is interesting to note that modern arguments regarding the status of embryos par-take of the sort of logic that Sterne here lays out. Laurence Sterne, *The Life and Opin-ions of Tristram Shandy, Gentleman* (New York: New American Library, 1960), p. 10.
39. Quoted in Meyer, op. cit., p. 51.
40. Quoted in K. J. Betteridge, "An Historical Look at Embryo Transfer," *Journal of Reproduction and Fertility* 62 (1981): 3.

41. Quoted in Adelmann, op. cit., vol. 2, p. 781.
42. Quoted in Betteridge, op. cit., p. 3.

CHAPTER 6

1. Jonathan Swift, "A Modest Proposal," in Martin Price, *The Restoration and the Eighteenth Century* (New York: Oxford University Press, 1973), p. 221.
2. Ibid., p. 223.
3. Ibid., p. 226.
4. Oliver Goldsmith, "The Deserted Village," in ibid., p. 696.
5. Joseph Spengler, in *French Predecessors of Malthus* (Durham, N.C.: Duke University Press, 1942), p. 28fn., cited Sebastien le Prestre, maréchal de Vauban's 1708 estimate "that one tenth of the French population was reduced to mendicancy, one half was on the verge of mendicancy, three tenths were badly off and in debt; only a part of the remaining one tenth could afford alms; not over ten thousand families (comprising about one fourth of one per cent of the population) were well off." See also p. 46.
6. Simon Schama, *Citizens* (New York: Alfred A. Knopf, 1989), p. 183.
7. See Lamont C. Cole, "Sketches of General and Comparative Demography," *Cold Spring Harbor Symposia of Quantitative Biology* 22 (1957): 3–5.
8. Michel Foucault, *The History of Sexuality*, trans. Robert Hurley (New York: Vintage Books, 1990), pp. 25ff.
9. A few critics disputed the importance of population, especially the Physiocrats of 1760s and 1770s France, who believed that wealth was the goal of nations, and that population growth depended entirely upon the growth of wealth; and the *philosophes*, including Condorcet, Diderot, Voltaire, and others who generally advanced more sophisticated arguments regarding the relationship of population to economy. See Spengler, op. cit., pp. 176, 212ff.
10. See William Godwin's translation of these observations in *Of Population* (New York: August M. Kelley Bookseller, 1964), pp. 36–40, and, for the impact on French thinkers, Spengler, op. cit., p. 77.
11. John Graunt, "Natural and Political Observations Mentioned in a Following Index, and Made upon the Bills of Mortality," reprinted in Kenneth Weiss and Paul Ballonoff, *Demographic Genetics* (Stroudsberg, Penn.: Dowden, Hutchinson & Ross, Inc., 1975), p. 43. This view was expressed repeatedly during this period. Frenchman M. de Belesbat, for example, wrote in 1706, "Princes are powerful only in consequence of the great number of men that they command, and not at all because of their wealth and the extent of their states. . . . The wealth of princes comes from the fertility of their States; fertility, from good cultivation; good cultivation, from the great number of men who are employed therein. Consequently princes who know their true interests will do everything to conserve and augment the number of their subjects, to enrich them and render them happy." Quoted in Spengler, op. cit., p. 39.
12. Quoted in Cole, op. cit., p. 4.
13. Marvin Harris, *Cannibals and Kings* (New York: Random House, 1977), p. 184.
14. Reported in John Brownlow's *The History and Design of the Foundling Hospital*, published in 1868 and quoted in William L. Langer, "Infanticide: A Historical Survey," *History of Childhood Quarterly* 1 (1974): 359. Harris, op. cit., p. 184, reported that of these 14,934 children, only some 4,400 survived to adolescence. However, infants consigned to the workhouse system fared even worse. According to Harris, "Parish officers assigned the [workhouse] infants to women who were nicknamed 'killing nurses' or 'she-butchers' because 'no child ever escaped their care alive.'"
15. Langer, op. cit., p. 358, gives the 5,000-per-year figure from the mid-1830s; David I. Kertzer, *Sacrificed for Honor* (New York: Beacon Press, 1993), p. 13, gives an annual average of 9,458 for the period 1880–89.
16. Kertzer, op. cit., p. 13.

17. Ibid., p. 138.
18. Peter C. Hoffer and N. E. H. Hull, *Murdering Mothers* (New York: New York University Press, 1981), pp. 42–49.
19. Ibid., p. x.
20. Ibid., p. 150.
21. Johann Wolfgang Goethe, *Faust,* ed. Cyrus Hamlin and trans. Walter Arndt (New York: W. W. Norton & Co., 1976), p. 98.
22. See Maria W. Piers, *Infanticide* (New York: W. W. Norton & Co., 1978), pp. 69–73, for an interesting discussion of contemporary campaigns against capital punishment for infanticide.
23. Spengler, op. cit., p. 30fn. Among others, Rousseau was an ardent supporter of breast-feeding by mothers; he tied its lack to a decline in morals.
24. We know that ancient Roman physicians employed this extracting tool because excavators unearthed finely forged forceps and a speculum at Pompeii. However, the invention had been lost by the Middle Ages.
25. Laurence Sterne, *The Life and Opinions of Tristram Shandy, Gentleman* (New York: New American Library, 1960), p. 46.
26. Ibid., p. 118.
27. Ibid., p. 122.
28. D'Alembert's mother, according to Francis Galton, was one Mademoiselle de Tencin, a nun who renounced her vows and became a novelist and an adventuress. Once d'Alembert, whose given name was that of the town in whose public market he was found, became famous, his mother sought him out and revealed her identity. See Francis Galton, *Hereditary Genius* (New York: St. Martin's Press, 1978), p. 208.
29. See John Bagnell Bury, *The Idea of Progress* (London: MacMillan & Co., 1924), pp. 76–228, for a lengthy discussion of late-seventeenth- and eighteenth-century views of science and progress, the roles of nature and nurture in shaping individuals, and the issue of perfectibility.
30. Quoted in ibid., p. 226.
31. Godwin read Wallace, and Malthus was later accused of having plagiarized this work.
32. In Garrett Hardin, *Population, Evolution, and Birth* (San Francisco: W. H. Freeman, 1964), p. 21. At the time this volume was published, Hardin was vice president of the American Eugenics Society.
33. During the seventeenth and eighteenth centuries, many French students of overpopulation, although also fearing the negative effects of luxury, tended more than the English to recognize the incendiary potential of intractable poverty, which they warned could lead to state-toppling uprisings. Thus, many argued in favor of state welfare programs. See Spengler, op. cit., pp. 16, 17. However, Spengler (p. 21) pointed out that it was also the case that powerful Frenchmen, including a finance minister for Louis XIV, shared the attitude that "the masses (whom kindness, well-being, and wealth tended to corrupt) made satisfactory beasts of burden only so long as wages were kept at a subsistence level and conditions of work were made onerous and unfavorable."
34. Wrangler honors were highly prestigious and nationally recognized. Any number of British scientists gained initial fame as Wranglers.
35. Beginning with the 1803 edition, the title was changed to *An Essay on the Principle of Population; or A View of its past and present Effects on Human Happiness; With an Inquiry into our Prospects respecting the future Removal or Mitigation of the Evils which it occasions.*
36. Godwin, op. cit., pp. iii–vii.
37. He described his visit to the St. Petersburg home in 1789 in *An Essay on the Principle of Population,* Cambridge variorum edition, ed. Patricia James (Cambridge, England: Cambridge University Press, 1989), vol. I, pp. 173–75. He was "particularly struck with the extraordinary degree of neatness, cleanliness, and sweetness which appeared to prevail in every department. The house itself had been a palace, and all the rooms were large, airy, and even elegant."

38. Quoted in Godwin, op. cit., p. 31. This passage appeared in the 1798 edition, but was removed from all subsequent ones.
39. Malthus, op. cit., vol. 1, p. 12.
40. Quoted in Spengler, op. cit., p. 79.
41. Malthus, op. cit., vol. 1, p. 338.
42. Ibid., p. 18. Malthus also held that vice itself could cause misery.
43. Ibid., p. 20.
44. Malthus, 1798 edition of the *Essay,* quoted in Hardin, op. cit., p. 8. Elsewhere, Malthus reiterated that poverty is not related to "forms of government, or the unequal division of property," and that the poor themselves should be blamed for their condition, because they lacked moral restraint. Malthus, op. cit., Cambridge variorum edition, vol. 1, pp. 129, 130.
45. Malthus, op. cit., Cambridge variorum edition, vol. 2, p. 127. Although supporters of Malthus (for example, Patricia James, on p. xiii of the variorum edition) have pointed out that he removed this passage and should not be held to it, Godwin's note on the matter bears repeating. He wrote, "This passage, which occurs in the Second Edition in quarto, p. 531, is not to be found in the Fifth Edition of the Essay. But I beg leave once for all to observe, that those sentences of our author, the sense of which he has never shewn the slightest inclination to retract, and the spirit of which on the contrary is of the essence of his system, I do not hold myself bound to pass over unnoticed, merely because he has afterwards expunged them, that he might not 'inflict an unnecessary violence on the feelings of his readers [*Quarterly Review* for July 1817],' or that he might 'soften some of the harshest conclusions of the first Essay [Malthus, Preface to the second edition].' " See Godwin, op. cit., p. 19fn.
46. Malthus, op. cit., Cambridge variorum edition, vol. 2, pp. 137–41. However, as a firm believer that the poor should be enabled to take individual responsibility, he insisted that a good living wage was necessary. See also vol. 2, p. 248.
47. Ibid., p. 162.
48. Ibid., p. 155.
49. Ibid.
50. Ibid., p. 250. See also a very similar passage on p. 102 of the same volume.
51. Francis Place, who was born to an alcoholic, physically abusive bailiff in London and worked his way up to become a master tailor and radical Parliamentarian, in 1823 wrote and distributed to parts of London and Manchester a broadside in three different versions titled "To the Married of Both Sexes," "To the Married of Both Sexes of the Working People," and "To the Married of Both Sexes in Genteel Life," which urged the working classes to use contraception, particularly a sponge inserted in the vagina, to avoid excess births. This would in part free them from having to send their children to work in mills and factories, or (what he only implied) from having to commit infanticide. See Place, in Hardin, op. cit., p. 192. Bradlaugh and Besant were tried and convicted in 1877 for having reprinted the birth control manual *The Fruits of Philosophy: or The Private Companion of Young Married People,* written by American physician Charles Knowlton. They served six months, until a high court overturned the conviction.
52. Cited in Allan Chase, *The Legacy of Malthus* (Urbana: University of Illinois Press, 1980), p. 83.

CHAPTER 7

1. J. V. Beckett, citing B. A. Holderness (1989) in *The Agricultural Revolution* (Cambridge, Mass.: Basil Blackwell, 1990), p. 55.
2. Harriet Ritvo, *The Animal Estate* (Cambridge, Mass.: Harvard University Press, 1987), p. 52.
3. William Godwin, *Of Population* (New York: August M. Kelley Bookseller, 1964), p. 21.

4. Ritvo, op. cit., p. 60.
5. Ibid.
6. Thomas Malthus, *An Essay on the Principle of Population,* Cambridge variorum edition, ed. Patricia James (Cambridge, England: Cambridge University Press, 1989), vol. I, p. 314.
7. Ibid., vol. 2, p. 344.
8. Cited in Joseph Spengler, *French Predecessors of Malthus* (Durham, N.C.: Duke University Press, 1942), pp. 128, 134.
9. See ibid., pp. 98ff.
10. Quoted in ibid., p. 168.
11. See ibid., p. 80, or Michel Foucault, *The History of Sexuality,* trans. Robert Hurley (New York: Vintage Books, 1990), p. 118.
12. Malthus, op. cit., vol. I, p. 25.
13. French linguist François Tant, quoted in Ashley Montagu, *Man's Most Dangerous Myth* (New York: Columbia University Press, 1945), p. 18.
14. Spengler, op. cit., p. 20fn., says attempts to differentiate serfs from landholders in this manner surfaced as early as the fifteenth century in France.
15. Montagu, op. cit., p. 18.
16. Ibid., p. 7.
17. Joseph-Arthur, comte de Gobineau, *The Inequality of Human Races,* trans. Adrian Collins (New York: Howard Fertig, 1967), p. xi.
18. Ibid., p. xiv.
19. Ibid., p. 25.
20. Ibid., p. 31.
21. Ibid., p. 210.
22. Ibid., p. 107.
23. Montagu, op. cit., p. 23.
24. Charles Darwin, *The Origin of Species and The Descent of Man* (New York: The Modern Library, 1936), p. 19.
25. Francis Galton, *Hereditary Genius* (New York: St. Martin's Press, 1978), p. 68.
26. Ibid.
27. Ibid., p. 84.
28. Cesare Lombroso, *Crime,* trans. Henry P. Horton (Boston: Little, Brown & Co., 1911), p. xviii.
29. Ibid., p. 174.
30. Francis Galton, *Inquiries into Human Faculty and Its Development* (New York: E. P. Dutton, 1911), p. 17.
31. Galton, *Hereditary Genius,* p. xx.
32. Ibid.
33. Ibid., p. x.
34. Galton, *Inquiries,* p. 220.
35. Ibid., pp. 150, 219.
36. Ibid., p. 155.
37. Ibid., p. 220.
38. See Daniel Kevles, *In the Name of Eugenics* (New York: Alfred A. Knopf, 1985), p. 37.
39. Ibid., p. 35.
40. Mark B. Adams, *The Wellborn Science* (New York: Oxford University Press, 1990), p. 5.
41. Kevles, op. cit., p. 59; see also Edward J. Larson, "The Rhetoric of Eugenics: Expert Authority and the Mental Deficiency Bill," *British Journal for the History of Science* 24 (1991): 47.
42. Kevles, op. cit., p. 63.
43. Walter Lippmann, "A Future for the Tests," *New Republic,* November 29, 1922, p. 10.
44. Foucault, op. cit., p. 125.
45. Quoted in Kevles, op. cit., p. 87.

CHAPTER 8

1. William H. Schneider, "The Eugenics Movement in France, 1890–1940," in Mark B. Adams, ed., *The Wellborn Science* (New York: Oxford University Press, 1990), p. 71.
2. Ibid., p. 72.
3. Ibid., p. 77.
4. This according to Daniel Kevles, *In the Name of Eugenics: Genetics and the Uses of Human Heredity* (New York: Alfred A. Knopf, 1985), p. 41.
5. Kenneth M. Ludmerer, *Genetics and American Society* (Baltimore: Johns Hopkins University Press, 1972), p. 39.
6. Ibid., p. 58.
7. Quoted in Ellsworth Huntington, *Tomorrow's Children* (New York: John Wiley & Sons, Inc., 1935), p. 3.
8. Ibid., p. 104.
9. Quoted in Kevles, op. cit., p. 62.
10. Ludmerer, op. cit., p. 42.
11. Ellen Chesler, *Woman of Valor* (New York: Simon & Schuster, 1992), pp. 14, 216.
12. Quoted in ibid., p. 60.
13. Quoted in J. B. S. Haldane, *Heredity and Politics* (New York: W. W. Norton & Co., 1938), p. 16. Haldane provided an incisive commentary on the notion that all such persons should be sterilized in the form of footnotes. To Laughlin's category "Blind" Haldane attached footnote 1, which reads, "E.g. Milton"; to the category "Deaf," footnote 2, which reads, "E.g. Beethoven"; to each of the words "homeless," "tramps," and "paupers," footnote 3, which reads, "E.g. Jesus."
14. Ibid., p. 18.
15. See Robert Proctor, *Racial Hygiene* (Cambridge, Mass.: Harvard University Press, 1988), p. 99.
16. Francis Galton, *Inquiries into Human Faculty* (New York: E. P. Dutton, 1911), p. 44.
17. Apparently, Goddard felt this argument was so righteous that it warranted distortions of the truth: Biologist Stephen J. Gould claimed that Goddard retouched photos of the "bad" side of the Kallikak family to make them appear more dull-witted and unregenerate. See Gould, *The Mismeasure of Man* (New York: W. W. Norton & Co., 1981), p. 171.
18. Two more states would add legislation before the movement was discredited.
19. Proctor, op. cit., p. 97.
20. Ibid., p. 15.
21. Ibid., p. 18.
22. Quoted in ibid., p. 51.
23. Sheila Faith Weiss, "The Race Hygiene Movement in Germany, 1904–1945," in Adams, op. cit., p. 24.
24. See Proctor, op. cit., p. 24.
25. Ibid., p. 79.
26. Weiss, in Adams, op. cit., p. 9.
27. This is essentially the description given by Lenz of the Nordic "type." See Erwin Baur, Eugen Fischer, and Fritz Lenz, *Human Heredity*, trans. Eden and Cedar Paul (New York: The MacMillan Co., 1931), pp. 659–62.
28. Ibid., p. 109.
29. This figure is given by Len Cooper, in "Aryan Nation: Germany's Cruel African Heritage," *Washington Post*, February 20, 1994; however, some sources suggest that the number of children was closer to four hundred.
30. Baur-Fischer-Lenz, op. cit., p. 629.
31. Ibid., p. 181.
32. Ibid., p. 629.
33. Ibid., p. 183.
34. Ibid.

35. Ibid., p. 561.
36. Quoted in Proctor, op. cit., p. 58.
37. Benno Müller-Hill, *Murderous Science,* trans. George R. Frasler (Oxford University Press, 1988), p. 111, quoted Lenz's son Widukind as reporting that his father had learned from someone at Lehmanns Verlag, the publishing house run by Julius Lehmann, that Hitler had read the book. Ludmerer, op. cit., p. 114, also cited two other sources for this information, a 1933 piece by Hermann Muller in *Birth Control Review* and one by Popenoe in the *Journal of Heredity.* Proctor, op. cit., p. 60, cited yet another source, Hans F. K. Gunther's *Mein Eindruck auf Hitler.*
38. Müller-Hill, op. cit., p. 111.
39. Adolf Hitler, *Mein Kampf,* trans. James Murphy (London: Hurst and Blackett, Ltd., 1942), p. 141.
40. Ibid., pp. 42, 78, 134, 172.
41. Ibid., pp. 78, 164, 184, 215, 227, 343.
42. Ibid., p. 186, 215.
43. Ibid., p. 106.
44. Ibid., p. 143.
45. See Proctor, op. cit., pp. 96, 101.
46. Hitler, op. cit., p. 227.
47. The precise number of people affected is in doubt. Weiss, in Adams, op. cit., p. 44, gave the 400,000 figure. Kevles, op. cit., p. 117, gave a figure of 225,000 sterilizations by 1936. Claudia Koonz, *Mothers in the Fatherland* (New York: St. Martin's Press, 1987), p. 189, cited in her text 56,200 operations through the end of 1934, but in a footnote cited Lewy as calculating total sterilizations of 32,000 in 1934, 73,000 in 1935, and 63,000 in 1936.
48. Quoted in Müller-Hill, op. cit., p. 29.
49. Ibid., p. 13.
50. Ibid.
51. Ibid., p. 14.
52. Ibid.
53. Shown in Proctor, op. cit., p. 50.
54. Quoted in Koonz, op. cit., p. 189.
55. Ibid., p. xxiii.
56. Baur-Fischer-Lenz, op. cit., p. 674.
57. Quoted in Müller-Hill, op. cit., p. 81. Streicher's views perhaps reflected his belief in telegony, the notion then current in animal husbandry that a female's first mating forever influences the quality of her offspring.
58. Ibid., p. 33; quoted from the *Allgemeine Zeitung* of June 5, 1936.
59. Cited in Ludmerer, op. cit., p. 116.
60. Quoted in Müller-Hill, op. cit., p. 61.
61. Quoted by Mildred Dickemann, "Infanticide in Humans: Ethnography, Demography, Sociobiology, and History," in Glenn Hausfater and Sarah Blaffer Hrdy, eds., *Infanticide: Comparative and Evolutionary Perspectives* (New York: Aldine Publishing Co., 1984), p. 429.
62. See Kevles, op. cit., p. 165.
63. Haldane, op. cit., p. 86.
64. Ibid., p. 95.
65. Ibid., p. 97.
66. "The Geneticists' Manifesto," *The Journal of Heredity* 30 (1939): 372. See also Frederick Osborn, "The American Concept of Eugenics," in the March 1939 issue of the magazine.
67. Ashley Montagu, "The Concept of Race in the Human Species in the Light of Genetics," *Journal of Heredity* 32 (1941): 243.
68. See Ludmerer, op. cit., p. 146, and H. J. Muller, "Human Evolution by Voluntary Choice of Germ Plasm," *Science* 134 (1961): 643.
69. Quoted in Ludmerer, op. cit., p. 177.

CHAPTER 9

1. Quoted in A. T. Gregoire and Robert C. Mayer, "The Impregnators," *Fertility and Sterility* 16 (1965): 133.
2. Ibid.
3. Hermann Rohleder, in *Test Tube Babies: A History of the Artificial Impregnation of Human Beings* (New York: Panurge Press, 1934), pp. 35ff., cites an Arab source describing this practice.
4. Ivanov's name is also sometimes translated as Elias Iwanoff in early reports on artificial insemination.
5. Translated from Ivanov's 1903 work, *"Artificial Impregnation of Mammals"*; quoted in *Dictionary of Scientific Biography*, ed. Charles Coulston Gillespie (New York: Scribner, 1970) p. 31.
6. H. Brewer estimate cited in Clair E. Folsome, "Artificial Insemination," *American Journal of Obstetrics and Gynecology* 45 (1943): 915.
7. Quoted in "Sir John Hammond, C.B.E., F.R.S.: An Interview," *Journal of Reproduction and Fertility* 3 (1962): 6.
8. Ibid.
9. Ibid., p. 7.
10. Ibid.
11. Ibid., p. 8.
12. Ibid.
13. *Law and Ethics of AID and Embryo Transfer*, CIBA Foundation Symposium 17 (New York: Elsevier, 1972), p. 4.
14. Cited in R. C. Jones, "Uses of Artificial Insemination," *Nature* 229 (1971): 534.
15. Jerome K. Sherman, "Synopsis of the Use of Frozen Human Semen Since 1964," *Fertility and Sterility* 24 (1973): 409.
16. Rohleder, op. cit., p. 59.
17. Ibid., pp. 97ff.
18. Ibid., p. 100.
19. Ibid., p. 167.
20. F. I. Seymour and A. Koerner quoted in Folsome, op. cit., pp. 917ff.
21. Ibid., p. 920.
22. Mary Barton, Kenneth Walker, and B. P. Wiesner, "Artificial Insemination," *British Medical Journal*, January 13, 1945, p. 40.
23. Ibid.
24. Ibid.
25. This occurred in 1980. Cited in Gena Corea, *The Mother Machine* (New York: Harper & Row, 1985), p. 42.
26. Feversham Committee, "Human Artificial Insemination," *British Medical Journal*, July 30, 1960, p. 379; U.S. figures from Alan F. Guttamacher, cited in G. Langer, E. Lemberg, and M. Sharf, "Artificial Insemination," *International Journal of Fertility* 14 (1969): 233.
27. Quoted in Robert Proctor, *Racial Hygiene* (Cambridge, Mass.: Harvard University Press, 1988), p. 244.
28. This was of sufficient concern that sperm banks actually adopted rules to forestall it. A 1972 *Time* magazine story on sperm banking reported of Manhattan's Idant that "there is a 45-day waiting period . . . before sperm can be withdrawn from the bank—long enough to prevent an already pregnant woman from using artificial insemination as a cover-up for an illegitimate baby." See "Frozen Assets," *Time*, January 3, 1972, p. 52.
29. Folsome, op. cit., p. 924.
30. Quoted in R. Snowden, G. D. Mitchell, and E. M. Snowden, *Artificial Reproduction* (London: George Allen & Unwin, 1983), p. 57.
31. Feversham Committee, op. cit., p. 379.

32. Ibid., p. 380.
33. Ibid.
34. "Report of Panel on Human Artificial Insemination," *British Medical Journal Supplement,* April 7, 1973, p. 4.
35. Ibid., p. 5.
36. The case was People v. Sorenson, [1968] 60 Cal. Rptr. 495, 437 P.2d 495.

CHAPTER 10

1. Hermann J. Muller, "Human Evolution by Voluntary Choice of Germ Plasm: This procedure should be more acceptable and effective than differential control over family size," *Science* 134 (1961): 643.
2. Jerome K. Sherman, "Research on Frozen Human Semen: Past, Present, and Future," *Fertility and Sterility* 15 (1964): 491.
3. Ibid.
4. Ibid.
5. Ibid., p. 495.
6. Ibid., p. 497.
7. John Maynard Smith, "Eugenics and Utopia," *Daedalus* 94 (1965): 487.
8. Cited in Robert T. Francoeur, *Utopian Motherhood* (Garden City, N.Y.: Doubleday & Co., 1970), p. 57.
9. Ibid., p. 19.
10. Alan Parkes and J. E. Meade, eds., *Biological Aspects of Social Problems* (New York: Plenum Press, 1965), p. 212.
11. Based on 1965 figures which put infertility at about 11.2 percent of the country's 26,454,000 married couples. See Anjani Chandra and William D. Mosher, "The Demography of Infertility and the Use of Medical Care for Infertility," *Infertility and Reproductive Medicine Clinics of North America* 5 (1994): 287.
12. Martin Curie-Cohen, Lesleigh Luttrell, and Sander Shapiro, "Current Practice of Artificial Insemination by Donor in the United States," *New England Journal of Medicine* 300 (1979): 588.
13. Ibid.
14. Clinical social worker Jean Benward has pointed out that this same attitude prevails regarding current IVF/embryo transfer practices.
15. Curie-Cohen et al., op. cit., p. 589.
16. Patricia Mahlstedt and Kris Probasco, "Sperm Donors: Their Attitudes Toward Providing Medical and Psychological Information for Recipient Couples and Donor Offspring," *Fertility and Sterility* 56 (1991): 748.
17. Ibid.
18. Quoted in Francoeur, op. cit., p. 37.
19. Jerome K. Sherman, "Synopsis of the Use of Frozen Human Semen Since 1964: State of the Art of Human Semen Banking," *Fertility and Sterility* 24 (1973): 404.
20. Quoted in Jan Zimmerman, ed., *The Technological Woman* (New York: Praeger, 1983), p. 211.
21. R. G. Snowden, G. D. Mitchell, and E. M. Snowden, *Artificial Reproduction: A Social Investigation* (London: George Allen & Unwin, 1983), p. 85.
22. Ibid., p. 137.
23. Ibid., p. 124.
24. R. S. Ledward, E. M. Symonds, and S. Eynon, "Social and Environmental Factors as Criteria for Success in Artificial Insemination by Donor (AID)," *Journal of Biosocial Science* 14 (1982): 270.
25. Snowden et al., op. cit., p. 97.
26. Ibid., p. 88.
27. Ibid., p. 98.

28. Quoted in "The Proxy Fathers: Sowing the Seeds of Despair," *Sunday Times* (London), April 11, 1982, p. 29.
29. Mahlstedt and Probasco, op. cit., p. 749.
30. Ibid., p. 753.
31. Patricia Mahlstedt, in remarks given during a workshop on Legal and Bioethical Challenges to Effective Infertility Counseling at the twenty-seventh annual postgraduate course at the 1994 ASRM meeting in San Antonio. Mahlstedt's presentation was titled "Disclosure Versus Non-Disclosure in Donor Conception: A Private Decision or a Public Concern?"

CHAPTER 11

1. Quoted in K. J. Betteridge, "An Historical Look at Embryo Transfer," *Journal of Reproduction and Fertility* 62 (1981): 1.
2. Walter Heape, "Preliminary Note on the Transplantation and Growth of Mammalian Ova within a Uterine Foster Mother," *Proceedings of the Royal Society of London,* November 27, 1890, p. 457.
3. Charles Thibault, "Citation for M.C. Chang," in Howard W. Jones, Jr., and Charlotte Schrader, eds., *In Vitro Fertilization and Other Assisted Reproduction* (New York: New York Academy of Sciences, 1988), p. xiv.
4. "Conception in a Watch Glass," *New England Journal of Medicine* 217 (1937): 678.
5. Yevgeny Zamyatin, *We,* trans. Mirra Ginsburg (New York: Bantam Books, 1972), p. 22.
6. Ibid., p. 14.
7. George Orwell, *1984* (New York: New American Library, 1961), p. 57.
8. M. C. Chang, "Fertilization of Rabbit Ova In Vitro," *Nature* 184 (1959): 467.
9. Quoted in Robert T. Francoeur, *Utopian Motherhood* (Garden City, N.Y.: Doubleday & Co., 1970), p. 58.
10. Gregory Pincus, *The Control of Fertility* (New York: Academic Press, 1965), pp. 6, 8.
11. Sir Julian Huxley, "Eugenics in Evolutionary Perspective," *Perspectives in Biology and Medicine,* Winter 1963, p. 157.
12. Ibid., p. 162.
13. Ibid., p. 171.
14. Julian Huxley, *The Human Crisis* (Seattle: University of Washington Press, 1963), p. 20.
15. Joshua Lederberg, "Experimental Genetics and Human Evolution," *Bulletin of the Atomic Scientists* 22 (1966): 4.
16. Ibid., p. 7.
17. Ibid., p. 8.
18. Ibid., p. 9.
19. Peter Medawar, "The Genetic Improvement of Man," *Australian Annals of Medicine* 4 (1969): 319.

CHAPTER 12

1. This and other quotes, as well as details of Edward's career, come from the autobiographical *A Matter of Life,* coauthored with Patrick Steptoe (New York: William Morrow, 1980). See p. 24.
2. Ibid., p. 27.
3. Ibid., p. 38.
4. Ibid., p. 47.
5. Ibid., p. 52.
6. Ibid., p. 61.

7. Ibid., pp. 79, 185.
8. R. G. Edwards and David Sharpe, "Social Values and Research in Human Embryology," *Nature* 231 (1971): 89.
9. Ibid., p. 90.
10. Ibid.
11. See R. G. Edwards, B. D. Bavister, and P. C. Steptoe, *Nature* 21 (1969): 982.
12. Edwards and Steptoe, op. cit., p. 86.
13. Ibid., p. 84.
14. Ibid., pp. 99ff.
15. Ibid., p. 99.
16. Ibid., p. 101.
17. R. G. Edwards and Ruth Fowler, "Human Embryos in the Laboratory," *Scientific American* 223 (1970): 53.
18. Ibid. Edwards also wrote about this, in almost identical language, in his 1971 *Nature* article with David Sharpe.
19. Ibid., p. 54.
20. Edwards and Steptoe, op. cit., p. 106.
21. In "In Vitro Fertilization of Human Ova and Blastocyst Transfer: An Invitational Symposium," *Journal of Reproductive Medicine* 11 (1973): 193.
22. Quoted in Paul Ramsey, "Shall We 'Reproduce'? I. The Medical Ethics of In Vitro Fertilization," *JAMA* 220 (1972): 1348.
23. Leon R. Kass, "Babies by Means of In Vitro Fertilization: Unethical Experiments on the Unborn," *New England Journal of Medicine* 285 (1971): 1177.
24. Ibid., p. 1175.
25. Ibid., p. 1176.
26. Ramsey, op. cit., p. 1349.
27. Ibid., p. 1481.
28. Edwards and Steptoe, op. cit., p. 113.
29. Ibid., p. 114. See also "Storm over Work on Test-Tube Babies," *Times* (London), October 18, 1971, p. 5.
30. Edwards and Steptoe, op. cit., p. 114.
31. Lesley and John Brown, as told to Sue Freeman, *Our Miracle Called Louise* (New York: Paddington Press, Ltd., 1979), p. 106.
32. Edwards and Steptoe, op. cit., p. 152.
33. Ibid., p. 154.
34. Brown and Brown, op. cit., p. 152.
35. Edwards and Steptoe, op. cit., p. 185.
36. Ibid., p. 1.

CHAPTER 13

1. See David Rorvik, "The Embryo Sweepstakes," *New York Times Magazine*, September 15, 1974, p. 17, and Vance Packard, *The People Shapers* (Boston: Little, Brown & Co., 1977), p. 193. Rorvik subsequently coauthored a book with Shettles, *How to Choose the Sex of Your Baby*, a consumer guide to gender selection techniques, which entered its fourth edition in 1989.
2. Quoted in Rorvik, op. cit., p. 17.
3. Similarly, two Indian researchers, Subhas Mukherjee and Saroj Kanti Bhattacharya, have been denied recognition for the 1978 birth of Durga Agarwal in Calcutta, because their documentation was deemed insufficient. If indeed the team did succeed, Agarwal would be the world's second IVF baby, and the first to have been produced from a frozen embryo.
4. Carl Wood and Ann Westmore, *Test-Tube Conception* (Englewood Cliffs, N.J.: Prentice-Hall, 1984), p. 42.

5. Cited in Clifford Grobstein, *From Chance to Purpose* (Reading, Mass.: Addison-Wesley Publishing Co., 1981), p. 109.
6. Cited in Ethics Advisory Board, "HEW Support of Research Involving Human In Vitro Fertilization and Embryo Transfer," reprinted in Grobstein, op. cit., p. 197.
7. Wood and Westmore, op. cit., p. 94.
8. See *New Scientist*, April 12, 1984, p. 3.
9. "Australia Dispute Arises on Embryos," *New York Times*, June 23, 1984, p. 30.
10. Interview with Alan Trounson, July 24, 1995.
11. Beverly Merz, "Stock Breeding Technique Applied to Human Infertility," *Medical News* 250 (1983): 1257.
12. Quoted in Diana Frank and Marta Vogel, *The Baby Makers* (New York: Carroll & Graf Publishers, Inc., 1988), p. 84.
13. Howard Jones, Jr., "The Ethics of In Vitro Fertilization—1982," *Fertility and Sterility* 37 (1982): 149.
14. Ibid., p. 147.
15. Ibid., p. 148.
16. Ibid.
17. See Steven York, M.D., and Risa Adler-York v. Howard W. Jones, Jr., M.D., Suheil Muasher, M.D., Medical College of Hampton Roads, t/a the Howard and Georgeanna Jones Institute for Reproductive Medicine, and Sentara Health System, t/a Sentara Norfolk General Hospital, Ed.VA. 717 F Suppl 421.
18. This was not an idle proviso. In 1990, a divorcing Tennessee couple, the Davises, staged a battle over seven embryos that they had consigned to storage two years earlier while undergoing IVF and were awarded joint custody of the "children in vitro" by a state appellate court. However, the Tennessee Supreme Court, in affirming the decision, declared that the embryos could neither be used by the former Mrs. Davis nor donated to another couple, since Mr. Davis objected. See Davis v. Davis, [Tenn. 1992] 842 S.W.2d 588.
19. See Sherman Elias and George J. Annas, *Reproductive Genetics and the Law* (Chicago: Year Book Medical Publishers, Inc., 1987), p. 237.
20. See the EAB report in Grobstein, op. cit., pp. 155–204.
21. See the dissenting opinion of J. Kennard in Johnson v. Calvert, [May 1993] 5 Cal.4th 84; 19 Cal.Rptr.2d, 851 P.2d 776.
22. EAB report in Grobstein, op. cit., pp. 188–90.
23. Ibid., p. 199.
24. Ibid., p. 203.
25. Grobstein, op. cit., p. 113.
26. See the table compiled by LeRoy Walters which appears in the Institute of Medicine, *Medically Assisted Conception: An Agenda for Research* (Washington, D.C.: National Academy Press, 1989), p. 73.
27. Quoted in Omar Sattaur, "New Conception Threatened by Old Morality," *New Scientist*, September 27, 1984, p. 13.
28. Quoted in Anne Marie Moulin's review of *L'oeuf transparent* in *Journal of Medicine and Philosophy* 14 (1989): 588.
29. Quoted in Gail Vines, "Test-Tube Pioneer Fears Rise of Eugenics," *New Scientist*, October 9, 1986, p. 17.
30. Ibid.
31. Ibid.
32. Quoted in Rita Arditti, Renate Duelli Klein, and Shelley Minden, eds., *Test-Tube Women* (Winchester, Mass.: Pandora Press, 1989), p. 345.
33. Quoted by Gena Corea in ibid., p. 29.
34. Grobstein, op. cit., p. 13.
35. Institute of Medicine, op. cit., p. 2.
36. Cited in Arditti et al., op. cit., p. xiii.
37. Robyn Rowland in Gena Corea et al., *Man-Made Women* (Bloomington: Indiana University Press, 1987), p. 75.

38. Kishwar, in ibid., p. 33.
39. Cited by Holmes in ibid., p. 18.
40. Rowland in ibid., pp. 74ff.
41. See Arditti et al., op. cit., p. xix.
42. Ibid., p. xx.
43. Elaine Hoffman Baruch, Amadeo F. D'Adamo, Jr., and Joni Seager, eds., *Embryos, Ethics, and Women's Rights* (New York: Harrington Park Press, 1988), p. 95.
44. Hubbard in Arditti et al., op. cit., p. 334.
45. Ibid., p. 340.
46. Quoted in Gail Vines, "Whose Baby Is It Anyway?" *New Scientist,* July 3, 1986, p. 27.
47. John Robertson, *Children of Choice: Freedom and the New Reproductive Technologies* (Princeton, N.J.: Princeton University Press, 1994), pp. 179ff.
48. Hubbard in Baruch et al., op. cit., p. 232.
49. Ibid., p. 203.
50. Quoted in Robert F. Howe, "Doctor Lied About Pregnancy Status, Fertility Patient Says as Trial Begins," *Washington Post,* February 11, 1992, B section, p. 1.
51. Quoted in Robert F. Howe, "Jacobson Misled Hundreds, Professor Testifies," *Washington Post,* February 19, 1992, D section, p. 3.

CHAPTER 14

1. Quoted in Rebecca Kolberg, "Human Embryo Cloning Reported," *Science* 262 (1993): 652.
2. Gina Kolata, "Cloning Human Embryos: Debate Erupts over Ethics," *New York Times,* October 26, 1993, p. A1.
3. Quoted in Kolberg, op. cit., p. 653.
4. Ethics Committee of the American Fertility Society, "Ethical Considerations of Assisted Reproductive Technologies," *Fertility and Sterility* 62, suppl. 1 (1994): 62S.
5. Reuters, "By Large Margin, Americans Oppose Cloning of Humans," *New York Times,* November 11, 1993, p. B9.
6. See Johnson v. Calvert, [May 1993] 5 Cal.4th 84; 19 Cal. Rptr.2d 494; 851 P.2d 776.
7. In Pennsylvania in 1995, there was a tragic case in which a twenty-six-year-old single man, James Alan Austin, obtained a child in this fashion. He then bludgeoned the five-week-old boy to death, was arrested, and confessed to the killing.
8. See Johnson V. Calvert, op. cit., p. 109.
9. Patricia Mahlstedt, "Disclosure Versus Non-Disclosure in Donor Conception: A Private Decision or a Public Concern," in *Legal and Bioethical Challenges to Effective Infertility Counseling,* notes from an American Fertility Society postgraduate course held November 5–6, 1994, p. 82.
10. Jean Benward case study, presented at an American Fertility Society postgraduate course held November 5–6, 1994.
11. Ethics Committee, op. cit., p. 49S.
12. Interview with Richard Paulson, USC, November 1, 1994.
13. Quoted in Edmund D. Pellegrino, John Collins Harvey, and John P. Langan, eds., *Gift of Life: Catholic Scholars Respond to the Vatican Instruction* (Washington, D.C.: Georgetown University Press, 1990), p. 13.
14. Ibid., p. 20.
15. Ethics Committee, op. cit., p. 66S.
16. Quoted in Edward J. Larson, "The Rhetoric of Eugenics," *British Journal for the History of Science* 24 (1991): 50.
17. Ethics Committee, op. cit., p. 65S.
18. Bentley Glass, "Racism and Eugenics in International Context," *Quarterly Review of Biology* 68 (1993): 61.
19. John Maddox, "Adventures in the Germ Line," *New York Times,* December 11, 1994, E section, p. 15.

20. Ibid.
21. John A. Robertson, *Children of Choice: Freedom and the New Reproductive Technologies* (Princeton, N.J.: Princeton University Press, 1994), p. 167.
22. W. French Anderson, "Genetics and Human Malleability," *Hastings Center Report*, January/February 1990, p. 23.
23. Clifford Grobstein, *From Chance to Purpose: An Appraisal of External Human Fertilization* (Reading, Mass.: Addison-Wesley Publishing Co., 1981), p. xi.
24. Hans Eysenck, in Francis Galton, *Hereditary Genius* (New York: St. Martin's Press, 1978), p. vi.
25. Don Peppers and Martha Rogers, "Let's Make a Deal," *WIRED* 2.02 (1994): 74.

BIBLIOGRAPHY

BOOKS

Adams, Mark B., ed. *The Wellborn Science: Eugenics in Germany, France, Brazil, and Russia.* New York: Oxford University Press, 1990.

Adelmann, Howard B. *Marcello Malpighi and the Evolution of Embryology.* 5 vols. Ithaca, N.Y.: Cornell University Press, 1966.

Allen, J., J. Golson, and R. Jones, eds. *Sunda and Sahul: Prehistoric Studies in Southeast Asia, Melanesia, and Australia.* New York: Academic Press, 1977.

Ammerman, Albert J., and L. L. Cavalli-Sforza. *The Neolithic Transition and the Genetics of Populations in Europe.* Princeton, N.J.: Princeton University Press, 1984.

Amt, Emilie, ed. *Women's Lives in Medieval Europe.* New York: Routledge, 1993.

Aptekar, Herbert. *Anjea: Infanticide, Abortion and Contraception in Savage Society.* New York: William Godwin, Inc., 1931.

Arditti, Rita, Renate Duelli Klein, and Shelley Minden, eds. *Test-Tube Women: What Future for Motherhood?* Winchester, Mass.: Pandora Press, 1989.

Ariès, Philippe, and Georges Duby, eds. *A History of Private Life.* Vols. 1–3. Cambridge, Mass.: The Belknap Press of Harvard University Press, 1989.

Aristotle. *Politics.* Trans. H. Rackham. Cambridge, Mass.: Harvard University Press, 1932.

Baruch, Elaine Hoffman, Amadeo F. D'Adamo, Jr., and Joni Seager, eds. *Embryos, Ethics, and Women's Rights: Exploring the New Reproductive Technologies.* New York: Harrington Park Press, 1988.

Baur, Erwin, Eugen Fischer, and Fritz Lenz. *Human Heredity.* Trans. Eden and Cedar Paul. New York: The MacMillan Co., 1931.

Bergmann, Martin S. *In the Shadow of Moloch: The Sacrifice of Children and Its Impact on Western Religions.* New York: Columbia University Press, 1992.

Bock, Gisella, and Pat Thane. *Maternity and Gender Policies: Women and the Rise of the European Welfare States, 1880s–1950s.* New York: Routledge, 1991.

Bodmer, Walter, and Robin McKie. *The Book of Man: The Human Genome Project and the Quest to Discover Our Genetic Heritage.* New York: Scribner, 1994.

British Medical Association. *Our Genetic Future: The Science and Ethics of Genetic Technology.* New York: Oxford University Press, 1992.

Brown, Lesley, John Brown, and Sue Freeman. *Our Miracle Called Louise.* New York: Paddington Press, 1979.

Bury, John Bagnell. *The Idea of Progress: An Inquiry into Its Origin and Growth.* London: MacMillan & Co., Ltd., 1924.

Carr-Saunders, Alexander Morris. *Population.* London: Oxford University Press, 1925.

———. *The Population Problem: A Study in Human Evolution.* London: Oxford University Press, 1922.

Carter, C. O. *Developments in Human Reproduction and Their Eugenic, Ethical Implications: Proceedings of the Nineteenth Annual Symposium of the Eugenics Society.* New York: Academic Press, 1983.

Cartledge, Paul. *Hellenistic and Roman Sparta.* London: Routledge, 1989.

Cave-Browne, John. *Indian Infanticide: Its Origin, Progress, and Suppression.* London: W. H. Allen & Co., 1857.

Chadwick, Ruth, ed. *Ethics, Reproduction and Genetic Control.* London: Croom Helm, 1987.

Chase, Allan. *The Legacy of Malthus: The Social Cost of the New Scientific Racism.* Urbana: University of Illinois Press, 1980.

Chesler, Ellen. *Woman of Valor: Margaret Sanger and the Birth Control Movement in America.* New York: Simon & Schuster, 1992.

Cipolla, Carlo M. *Miasmas and Disease: Public Health and the Environment in the Pre-Industrial Age.* Trans. Elizabeth Potter. New Haven: Yale University Press, 1992.

Corea, Gena. *The Mother Machine: Reproductive Technologies from Artificial Insemination to Artificial Wombs.* New York: Harper & Row, 1985.

Corea, Gena, et al. *Man-Made Women: How New Reproductive Technologies Affect Women.* Bloomington: Indiana University Press, 1987.

Coughlan, Michael J. *The Vatican, the Law and the Human Embryo.* Iowa City: University of Iowa Press, 1990.

Darwin, Sir Francis. *Rustic Sounds and Other Studies in Literature and Natural History.* 1917. Reprint, Freeport, N.Y.: Books for Libraries Press, 1969.

Davis, Bernard D. *The Genetic Revolution: Scientific Prospects and Public Perceptions.* Baltimore: The Johns Hopkins University Press, 1991.

Duster, Troy. *Backdoor to Eugenics.* New York: Routledge, 1990.

Eaves, L. J., H. J. Eysenck, and N. G. Martin. *Genes, Culture and Personality: An Empirical Approach.* San Diego: Academic Press, 1989.

Edwards, Robert, and Patrick Steptoe. *A Matter of Life: The Story of a Medical Breakthrough.* New York: William Morrow, 1980.

Elias, Sherman, and George Annas. *Reproductive Genetics and the Law.* Chicago: Year Book Medical Publishers, 1987.

Fletcher, Joseph. *The Ethics of Genetic Control: Ending Reproductive Roulette.* Buffalo, N.Y.: Prometheus Books, 1988.

Forrest, W. G. *A History of Sparta, 950–192 B.C.* New York: W. W. Norton & Co., 1968.

Foucault, Michel. *The History of Sexuality, an Introduction.* Vol. 1. Trans. Robert Hurley. New York: Vintage Books, 1978.

Francoeur, Robert T. *Utopian Motherhood: New Trends in Human Reproduction.* Garden City, N.Y.: Doubleday & Co., 1970.

Frank, Diana, and Marta Vogel. *The Baby Makers.* New York: Carroll & Graf Publishers, 1988.

Galton, Francis. *Hereditary Genius: An Inquiry into Its Laws and Consequences.* Introduction by H. J. Eysenck. London: Julian Friedmann Publishers, 1978.

———. *Inquiries into Human Faculty and Its Development.* London: J. M. Dent & Sons, 1911.

Gamble, Clive. *The Paleolithic Settlement of Europe.* Cambridge, England: Cambridge University Press, 1986.

Gardner, Jane F., and Thomas Wiedemann. *The Roman Household.* New York: Routledge, 1991.

Glover, Jonathan, et al. *Ethics of the New Reproductive Technologies: The Glover Report to the European Commission.* De Kalb: Northern Illinois University Press, 1989.

Gobineau, Joseph Arthur, comte de. *The Inequality of Human Races.* Trans. Adrian Collins. New York: Howard Fertig, 1967.

Godwin, William. *Of Population: An Enquiry Concerning the Power of Increase in the Numbers of Mankind, Being an Answer to Mr. Malthus's Essay on That Subject (1820).* New York: August M. Kelley Bookseller, 1964.

Goethe, Johann Wolfgang. *Faust.* Ed. Cyrus Hamlin and trans. Walter Arndt. New York: W. W. Norton & Co., 1976.

Gould, Stephen Jay. *The Mismeasure of Man.* New York: W. W. Norton & Co., 1981.

Graham, Robert Klark. *The Future of Man.* Introduction by Sir Cyril Burt. North Quincy, Mass.: The Christopher Publishing House, 1970.

Grinder, Robert E. *A History of Genetic Psychology: The First Science of Human Development.* New York: John Wiley & Sons, 1967.

Grobstein, Clifford. *From Chance to Purpose: An Appraisal of External Human Fertilization.* Reading, Mass.: Addison-Wesley Publishing Co., 1981.

Guha, Sujoy K. *Bioengineering in Reproductive Medicine.* Boca Raton, Fla.: CRC Press, 1990.

Gustafson, James M., and W. French Anderson. *Genetic Engineering and Humanness: A Revolutionary Prospect.* Washington, D.C.: Washington National Cathedral, 1992.

Hadingham, Evan. *Secrets of the Ice Age: The World of the Cave Artists.* New York: Walker & Co., 1979.

Haldane, J. B. S. *Heredity and Politics.* New York: W. W. Norton & Co., 1938.

Hardin, Garrett. *Population, Evolution, and Birth Control: A Collage of Controversial Ideas.* San Francisco: W. H. Freeman & Co., 1969.

Harris, Marvin. *Cannibals and Kings.* New York: Random House, 1977.

Harvey, William. *On the Generation of Animals.* Trans. Robert Willis. Vol. 28. Great Books of the Western World Series. Ed. Robert Maynard Hutchins. Chicago: Encyclopaedia Britannica, 1952.

Hausfater, Glenn, and Sarah Blaffer Hrdy, eds. *Infanticide: Comparative and Evolutionary Perspectives.* New York: Aldine Publishing Co., 1984.

Herlihy, David, and Christiane Klapisch-Zuber. *Tuscans and Their Families: A Study of the Florentine Catasto of 1427.* New Haven: Yale University Press, 1985.

Himmelfarb, Gertrude. *Poverty and Compassion: The Moral Imagination of the Late Victorians.* New York: Vintage Books, 1992.

Hitler, Adolf. *Mein Kampf.* Trans. James Murphy. London: Hurst and Blackett, 1942.

Hoffer, Peter C., and N. E. H. Hull. *Murdering Mothers: Infanticide in England and New England, 1558–1803.* New York: New York University Press, 1981.

Holmes, Helen B., Betty B. Hoskins, and Michael Gross, eds. *The Custom-Made Child? Women-Centered Perspectives.* Clifton, N.J.: The Humana Press, 1981.

Hubbard, Ruth, and Elijah Wald. *Exploding the Gene Myth.* Boston: Beacon Press, 1993.

Hunt, David. *Parents and Children in History: The Psychology of Family Life in Early Modern France.* New York: Basic Books, 1970.

Huntington, Ellsworth. *Tomorrow's Children: The Goal of Eugenics.* New York: John Wiley & Sons, 1935.

Huxley, Julian. *The Human Crisis.* Seattle: University of Washington Press, 1963.

Institute of Medicine. *Assessing Medical Technologies.* Washington, D.C.: National Academy Press, 1985.

Isaac, Glynn. *The Archaeology of Human Origins.* Cambridge, England: Cambridge University Press, 1989.

Kay, Margarita Artschwager. *Anthropology of Human Birth.* Philadelphia: F. A. Davis Co., 1982.

Kertzer, David I. *Sacrificed for Honor: Italian Infant Abandonment and the Politics of Reproductive Control.* Boston: Beacon Press, 1993.

Kevles, Daniel. *In the Name of Eugenics: Genetics and the Uses of Human Heredity.* New York: Alfred A. Knopf, 1985.

Knecht, Heidi, Anne Pike-Tay, and Randall White, eds. *Before Lascaux: The Complex Record of the Early Upper Paleolithic.* Boca Raton, Fla.: CRC Press, 1993.

Kohl, Marvin, ed. *Infanticide and the Value of Life.* Buffalo, N.Y.: Prometheus Books, 1978.

Kraut, Alan M. *Silent Travelers: Germs, Genes, and the "Immigrant Menace."* New York: Basic Books, 1994.

Lacey, W. K. *The Family in Classical Greece.* Ithaca, N.Y.: Cornell University Press, 1968.

Ladurie, LeRoy. *Montaillou: The Promised Land of Error.* Trans. Barbara Bray. New York: George Braziller, 1978.

Law and Ethics of AID and Embryo Transfer. Ciba Foundation Symposium 17. New York: Elsevier, 1972.

Lecky, William E. H. *History of European Morals, from Augustus to Charlemagne.* Vol. 2. New York: D. Appleton & Co., 1900.

Lee, Richard, and Irven DeVore, eds. *Man the Hunter.* Chicago: Aldine Publishing Co., 1968.

Lee, Simon. *Laws and Morals: Warnock, Gillick and Beyond.* Oxford: Oxford University Press, 1986.

Lefkowitz, Mary R., and Maureen B. Fant. *Women's Life in Greece and Rome: A Source Book in Translation.* 2d ed. Baltimore: The Johns Hopkins University Press, 1992.

Lerner, Richard. *Final Solutions: Biology, Prejudice, and Genocide.* Forewords by R. C. Lewontin and Benno Müller-Hill. University Park, Penn.: The Pennsylvania State University Press, 1992.

Levine, Robert J. *Ethics and Regulation of Clinical Research.* 2d ed. New Haven: Yale University Press, 1988.

Lewontin, Richard, Steven Rose, and Leon J. Kamin. *Not in Our Genes: Biology, Ideology, and Human Nature.* New York: Pantheon Books, 1984.

Lombroso, Cesare. *Crime: Its Causes and Remedies.* Trans. Henry P. Horton. Boston: Little, Brown & Co., 1911.

Ludmerer, Kenneth M. *Genetics and American Society: A Historical Appraisal.* Baltimore: The Johns Hopkins University Press, 1972.

Lyon, Jeff. *Playing God in the Nursery.* New York: W. W. Norton & Co., 1985.

Malthus, T. R. *An Essay on the Principle of Population; or A View of its past and present Effects on Human Happiness; With an Inquiry into our Prospects respecting the future Removal or Mitigation of the Evils which it occasions.* 1803 ed. Ed. Patricia James. Reprint, with 1806, 1807, 1817, and 1826 variora, Cambridge, England: Cambridge University Press, 1989.

Mazumdar, Pauline M. H. *Eugenics, Human Genetics and Human Failings: The Eugenics Society, Its Sources and Its Critics in Britain.* London: Routledge, 1992.

McDaniel, Walton Brooks. *Conception, Birth and Infancy in Ancient Rome and Modern Italy.* Lancaster, Penn.: Business Press, 1948.

McKeown, Thomas. *The Modern Rise of Population.* New York: Academic Press, 1976.

Meyer, Arthur William. *The Rise of Embryology.* Stanford Calif.: Stanford University Press, 1939.

Miller, Barbara. *The Endangered Sex: Neglect of Female Children in Rural North India.* Ithaca, N.Y.: Cornell University Press, 1981.

Montagu, M. F. Ashley. *Man's Most Dangerous Myth: The Fallacy of Race.* Foreword by Aldous Huxley. New York: Columbia University Press, 1945.

Müller-Hill, Benno. *Murderous Science: Elimination by Scientific Selection of Jews, Gypsies, and Others, Germany, 1933–1945.* Trans. George R. Fraser. Oxford: Oxford University Press, 1988.

Nisbet, Robert. *History of the Idea of Progress.* New York: Basic Books, 1980.

Parkes, Alan, and J. E. Meade, eds. *Biological Aspects of Social Problems.* New York: Plenum Press, 1965.

Pearl, Raymond. *The Biology of Population.* New York: Alfred A. Knopf, 1925.

Pellegrino, Edmund D., John Collins Harvey, and John P. Langan, eds. *Gift of Life: Catholic Scholars Respond to the Vatican Instruction.* Washington, D.C.: Georgetown University Press, 1990.

Piers, Maria. *Infanticide.* New York: W. W. Norton & Co., 1978.

Pincus, Gregory. *The Control of Fertility.* New York: Academic Press, 1965.

———. *The Eggs of Mammals.* New York: The MacMillan Co., 1936.

Place, Francis. *Illustrations and Proofs of the Principle of Population.* London: George Allen & Unwin, 1930. Reprint, with critical and textual notes by Norman E. Hines, New York: Augustus M. Kelley Bookseller, 1967.

Plato. *Great Dialogues of Plato.* Ed. Eric H. Warmington and Philip G. Rouse and trans. W. H. D. Rouse. New York: The New American Library, 1956.

Pliny. *Natural History.* Trans. H. A. Rackham. Cambridge, Mass.: Harvard University Press, 1938.

Plutarch's Moralia. Trans. Frank Cole Babbit. Cambridge, Mass.: Harvard University Press, 1936.

Poirier, Frank. *Understanding Human Evolution.* 3d ed. Englewood Cliffs, N.J.: Prentice-Hall, 1993.

Pomeroy, Sarah. *Women in Hellenistic Egypt.* New York: Schocken Books, 1984.

Price, Martin. *The Restoration and the Eighteenth Century.* New York: Oxford University Press, 1973.

Proctor, Robert. *Racial Hygiene: Medicine Under the Nazis.* Cambridge, Mass.: Harvard University Press, 1988.

Radin, Paul. *The Racial Myth.* New York: Whittlesey House, 1934.

Rawson, Beryl. *The Family in Ancient Rome: New Perspectives.* Ithaca, N.Y.: Cornell University Press, 1986.

Raymond, Janice. *Women as Wombs: Reproductive Technologies and the Battle over Women's Freedom.* New York: Harper San Francisco, 1993.

Rich, Adrienne. *Of Woman Born: Motherhood as Experience and Institution.* New York: W. W. Norton & Co., 1986.

Riddle, John M. *Contraception and Abortion from the Ancient World to the Renaissance.* Cambridge, Mass.: Harvard University Press, 1992.

Robinson, Daniel N., ed. *Significant Contributions to the History of Psychology, 1750–1920.* Comparative Psychology, Dser., vol. 3. Washington, D.C.: University Publications of America, 1977.

Roe, Shirley A. *Matter, Life, and Generation: Eighteenth-Century Embryology and the Haller-Wolff Debate.* Cambridge, England: Cambridge University Press, 1981.

Rosenthal, Joel T., ed. *Medieval Women and the Sources of Medieval History.* Athens: The University of Georgia Press, 1990.

Schama, Simon. *Citizens: A Chronicle of the French Revolution.* New York: Alfred A. Knopf, 1989.

Scheper-Hughes, Nancy. *Death Without Weeping: The Violence of Everyday Life in Brazil.* Berkeley: University of California Press, 1992.

Schwartz, Jeffrey. *What the Bones Tell Us.* New York: Henry Holt & Co., 1993.

Seneca. *Moral Essays.* Trans. John W. Basore. Cambridge, Mass.: Harvard University Press, 1970.

Shipman, Pat, and Erik Trinkhaus. *The Neanderthals: Changing the Image of Mankind.* New York: Alfred A. Knopf, 1993.

Snowden, R. G., G. D. Mitchell, and E. M. Snowden. *Artificial Reproduction: A Social Investigation.* London: George Allen & Unwin, 1983.

Sophocles, *Oedipus Rex,* in *Three Theban Plays.* Trans. Henry H. Stevens. Yarmouthport, Mass.: The Register Press, 1958.

Spengler, Joseph. *French Predecessors of Malthus: A Study in Eighteenth-Century Wage and Population Theory.* Durham, N.C.: Duke University Press, 1942.

Stent, Gunther. *The Coming of the Golden Age: A View of the End of Progress.* Garden City, N.Y.: The Natural History Press, 1969.

Sterne, Laurence. *The Life and Opinions of Tristram Shandy, Gentleman.* New York: New American Library, 1960.

Stone, Lawrence. *The Family, Sex and Marriage in England, 1500–1800.* New York: Harper & Row, 1977.

Strathern, Marilyn. *Reproducing the Future: Anthropology, Kinship, and the New Reproductive Technologies.* New York: Routledge, 1992.

Stuard, Susan Mosher, ed. *Women in Medieval Society.* Philadelphia: University of Pennsylvania Press, 1976.

Suetonius. *The Lives of the Caesars.* Trans. J. C. Rolfe. New York: G. P. Putnam's Sons, 1914.

Sumner, William Graham. *Folkways: A Study of the Sociological Importance of Usages, Manners, Customs, Mores, and Morals.* Boston: Ginn & Co., 1911.

Terence. *The Lady of Andros.* Trans. John Sargeaunt. New York: G. P. Putnam's Sons, 1912.

——. *The Mother-in-Law.* Trans. John Sargeaunt. New York: G. P. Putnam's Sons, 1912.

Thomson, Sir Godfrey. *The Factorial Analysis of Human Ability.* London: University of London Press, 1951.

Tooley, Michael. *Abortion and Infanticide.* Oxford: Clarendon Press, 1983.

Tuchman, Barbara. *A Distant Mirror: The Calamitous Fourteenth Century.* New York: Alfred A. Knopf, 1978.

Warren, Mary Anne. *Gendercide.* Totowa, N.J.: Rowman and Allanhead, 1985.

Washburn, Sherwood, ed. *Social Life of Early Man.* Chicago: Aldine Publishing Co., 1961.

Weindling, Paul. *Health, Race, and German Politics Between National Unification and Nazism, 1870–1945.* Cambridge, England: Cambridge University Press, 1989.

Weiss, Kenneth M., and Paul A. Ballonoff. *Demographic Genetics.* Benchmark Papers in Genetics, vol. 3. Stroudsburg, Penn.: Dowden, Hutchinson & Ross, 1975.

Weiss, Kenneth M., and R. H. Ward. *The Demographic Evolution of Human Populations.* London: Academic Press, 1976.

Wisot, Arthur L., and David R. Meldrum. *New Options for Fertility: A Guide to In Vitro Fertilization and Other Assisted Reproduction Methods.* New York: Pharos Books, 1990.

Wood, Carl, and Ann Westmore. *Test-Tube Conception.* Englewood Cliffs, N.J.: Prentice-Hall, 1984.

Wymelenberg, Suzanne. *Science and Babies: Private Decisions, Public Dilemmas.* Washington, D.C.: National Academy Press, 1990.

Wymer, John. *The Paleolithic Age.* New York: St. Martin's Press, 1982.

Zimmerman, Jan, ed. *The Technological Woman: Interfacing with Tomorrow.* New York: Praeger, 1983.

SCIENTIFIC PAPERS, JOURNAL ARTICLES, AND PROCEEDINGS

Adelman, Stuart, and Saul Ronsenzweig. "Prenatal Predetermination of the Sex of Offspring: II. The Attitudes of Young Married Couples with High School and with College Education." *Journal of Biosocial Science* 10 (1978):235.

Anderson, W. French. "Gene Therapy in Human Beings: When Is It Ethical to Begin?" *New England Journal of Medicine* 303 (1980):1293.

———. "Prospects for Human Gene Therapy." *Science* 226 (1984):401.

———. "Human Gene Therapy: Scientific and Ethical Considerations." *Journal of Medicine and Philosophy* 10 (1985):275.

———. "Human Gene Therapy: Why Draw a Line?" *Journal of Medicine and Philosophy* 14 (1989):681.

———. "Genetics and Human Malleability." *Hastings Center Report,* January/February 1990, p. 21.

Andrews, William J. "Eugenics Revisited." *Mankind Quarterly* 30 (1990):235.

Annas, George J. "Artificial Insemination: Beyond the Best Interests of the Donor." *Hastings Center Report,* August 1979, p. 14.

———. "Contracts to Bear a Child: Compassion or Commercialism?" *Hastings Center Report,* April 1981, p. 23.

Annas, George J., and Sherman Elias. "Legal and Ethical Implications of Fetal Diagnosis and Gene Therapy." *American Journal of Medical Genetic* 35 (1990):215.

Apperley, Jane F., and David A. Williams. "Gene Therapy: Current Status and Future Directions." *British Journal of Haematology* 75 (1990):148.

Asch, Ricardo H., Jose P. Balmaceda, Linda R. Ellsworth, and P. C. Wong. "Gamete Intra-Fallopian Transfer (GIFT): A New Treatment for Infertility." *International Journal of Fertility* 30 (1985):41.

Barton, Mary, Kenneth Walker, and B. P. Wiesner. "Artificial Insemination." *British Medical Journal,* January 13, 1945, p. 40.

Betteridge, K. J. "An Historical Look at Embryo Transfer." *Journal of Reproduction and Fertility* 62 (1981):1.

Bhattacharya, Bhairab C. "A Convectional Counter-Stream Sedimentation Process for Separating Biological Isotopes." *IEEE Transactions on Biomedical Engineering* BME-26 (1979):160.

Blandau, Richard J. "In Vitro Fertilization and Embryo Transfer." *Fertility and Sterility* 33 (1980):3.

Blumberg, Bruce D., Mitchell S. Golbus, and Karl Hanson. "The Psychological Sequelae of Abortion Performed for a Genetic Indication." *American Journal of Obstetrics and Gynecology* 122 (1975):799.

Briggs, Robert, and Thomas J. King. "Transplantation of Living Nuclei from Blastula Cells into Enucleated Frogs' Eggs." *Proceedings of the National Academy of Science* 38 (1952):455.

Buster, John, Maria Bustillo, Ian Thorneycroft, James Simon, Stephen Boyers, and John Marshall. "Non-Surgical Transfer of In Vivo Fertilised Donated Ova to Five Infertile Women: Report of Two Pregnancies." *Lancet*, July 23, 1983, p. 223.

Cavalli-Sforza, L. L., A. C. Wilson, C. R. Cantor, R. M. Cook-Deegan, and M. C. King. "Call for a Worldwide Survey of Human Genetic Diversity: A Vanishing Opportunity for the Human Genome Project." *Genomics* 11 (1991):490.

Cederqvuist, Lars L., and Fritz Fuchs. "Antenatal Sex Determination: A Historical Review." *Clinical Obstetrics and Gynecology* 13 (1970):159.

Chang, M. C. "Fertilizing Capacity of Spermatozoa Deposited into the Fallopian Tubes," *Nature* 168 (1951):697.

———. "Fertilization of Rabbit Ova In Vitro," *Nature* 184 (1959):466.

Chen, Christopher. "Pregnancy After Human Oocyte Cryopreservation." *Lancet*, April 19, 1986, p. 884.

Cole, Lamont C. "Sketches of General and Comparative Demography." *Symposia of Quantitative Biology* 22 (1957):1.

"Conception in a Watch Glass." *New England Journal of Medicine* 217 (1937):678.

Croxatto, H. B., B. Fuentealba, S. Diaz, L. Pastene, and H. J. Tatum. "A Simple Nonsurgical Technique to Obtain Unimplanted Eggs from Human Uteri." *American Journal of Obstetrics and Gynecology* 112 (1972):662.

Curie-Cohen, Marie, Lesleigh Luttrell, and Sander Shapiro. "Current Practice of Artificial Insemination by Donor in the United States." *New England Journal of Medicine* 300 (1979):585.

Custis, Donald L. "An Undergraduate Views Eugenics." *Journal of Heredity* 30 (1939):412.

Diasio, Robert B., and Robert H. Glass. "Effects of pH on the Migration of X and Y Sperm." *Fertility and Sterility* 22 (1971):303.

Donchin, Anne. "The Growing Feminist Debate over the New Reproductive Technologies." *Hypatia* 4 (1989):136.

Edwards, R. G. "Maturation In Vitro of Human Ovarian Oocytes." *Lancet*, November 6, 1965, p. 926.

———. "Preliminary Attempts to Fertilize Human Oocytes Matured In Vitro." *American Journal of Obstetrics and Gynecology* 96 (1966):192.

Edwards, R. G., B. D. Bavister, and P. C. Steptoe. "Early Stages of Fertilization In Vitro of Human Oocytes Matured In Vitro." *Nature* 221 (1969):632.

Edwards, R. G., and Ruth E. Fowler. "Human Embryos in the Laboratory." *Scientific American* 223 (1970):45.

Edwards, R. G., and David J. Sharpe. "Social Values and Research in Human Embryology." *Nature* 231 (1971):87.

Edwards, R. G., P. C. Steptoe, and J. M. Purdy. "Fertilization and Cleavage In Vitro of Preovulator Human Oocytes." *Nature* 227 (1970):1307.

Edwards, R. G., P. C. Steptoe, and J. M. Purdy. "Establishing Full-Term Human Pregnancies Using Cleaving Embryos Grown In Vitro." *British Journal of Obstetrics and Gynaecology* 87 (1980):737.

Folsome, Clair E. "The Status of Artificial Insemination: A Critical Review." *American Journal of Obstetrics and Gynecology* 45 (1943):915.

Foss, G. L., "Artificial Insemination by Donor: A Review of Twelve Years' Experience." *Journal of Biosocial Science* 14 (1982):253.

Fowler, R. E., R. G. Edwards, D. E. Walters, S. T. H. Chan, and P. C. Steptoe. "Steroidogenesis in Preovulatory Follicles of Patients Given Human Menopausal and Chorionic Gonadotrophins as Judged by the Radioimmunoassay of Steroids in Follicular Fluid." *Journal of Endocrinology* 77 (1978):161.

Franklin, Sarah, and Maureen McNeil. "Reproductive Futures: Recent Literature and Current Feminist Debates on Reproductive Technologies." *Feminist Studies* 14 (1988):545.

Freidmann, Theodore. "Opinion: The Human Genome Project—Some Implications of

Extensive 'Reverse Genetic' Medicine." *American Journal of Human Genetics* 46 (1990):407.

"Genetic Engineering in Man: Ethical Considerations." *JAMA* 22 (1972):721.

"German Genetic Literature." *Journal of Heredity* 30 (1939):445.

Gerstel, Gerda. "A Psychoanalytic View of Artificial Donor Insemination." *American Journal of Psychotherapy* 17 (1963):64.

Glass, Bentley. "Racisim and Eugenics in International Context." *Quarterly Review of Biology* 68 (1993):61.

Gregoire, A. T., and Robert C. Mayer. "The Impregnators," *Fertility and Sterility* 16 (1965):130.

Guttmacher, Alan F. "Artificial Insemination." *Annals of the New York Academy of Sciences* 97 (1962):623.

Harper, Peter S. "The Human Genome Project and Medical Genetics." *Journal of Medical Genetics* 29 (1992):1.

Harris, G. W. "Humours and Hormones." *Proceedings of the Society for Endocrinology* 53 (1972):suppl., ii.

Hartman, Carl G. "A Half Century of Research in Reproductive Physiology." *Fertility and Sterility* 12 (1960):1.

Hirschfeld, Lawrence A., James Howe, and Bruce Levin. "Warfare, Infanticide, and Statistical Inference: A Comment on Divale and Harris." *American Anthropologist* 80 (1978):110.

Huxley, Sir Julian. "Eugenics in Evolutionary Perspective." *Perspectives in Biology and Medicine,* Winter 1963, p. 155.

Illmensee, Karl, and Peter C. Hoppe. "Nuclear Transplantation in *Mus musculus:* Developmental Potential of Nuclei from Preimplantation Embryos." *Cell* 23 (1981):9.

Izuka, Rihachi, Yoshiaki Sawada, Nobuhiro Nishina, and Michie Ohi. "The Physical and Mental Development of Children Born Following Artificial Insemination." *International Journal of Fertility* 13 (1968):24.

Jacobson, Cecil B., James G. Sites, and Luis F. Arias-Bernal. "In Vitro Maturation and Fertilization of Human Follicular Oocytes." *International Journal of Fertility* 15 (1970):103.

Jones, Howard W., Jr. "The Ethics of In Vitro Fertilization—1982." *Fertility and Sterility* 37 (1982):146.

———. "The Norfolk Experience: In Vitro Fertilization Comes to the States." Excerpt of a speech given at Oss, The Netherlands, November 1993.

Jones, Howard W., Jr., and Charlotte Schrader. *In Vitro Fertilization and Other Assisted Reproduction. Annals of the New York Academy of Sciences.* Vol. 541. New York: The New York Academy of Sciences, 1988.

Jones, R. C. "Uses of Artificial Insemination." *Nature* 229 (1971):534.

Jordan, Elke. "The Human Genome Project: Where Did It Come From, Where Is It Going?" *American Journal of Human Genetics* 51 (1992):1.

Kasai, M., K. Niwa, and A. Iritani. "Effects of Various Cryoprotective Agents on the Survival of Unfrozen and Frozen Mouse Embryos." *Journal of Reproduction and Fertility* 63 (1981):175.

Kass, Leon R. "Babies by Means of In Vitro Fertilization: Unethical Experiments on the Unborn?" *New England Journal of Medicine* 285 (1971):1174.

Kaye, Howard L. "Are We the Sum of Our Genes?" *Wilson Quarterly,* Spring 1992, p. 77.

Kellum, Barbara A. "Infanticide in England in the Later Middle Ages." *History of Childhood Quarterly* 1 (1974):367.

Kennedy, Joseph F., and Roger P. Donahue. "Human Oocytes: Maturation in Chemically Defined Media." *Science* 164 (1969):1292.

Ketzer, D., de, P. Dennis, B. Hudson, J. Leeton, A. Lopata, K. Outch, J. Talbot, and C. Wood. "Transfer of a Human Zygote." *Lancet,* September 29, 1973, p. 728.

Kevles, Daniel J. "Controlling the Genetic Arsenal." *Wilson Quarterly,* Spring 1992, p. 68.

Klawiter, Maren. "Using Arendt and Heidegger to Consider Feminist Thinking on Women and Reproductive/Infertility Technologies." *Hypatia* 5 (1990):65.

Langer, G., E. Lemberg, and M. Sharf. "Artificial Insemination." *International Journal of Fertility* 14 (1969):232.

Langer, William L. "Infanticide: A Historical Survey." *History of Childhood Quarterly* 1 (1974):353.

Lawn, L., and R. A. McCance. "Ventures with an Artificial Placenta: I. Principles and Preliminary Results." *Proceedings of the Royal Society* 155 (1961):500.

Lederberg, Joshua. "Experimental Genetics and Human Evolution." *Bulletin of the Atomic Scientists* 22 (1966):4.

Ledward, R. S., E. M. Symonds, and S. Eynon. "Social and Environmental Factors as Criteria for Success in Artificial Insemination by Donor (AID)." *Journal of Biosocial Science* 14 (1982):263.

Lopata, A., Ian W. H. Johnston, Ian J. Hoult, and Andrew I. Speirs. "Pregnancy Following Intrauterine Implantation of an Embryo Obtained by In Vitro Fertilization of a Preovulatory Egg." *Fertility and Sterility* 33 (1980):117.

Lopata, A., R. McMaster, J. C. McBain, and W. I. H. Johnston. "In Vitro Fertilization of Preovulatory Human Eggs." *Journal of Reproduction and Fertility* 52 (1978):339.

Lopata, A., A. Henry Sathananthan, John C. McBain, W. Ian H. Johnston, and Andrew I. Speirs. "The Ultrastructure of the Preovulatory Human Egg Fertilized In Vitro." *Fertility and Sterility* 33 (1980):12.

Lorber, Judith, and Lakshmi Bandlamudi. "The Dynamics of Marital Bargaining in Male Infertility." *Gender and Society* 7 (1993):32.

Mahlstedt, Patricia P. "The Psychological Component of Infertility." *Fertility and Sterility* 43 (1985):335.

————. "Psychological Issues of Infertility and Assisted Reproductive Technology." *Urologic Clinics of North America* 21 (1994):557.

Mahlstedt, Patricia P., and Kris A. Probasco. "Sperm Donors: Their Attitudes Toward Providing Medical and Psychosocial Information for Recipient Couples and Donor Offspring." *Fertility and Sterility* 56 (1991):747.

Marks, Jonathan. "Historiography of Eugenics." *American Journal of Human Genetics* 52 (1993):650.

McMaster, R., R. Yanagimachi, and A. Lopata. "Penetration of Human Eggs by Human Spermatozoa In Vitro." *Biology of Reproduction* 19 (1978):212.

Medawar, P. B. "The Genetic Improvement of Man." *Australian Annals of Medicine* 4 (1969):317.

Menkin, Miriam F., and John Rock. "In Vitro Fertilization and Cleavage of Human Ovarian Eggs." *American Journal of Obstetrics and Gynecology* 55 (1948):440.

Mennuti, Michael T. "Prenatal Diagnosis—Advances Bring New Challenges." *New England Journal of Medicine* 320 (1989):661.

Montagu, M. F. Ashley. "The Concept of Race in the Human Species in the Light of Genetics." *Journal of Heredity* 32 (1941):243.

Moulin, Anne Marie. Review of Jacques Testart, *L'oeuf transparent*. *Journal of Medicine and Philosophy* 14 (1989):587.

Parkes, A. S. "Preservation of Human Spermatozoa at Low Temperatures." *British Medical Journal*, August 18, 1945, p. 212.

Paulson, Richard J., Mark V. Sauer, Mary M. Francis, Thelma M. Macaso, and Rogerio A. Lobo. "In Vitro Fertilization in Unstimulated Cycles: The University of Southern California Experience." *Fertility and Sterility* 57 (1992):290.

————. "A Prospective Controlled Evaluation of TEST-Yolk Buffer in the Preparation of Sperm for Human In Vitro Fertilization in Suspected Cases of Male Infertility." *Fertility and Sterility* 58 (1992):551.

Pincus, G., and E. V. Enzmann. "Can Mammalian Eggs Undergo Normal Development In Vitro?" *Proceedings of the National Academy of Science* 20 (1934):121.

Polge, C., A. U. Smith, and A. S. Parkes. "Revival of Spermatozoa After Vitrification and Dehydration at Low Temperatures." *Nature* 164 (1949):666.

Population Studies: Animal Ecology and Demography. Symposia on Quantitative Biology. Vol. 22. Cold Spring Harbor, N.Y.: Cold Spring Harbor Biological Laboratory, 1957.

Rafter, Nichole Hahn. "White Trash: Eugenics as Social Ideology." *Society,* November/December 1988, p. 43.

Ramsey, Paul. "Shall We 'Reproduce'? I. The Medical Ethics of In Vitro Fertilization." *JAMA* 220 (1972):346.

———. "Shall We 'Reproduce'? II. Rejoinders and Future Forecast." *JAMA* 220 (1972):1480.

Revelle, Roger. "Can Man Domesticate Himself?" *Bulletin of the Atomic Scientists* 22 (1966):2.

Robertson, John A. "Ethical and Legal Issues in Preimplantation Genetic Screening." *Fertility and Sterility* 57 (1992):1.

Roper, A. G. "Ancient Eugenics." *Mankind Quarterly* 32 (1992):383.

Rothschild, Lord. "Did Fertilization Occur?" *Nature* 221 (1969):981.

Schumacher, Gebhard F. B., et al. "In Vitro Fertilization of Human Ova and Blastocyst Transfer: An Invitational Symposium." *Journal of Reproductive Medicine* 11 (1973):192.

Selden, Steven. "Resistance in School and Society: Public and Pedagogical Debates about Eugenics, 1900–1947." *Teachers College Record* 90 (1988):61.

———. "Selective Traditions and the Science Curriculum: Eugenics and the Biology Textbook, 1914–1949." *Science Education* 75 (1991):493.

Seppala, Mark, and R. G. Edwards, eds. *In Vitro Fertilization and Embryo Transfer. Annals of the New York Academy of Sciences.* Vol. 442. New York: The New York Academy of Sciences, 1985.

Sherman, J. K. "Research on Frozen Human Semen: Past, Present, and Future." *Fertility and Sterility* 15 (1964):485.

———. "Synopsis of the Use of Frozen Human Semen Since 1964: State of the Art of Human Semen Banking." *Fertility and Sterility* 24 (1973):397.

Simpson, Nancy E. "Experiences of a Genetic Counselling Clinic in Kingston, Ontario." *Canadian Journal of Public Health* 66 (1975):375.

"Sir John Hammond: An Interview," *Journal of Reproduction and Fertility* 3 (1962):2.

Smith, John Maynard. "Eugenics and Utopia." *Daedalus* 94 (1965):487.

Soupart, Pierre, and Patricia Ann Strong. "Ultrastructural Observations on Human Oocytes Fertilized In Vitro." *Fertility and Sterility* 25 (1974):11.

Stephens, J. Claiborne, Mark L. Cavanaugh, Margaret I. Gradie, Martin L. Mador, and Kenneth D. Kidd. "Mapping the Human Genome: Current Status." *Science* 250 (1990):237.

Steptoe, P. C., and R. G. Edwards. "Birth After the Reimplantation of a Human Embryo." *Lancet,* August 12, 1978, p. 366.

Steptoe, P. C., R. G. Edwards, and J. M. Purdy. "Human Blastocysts Grown in Culture." *Nature* 229 (1971):133.

Swerdlow, Joel L. "The Double-Edged Helix." *Wilson Quarterly,* Spring 1992, p. 60.

Villa, Paolo, Claude Bouville, Jean Courtin, Daniel Helmer, Eric Mahieu, Pat Shipman, Giorgio Belluomini, and Marilí Bianca. "Cannibalism in the Neolithic." *Science* 233 (1986):431.

Walters, LeRoy. "Human In Vitro Fertilization: A Review of the Ethical Literature." *Hastings Center Report,* August 1979, p. 23.

Watson, James D. "The Human Genome Project: Past, Present, and Future." *Science* 248 (1990):44.

Weindling, Paul. "The 'Sonderweg' of German Eugenics: Nationalism and Scientific Internationalism." *British Journal for the History of Science* 22 (1989):321.

———. "The Survival of Eugenics in Twentieth-Century Germany." *American Journal of Human Genetics* 52 (1993):643.

Weiss, Kenneth M. "Demographic Models for Anthropology." *American Antiquity* 38 (1973):1.

Wertz, Dorothy C., John C. Fletcher, and John J. Mulvihill. "Medical Geneticists Confront Ethical Dilemmas: Cross-cultural Comparisons Among Eighteen Nations." *American Journal of Human Genetics* 46 (1990):1200.

Whittingham, D. G. "Fertilization of Mouse Eggs In Vitro." *Nature* 220 (1968):592.

Whittingham, D. G., and W. K. Whitten. "Long-Term Storage and Aerial Transport of Frozen Mouse Embryos." *Journal of Reproduction and Fertility* 36 (1974):433.

Williamson, Nancy E., T. H. Lean, and D. Vengadasalam. "Evaluation of an Unsuccessful Sex Preselection Clinic in Singapore." *Journal of Biosocial Science* 10 (1978):375.

Wood, Carl, Alan Trounson, John Leeton, J. McKenzie Talbot, Beresford Buttery, Janice Webb, Jillian Wood, and David Jessup. "A Clinical Assessment of Nine Pregnancies Obtained by In Vitro Fertilization and Embryo Transfer." *Fertility and Sterility* 35 (1981):502.

NEWSPAPERS AND PERIODICALS

Angell, Marcia. "New Ways to Get Pregnant." *New England Journal of Medicine* 323 (1990):1200.

"Australia Dispute Arises on Embryos." *New York Times,* June 23, 1984, Section I, p. 30.

Bohlen, Celestine. "Almost Anything Goes in Birth Science in Italy." *New York Times,* April 4, 1995, p. A14.

Bragg, Rick. "Husband's Legacy Is Child Conceived After Death, but the Law Resists." *New York Times,* December 22, 1994, p. D22.

Brownlee, Shannon, Betsy Wagner, Monika Guttman, and Missy Daniel. "The Baby Chase." *U.S. News and World Report,* December 5, 1994, p. 84.

Burns, John F. "India Fights Abortion of Female Fetuses." *New York Times,* August 27, 1994, Section I, p. 5.

"By Large Margin, Americans Oppose Cloning of Humans." *New York Times,* November 1, 1993, p. B9.

Chira, Susan. "Of a Certain Age, and in a Family Way." *New York Times,* January 2, 1994, Section IV, p. 5.

Connor, Steve. "Scientists Licensed to Work on 'Pre-Embryos.' " *New Scientist,* November 21, 1985, p. 21.

Cooper, Len. "Aryan Nation: Germany's Cruel African Heritage." *Washington Post,* February 20, 1994, p. C3.

"Cutting Out the Middle Man." *Economist,* January 4, 1992, p. 27.

Dickson, David. "Europe Split on Embryo Research." *Science* 242 (1988):117.

Direcks, Anita, and Helen Bequaert Holmes. "Miracle Drug, Miracle Baby." *New Scientist,* November 6, 1986, p. 53.

"Doubts Grow About India's 'Cryogenic' Infant." *Medical World News,* December 25, 1978, p. 11.

"Egg Swapping at Birth Clinic Brings Charge." *New York Times,* July 9, 1995, Section I, p. 9.

Elmer-Dewitt, Philip. "The Genetic Revolution." *Time,* January 17, 1994, p. 46.

"Fertile Ground." *New Scientist,* July 3, 1986, p. 17.

"Fertility Clinic Is Sued over the Loss of Embryos." *New York Times,* October 1, 1995, Section I, p. 26.

"Fertility Doctors' Data Is Seized." *New York Times,* September 22, 1995, p. A25.

"The Fertility Doctor's Private Practices." *Newsweek,* March 16, 1992, p. 62.

Fishman, Steve. "Inconceivable Conception." *Vogue,* December 1994, p. 306.

Ford, Jane. "Test-Tube Baby Pioneer Quits." *New Scientist,* May 31, 1984, p. 5.

"Frozen Assets." *Time,* January 3, 1972, p. 52.

"Frozen Eggs Find Ethical Favour in Australia . . . but Iced Embryos Make Money in the US." *New Scientist,* April 24, 1986.

"The Glass Womb." *Time,* January 27, 1961, p. 32.

Gould, Donald. "Other People's Pregnancies." *New Scientist,* August 28, 1993, p. 44.

Gould, Stephen Jay. "The Smoking Gun of Eugenics." *Natural History,* December 1991, p. 8.

Gwynne, Peter, et al. "All About That Baby." *Newsweek,* August 7, 1978, p. 66.

————. "A Happy Accident? More Test-Tube Babies?" *Science Digest,* October 1978, p. 7.

Howe, Robert F. "Va. Fertility Doctor Goes on Trial Today." *Washington Post,* February 10, 1992, p. D3.

———. "Witness Tells of Jacobson's Restroom Visits." *Washington Post*, February 15, 1992, p. D1.

———. "Citing Cruel Lies by Jacobson, Judge Gives Him Five Years, Fine." *Washington Post*, May 9, 1992, p. D1.

Kolata, Gina. "In Vitro Fertilization Goes Commercial." *Science* 221 (1983):1160.

———. "Researcher Clones Embryos of Human Infertility Effort." *New York Times*, October 24, 1993, Section I, p. 1.

———. "Cloning Human Embryos: Debate Erupts over Ethics." *New York Times*, October 26, 1993, p. A1.

———. "Nightmare or the Dream of a New Era in Genetics?" *New York Times*, December 7, 1993, p. A1.

———. "Reproductive Revolution Is Jolting Old Views." *New York Times*, January 11, 1994, p. A1.

———. "Gene Technique Can Shape Future Generations." *New York Times*, November 22, 1994, p. C10.

Kolberg, Rebecca. "Human Embryo Cloning Reported." *Science* 262 (1993):652.

Lawson, Carol. "Celebrated Birth Aside, Teen Has a Typical Life." *New York Times*, October 4, 1993, p. A18.

Maddox, John. "Adventures in the Germ Line." *New York Times*, December 11, 1994, Section IV, p. 15.

Merz, Beverly. "Stock Breeding Technique Applied to Human Infertility." *JAMA* 250 (1983):1257.

"My Brother the Clone." *New York Times*, November 6, 1993, Section I, p. 22.

Mydans, Seth. "Fertility Clinic Told to Close Amid Complaints." *New York Times*, May 29, 1995, Section I, p. 7.

"Now It's Ice-Cube Babies." *Newsweek*, July 12, 1971, p. 83.

Ornstein, Peggy. "Looking for a Donor to Call Dad." *New York Times Magazine*, June 18, 1995, p. 28.

"Preventing 'Inferior' People in China." *New York Times*, December 27, 1993, p. A8.

Price, Frances. "Too Much of a Good Thing." *New Scientist*, August 18, 1990, p. 29.

"The Proxy Fathers: Sowing the Seeds of Despair?" *Sunday Times Magazine* (London), April 11, 1982, p. 28.

Rubin, Bernard. "Psychological Aspects of Human Artificial Insemination." *Archives of General Psychiatry* 13 (1965):121.

Sattaur, Omar. "New Conception Threatened by Old Morality." *New Scientist*, September 27, 1984, p. 12.

Simons, Marlise. "Uproar over Twins, and a Dutch Couple's Anguish." *New York Times*, June 28, 1995, p. A3.

Singer, Peter. "Technology and Procreation: How Far Should We Go?" *Technology Review*, February 1985, p. 23.

"Sperm-Bank Baby Gets Federal Benefits." *New York Times*, May 31, 1995, p. B7.

"Sperm on Deposit." *Newsweek*, August 30, 1971, p. 58.

"Storm over Work on Test Tube Babies." (London) *Times*, October 18, 1971, p. 5.

Sullivan, Walter. "Implant of Human Embryo Appears Near." *New York Times*, October 29, 1970, p. A1.

Tagliabue, John. "In Italy, a Child Is Born, and So Is a Lively Debate." *New York Times*, January 13, 1995, p. A8.

"Test-Tube Tempest." *Newsweek*, February 6, 1961, p. 78.

Tyler, Patrick E. "China Weighs Using Sterilization and Abortions to Stop 'Abnormal' Births." *New York Times*, December 22, 1993, p. A8.

———. "Chinese Start a Vitamin Program to Eliminate a Birth Defect." *New York Times*, January 11, 1994, p. C3.

———. "Population Control in China Falls to Coercion and Evasion." *New York Times*, June 25, 1995, Section I, p. 1.

"U.S. Teams Tool Up for Test-Tube Baby Production." *Medical World News*, December 25, 1978, p. 10.

Vines, Gail. "Whose Baby Is It Anyway?" *New Scientist,* July 3, 1986, p. 26.

————. "Test-Tube Pioneer Fears Rise of Eugenics." *New Scientist,* October 9, 1986, p. 17.

————. "Government Seeks Views on 'Test-Tube' Babies." *New Scientist,* December 18, 1986.

————. "New Shackles Hamper Test-Tube Baby Researchers." *New Scientist,* May 14, 1987, p. 22.

Walgate, Robert. "French Scientist Makes a Stand." *Nature* 323 (1986):385.

Westmore, Ann. "First Freeze-Thaw Baby Is Born." *New Scientist,* April 12, 1984, p. 3.

"What's Up, Doc?" *Time,* March 16, 1992, p. 33.

"Widow Hopes to Bear Children with Sperm from Dead Husband." *Miami Herald,* January 20, 1995.

"Woman Gives Birth to Triplets at Age Fifty." *New York Times,* November 12, 1992, p. A21.

OFFICIAL REPORTS AND GOVERNMENT DOCUMENTS

Ethics Committee of the American Fertility Society. "Ethical Considerations of the New Reproductive Technologies." *Fertility and Sterility* 46, suppl. 1 (1986).

————. "Ethical Considerations of the New Reproductive Technologies in Light of Instruction on Respect for Human Life in Its Origin and the Dignity of Procreation Issued by the Congregation for the Doctrine of the Faith." *Fertility and Sterility* 49, suppl. 1 (1988).

————. "Ethical Considerations of Assisted Reproductive Technologies." *Fertility and Sterility* 62, suppl. 1 (1994).

Feversham Committee. "Human Artificial Insemination." *British Medical Journal,* July 30, 1960, p. 379.

Medically Assisted Conception: An Agenda for Research. Report of a Study by a Committee of the Institute of Medicine. Washington, D.C.: National Academy Press, 1989.

Peel Report. "Report of Panel on Human Artificial Insemination." *British Medical Journal Supplement,* Appendix V, April 7, 1973, p. 3.

INDEX

Jordan, David Starr, 138
Journal of Heredity, 19, 151–52
Journal of Physical Anthropology, 136
Journal of the American Medical Association (JAMA), 163–64
Jukes family, 139
Jutras, Mark, 11

Kaiser Wilhelm Institute, 143, 151–52
Kallikak family, 139
Kansas Free Fair (1924), 137
Kass, Leon, 226–27, 243
Katz, Barbara, 247
Kennedy, Edward, 249
Kennedy, Joseph, 223
Kertzer, David, 106
Kishwar, Madhu, 255
Klapisch-Zuber, Christiane, 82
Klein, Renate Duelli, 254, 255
Kolata, Gina, 261
Koonz, Claudia, 147
Krebs-Ringer medium, 223

Lacedaemonian leap, 55
Ladurie, LeRoy, 76–77
Lady of Andros, The (Terence), 64
Lamarck, Jean-Baptiste, 99
Lancet, 216, 232, 241
Langer, William, 37
laparoscopy, 217–18
    harvesting eggs with, 198, 220–21, 229
Lapouge, Count Georges Vacher de, 133, 134–35
Laughlin, Harry, 138–39, 145, 152
Laurentian Hormone Conference, 197
Law for the Prevention of Genetically Diseased Offspring (German), 145
law of deviation from the mean, 127
Lebensborn (Well of life), 147
Lecky, William, 64
Lederberg, Joshua, 19–20, 152, 208–9, 222
Leeuwenhoek, Antonie van, 95–96
legitimacy issues, *see* illegitimacy
Lehmann, Julius Friedrich, 142
Leibniz, Gottfried Wilhelm, 93
Lenz, Fritz, 141, 142–43, 145, 147–48, 152
Lenz, Widukind, 143
Leonardo da Vinci, 87
lesbians, AID and, 181
leuprolide, 29
Lewis, Warren, 198, 200
Lewontin, Richard, 275
Leyland, Zoe, 239
life expectancy, 102–3
Linacre, Thomas, 85

lineage:
    female sexual purity and, 59, 72–73
    obsession with, in Middle Ages, 78–79
line breeding, 210
Lippmann, Walter, 131
Little, Clarence, 137
Locke, John, 87
Lombroso, Cesare, 128
London Foundling Hospital, 105
Lopata, Alex, 235, 237, 242
Lorber, Judith, 30
Lorenz, Konrad, 146
luteal phase, 205
luteinizing hormone (LH), 204, 205, 206, 223
Luttrell, Lesleigh, 179
luxury, degeneration associated with, 119
Lycurgus, 50–51, 52

Machigenga, 48
McKeown, Thomas, 40
McKusick, Victor, 215–16
McLaren, Anne, 252
Maclennan v. *Maclennan*, 170
MacMillan's, 126
Maddox, John, 272–73
magnifying equipment, 94–97
Mahlstedt, Patricia, 27, 28, 29, 180, 189–91, 263–64
Maimonides, 80
Maiscki, Ivan, 203
Malebranche, Nicolas, 95
Malpighi, Marcello, 94–95, 98
Malthus, Daniel, 111
Malthus, Thomas Robert, 111–15, 116, 118, 120
mammals:
    infanticide among, 37
    search for eggs of, 89, 92–93, 98–99
Man-Made Women, 254
Mantegazza, Paolo, 158–59
Man the Hunter, 43
March of Dimes Birth Defects Foundation, 33–34
Marcus Aurelius, 75
Maresca, Manny, 11
Maresca, Pam Williams, 11, 15
Marrs, Richard, 27, 245
Mars, 63
Marshall, F. H. A., 156, 192, 196, 204
Marshall, John, 241
Marx, Karl, 115
Mary, 71
Mastroianni, Luigi, 225, 234–35
matrilineal societies, 187